THE

HERMETIC AND ALCHEMICAL WRITINGS

OF

AUREOLUS PHILIPPUS THEOPHRASTUS BOMBAST,

OF HOHENHEIM, CALLED

PARACELSUS THE GREAT.

NOW FOR THE FIRST TIME FAITHFULLY TRANSLATED INTO ENGLISH.

EDITED WITH A BIOGRAPHICAL PREFACE, ELUCIDATORY NOTES, A COPIOUS HERMETIC
VOCABULARY, AND INDEX,

BY ARTHUR EDWARD WAITE.

IN TWO VOLUMES.

VOL. II.

HERMETIC MEDICINE AND HERMETIC PHILOSOPHY.

THE

HERMETIC and ALCHEMICAL
WRITINGS

OF

AUREOLUS PHILIPPUS THEOPHRASTUS BOMBAST,

OF HOHENHEIM, CALLED

PARACELSUS THE GREAT.

NOW FOR THE FIRST TIME FAITHFULLY TRANSLATED INTO ENGLISH.

EDITED WITH A BIOGRAPHICAL PREFACE, ELUCIDATORY NOTES, A COPIOUS HERMETIC
VOCABULARY, AND INDEX,

BY ARTHUR EDWARD WAITE.

IN TWO VOLUMES.

VOL. II.

HERMETIC MEDICINE AND HERMETIC PHILOSOPHY.

London :
JAMES ELLIOTT AND CO.,
TEMPLE CHAMBERS, FALCON COURT, FLEET STREET, E.C.
—
1894.

TABLE OF CONTENTS

VOLUME II

PART II

HERMETIC MEDICINE

APPENDICES

PART II.

HERMETIC MEDICINE.

THE ARCHIDOXIES OF THEOPHRASTUS PARACELSUS *

BOOK I

CONCERNING THE MYSTERY OF THE MICROCOSM †

IF, my dearest sons, we consider the misery by which we are detained in a gross and gloomy dwelling, exposed to hunger and to many and various accidents, from all sources, by which we are overwhelmed and surrounded, we see that we could scarcely flourish, or even live, so long as we followed the medicine prescribed by the ancients. For we were continually hedged in by calamities and bitter conditions, and were bound with terrible chains. Every day things became worse with us, as with others who were weighed in the same balance, whom also the ancients have not so far been able to help or to heal by means of their books. We do not in this place advance the different causes of this misfortune. This only we say, that many teachers by following the ancient methods have acquired for themselves much wealth, credit, and renown, though they did not deserve it, but got together such great resources by simple lies. From which consideration we have wished to elaborate and write this memorial work of ours, that we might arrive at a more complete and happier method of practice, since there are presented to us those mysteries of Nature which are too wonderful to be ever thoroughly investigated.

---

* The ten books of the *Archidoxies* stand in the same relation to Hermetic Medicine as the nine books *Concerning the Nature of Things* stand to Hermetic Chemistry and the science of metallic transmutation. They appear to have been reckoned among the most important works of Paracelsus, and the editions are exceedingly numerous. That which has been selected for translation is derived from the Geneva folio.

† The Microcosmos itself is to be understood thus, namely, as consisting of the four elements, and it is these invisibly. It is formed after the image of Him who created all things, and yet it has remained a creature. Therefore, it is partially one with the earth, because like the earth it has need of the other elements, heaven, air, and fire.—*De Hydropisi.* This, therefore, is the condition of the Microcosmos, or smaller world. It contains in its body all the minerals of the world. Consequently the body acquires its own medicine from the world. Hence it is clear that all minerals are useful to man, if any one of these be joined to its corresponding mineral in the body of the Microcosm. He who lacks this knowledge is by no means a philosopher or physician, for if a physician affirms that a certain marcasite is useful for this or that, it is first of all needful that he should know what is the marcasite of the world and what is the marcasite of the human body.—*Paramirum,* Lib. IV. *De Matrice.* There is a vast variety of things contained in the body of the Microcosm which elude the observation of the senses, though God, the Creator, has willed them to exist in that structure. There are, for example, more than a thousand species of trees, stones, minerals, manna, and metals. Even as the incredible magnitude of the Sun appears small to us by reason of its distance, so other things may be made to look small which are placed nearer at hand, and thus we come to understand that He who filled the shell of the heaven with so many and such great bodies, was able to include as many and as great wonders in the body of the Microcosm. Accordingly, know that the mysteries of the Microcosm are to be mystically understood--that is, we are no to measure local things according to proportion or substance, but must all the more arduously enquire in what matter we are to expect the effect of any agent.—*De Causis et Origine Luis Gallicæ*, Lib. V., c. 10.

Wherefore we have come to consider how that art can be reconciled with the mysteries of Nature, in opposition to those who, so far, have not been able to arrive at the art at all.

The strength of this mystery of Nature is hindered by the bodily structure, just as if one were bound in a prison with chains and fetters. From this the mind is free. For in its operation this mystery is like fire in green wood, which seeks to burn, but cannot on account of the moisture.

Since, then, hindrance arises from this source, one had to see how to get free from it. For such freedom being secured, this art of separation can only be compared to the art of the apothecaries, as light is compared to darkness. And this we say not in mere arrogance, but on account of the great frauds practised by apothecaries and physicians. Wherefore, not undeservedly, we call them darkness, or caves of robbers and impostors, since in them many persons are treated for gain by ignorant men ; persons who, if they were not rich, would at once be pronounced healthy, since the practitioners know that there is no remedy or help for these people in their consultations.

This, then, is worthy to be called an art, which teaches the mysteries of Nature ; which, by means of the quintessence, can cure a contraction and bring about health in the space of four days, whereas otherwise death would be the result. A wound, too, can be healed in twenty-four hours, which would scarcely yield to bodily treatment in as many days. Let us, therefore, readily approach by experiment this separation of the mysteries of Nature from the hindrances of the body.

First, then, we have to consider what is of all things most useful to man and most excellent. It is to learn the mysteries of Nature, by which we can discover what God is and what man is, and what avails a knowledge of heavenly eternity and earthly weakness. Hence arises a knowledge of theology, of justice, of truth, since the mysteries of Nature are the only true life of man, and those things are to be imitated which can be known and obtained from God as the Eternal Good. For although many things are gained in medicine, and many more in the mysteries of Nature, nevertheless after this life the Eternal Mystery remains, and what it is we have no foundation for asserting, save that which has been revealed to us by Christ. And hence arises the ignorant stupidity of theologians, who try to interpret the mysteries of God, whereof they know not the least jot ; and what it is not possible for man to formulate, namely, the will of Him who gave the mystery. But that word of His they twist to their own pride and avarice ; from whence arise misleading statements, which every day increase more and more. Hence it comes that we lightly value, nay, think nothing at all of that reason which is not evidently founded on the mysteries. In like manner the jurists have sanctioned laws according to their own opinions, which shall secure themselves against loss, though the safety of the State be imperilled.

Seeing, then, that in these faculties so many practices have come into vogue which are contrary to equity, let us dismiss the same to their proper time.

Nor do we care much for the vain talk of those who say more about God than He has revealed to them, and pretend to understand Him so thoroughly as if they had been in his counsels; in the meantime abusing us and depreciating the mysteries of Nature and of philosophy, about all of which they are utterly ignorant. The dishonest cry of these men is their principal knowledge, whereby they give themselves out to be those on whom our faith depends, and without whom heaven and earth would perish.

O consummate madness and imposture on the part of human creatures, in place whereof it would be more just that they should esteem themselves to be nought but unprofitable servants! Yet we, by custom imitating them, easily learn, together with them, to bend the word of our Teacher and Creator to our own pride. But since this word is not exactly known to us, can only be apprehended by faith, and is founded on no human reason, however specious, let us rather cast off this yoke, and investigate the mysteries of Nature, the end whereof approves the foundation of truth; and not only let us investigate these, but also the mysteries which teach us to fulfil the highest charity. And that is the treasure of the chief good which in this writing of our Archidoxies we understand in a material way.

From the aforesaid foundation we have drawn our medicine by experiment, wherein it is made clear to the eye that things are so. Then, coming to its practice, we divide this our book of Archidoxies into ten, as a sort of aid to the memory, so that we may not forget these matters, and at the same time may speak of them so far openly that we may be understood by our disciples, but not by the common people, for whom we do not wish these matters to be made too clear. We do not care to open our mind and thoughts and heart to those deaf ears, just as we do not wish to disclose them to impious men; but we shall endeavour to shut off our secrets from them by a strong wall and a key. And if by chance this our labour shall not be sufficiently safeguarded from those idiots who are enemies of all true arts, we shall forbear writing the tenth book, concerning the uses of those which precede it, so that we may not give the children's meat to the dogs. Nevertheless, the other nine will be sufficiently understood by our own disciples.

And, to speak more plainly of these matters, it must be known that in this treatise on the microcosm are proved and demonstrated all those points which it contains, which also embrace medicine, as well as those matters which are interconnected therewith. The subject of the microcosm is bound up with medicine and ruled by it, following it none otherwise than a bridled horse follows him who leads it, or a mad dog bound with chains. In this way I understand that medicine attracts Nature and everything that has life. Herein three things meet us, which shew by what forces they are filled and produced. Firstly, in what way the five senses are assisted by the mysteries of Nature, though those senses do not proceed from Nature, nor spring naturally as a herb from its seed, since there is no material which produces them. Secondly, the mobility of the body is to be considered : whence it proceeds, by what power it is moved and exercised, and in what manner it is ministered to. Thirdly, there must be a knowledge of all the forces in the body, and what forces apply to each member, and are transmuted according to the same nature as the particular limb, when originally they are identical in Nature.

First, then, we will speak of these senses : sight, hearing, touch, taste, and smell. The following example teaches us. The eyes have a material substance, of which they are composed, as it is handed down in the composition of the body. So of the other senses. But vision itself does not proceed from the same source as the eye ; nor the hearing from sound, or from the same source as the ears ; nor touch from flesh, nor taste from the tongue, nor smell from the nostrils, any more than reason proceeds from the brain ; but these are the bodily instruments, or rather the envelopes in which the senses are born. For it must not be understood that these senses depend solely on the favour of God, in the sense that they do not belong to the nature of man, but are infused solely by the grace of God above and beyond all Nature, to the end that, if one were born blind, the mighty works of God might be made known to us. We must not think so in this case. For the abovementioned senses have each their own body, imperceptible, impalpable, just as the root of the body, on the other hand, exists in a tangible form. For man is made up of two portions, that is to say, of a material and a spiritual body. Matter gives the body, the blood, the flesh ; but spirit gives hearing, sight, feeling, touch, and taste. When, therefore, a man is born deaf, this happens from a defect of the domicile in which hearing should be quartered. For the spiritual body does not complete its work in a situation which is badly disposed.

Herein, then, are recognised the mighty things of God, that there are two bodies, an eternal and a corporeal, enclosed in one, as is made clear in the Generation of Men.* Medicine acts upon the house by purging it, so that the spiritual body may be able to perfect its actions therein, like civet in a pure and uncontaminated casket.

Coming next to the power of motion in the body, let us inquire whence it is produced and has its origin, that is, how the body unites itself to the medicine so that the faculty of motion is increased. The matter is thus to be understood. Everything that lives has its own motion from Nature. This is

---

* The treatise *De Generatione Hominis*, to which reference is here made, appears to be fragmentary in character. It regards the generation of all growing things as twofold—one where the nature and the semen are contained in a single essence, the other where the essence of the nature exists without the semen. In man it recognises the existence of four complexions, and it distinguishes between the outward and the interior man. Men, as regards their mortal part, are nothing but mere cattle. There is, however, an internal man wherein there is no nature of the animal, and to consider him in this his true nature must be counted among the highest branches of philosophy. There is an immortal as well as a corruptible body of man, and it is in this, by the infusion of God's power, that reason, discernment, wisdom, doctrine, art, and generally whatsoever is above mortality, do alone inhere. Man, therefore, endowed with wisdom and subtlety, can emerge from his external body. All wisdom and intelligence which man enjoys is eternal with this body, and man the interior can live after another manner than can man the outer. This internal man is illustrated and clothed upon with truth for ever.

sufficiently proved of itself so far as natural motion is concerned. But the motion of which we think may be described as that which springs from the will, as, for example, in lifting the arm one may ask how this is done, when I do not see any instrument by which I influence it; but that takes place which I desire to take place. So one must judge with leaping, walking, running, and other matters which occur in opposition to, or outside of, natural motion. They have their origin in this, that intention, a powerful mistress, exists above my notions in the following manner. The intention or imagination kindles the vegetative faculty as the fire kindles wood—as we describe more particularly in our treatise on the Imagination.* Nowhere is it more powerful to fulfil its operations than in its own body where it exists and lives. So, in every body, nothing is more easily kindled than the vegetative soul, because it runs and walks by itself and is disposed for this very purpose. For, even as a hidden or buried fire blazes forth so soon as it is exposed and catches the air, so my mind is intent upon seeing something. I cannot with my hands direct my eyes whither I will; but my imagination turns them whithersoever it is my pleasure to look. So, too, as to my motion must it be judged. If I desire to advance and arbitrarily propose this to myself, at once my body is directed to one or the other place fixed on by myself. And the more this is impressed on me by my imagination and thought, the more quickly I run. In this way, Imagination is the motive-power of my running. None otherwise does medicine purify those bodies in which there is a spiritual element, whence it happens that their motion is more easily perfected.

Thirdly, it must be understood that in the body a distribution is made over all the members of everything which is presented to it, either without or within. In this distribution a change takes place by which things are modified, so that one part subserves the constitution of the heart, another accommodates itself to the nature of the brain; and of the rest in like manner. For the body attracts to itself in two ways, from within and from without. Within, it attracts whatever is taken through the mouth. Externally, it attracts air,

---

* *De Virtute Imaginativa* is another treatise which has survived only in a mutilated state. It further insists on the division of man into two bodies, the one visible and the other invisible, but is devoted to the consideration of the second only. The imagination is the mouth of the body which is not visible. It is also the sun of man which acts within its own sphere after the manner of the celestial luminary. It irradiates the earth, which is man, just as the material sun shines upon the material world. As the one operates corporeally, so works the other, after a parallel manner, spiritually. And as the sun sends its force on a spot which it shines upon, so also the imagination, like a star, bursts upon the thing which it affects. Nor are all things posited in heat and cold only, but in every operation. As the sun works corporeally and effects this or that, so also the imagination, by giving fire and fuel, effects all things which the sun effects, not that it has need of instruments, but that it makes those things with which it burns. Consider the matter as follows: He who wishes to burn anything needs flint, fire, fuel, brimstone, a candle, etc., and so he obtains fire; but if the sun seeks to burn, it requires none of these things, doing all things together and at once, no one beholding its steel. Such also is the imagination. It tinges and paints its own surface, but no one sees its pencil, ceruse, or pigments; all things take place with it at once, just as fire from the sun bursts forth without any corporeal instrument. Let no one, therefore, be surprised that from the imagination corporeal works should proceed, since similar results are manifest with other things. The whole heaven, indeed, is nothing else but an imagination. Heaven works in man, stirs up pests, fevers, and other things, but it does not produce these by corporeal instruments, but after the same manner that the sun burns. The sun, indeed, is of one power only, the moon of one power only, and every separate star is of one power only. Man, however, is altogether a star. Even as he imagines himself to be, such he is, and he is that also which he imagines. If he imagines fire, there results fire; if war, there ensues war; and so on in like manner. This is the whole reason why the imagination is in itself a complete sun.

earth, water, and fire.   Thus, then, the subject is to be arranged and defined. Those matters which are received from within need not be described.   They are known by the foundation of our nature, what they are which are distributed, and we shall speak subsequently as to their division.   But, externally, one must understand whatever is necessary to itself the body attracts from the four elements.   Unless this were done the internal nutriment would not suffice to sustain the life of man.   For instance, moisture, not existing constitutionally in the body, is extracted by the body itself from water, whence it happens that if one stands or sits in water, it is not necessary that he satisfies his thirst from without.   It does not, indeed, take place in the same way that heat is extinguished by water, like fire ; but the internal heat attracts to itself the moisture from without, and imbibes it just as though it were from within. Hence it happens that in the Alps cattle are able to remain the whole summer without drinking ; the air is drink for them, or supplies its place : and the same should be judged with regard to man.

The nature of man, too, may be sustained in the absence of food, if the feet are planted in the earth.   Thus we have seen a man who lived six months without food and was sustained only by this method : he wore a clod of earth on his stomach, and, when it got dry, took a new and fresh one. He declared that during the whole of that time he never felt hungry.   The cause of this we shew in the treatise on the Appetite of Nature.*

So, in the matter of medicine, we have seen a man sustain himself for many years by the quintessence of gold, taking each day scarcely half a scruple of it.   In the same way, there have been many others who for so long as twenty years ate nothing, as I remember to have seen in our times.   This was by some attributed to the piety and goodness of the persons themselves, or even to God, which idea we would be the last to impugn or to criticise. But this, nevertheless, is an operation of Nature ; insomuch that sorrow and mental despondency take away hunger and thirst to such an extent that the body can sustain itself for many years by its own power of attraction.   So, then, food and drink are not thus arranged that it is absolutely necessary we should eat bread or meats, or drink wine or water, but we are able to sustain our life on air and on clods of earth ; and whatever is appointed for food, we should believe is so appointed that we should taste and try it, as we shall shew more at length in the " Monarchy of God."†   Let us, however, concede this point—that on account of our labours and such things, it cannot be that we do without temporal and bodily food, and that for many causes.   Wherefore food was ordained for this purpose, just as medicine was against diseases.

We will make a distinction of things entering into the body after the following fashion, that they are distributed through every part of it none otherwise than as if ardent wine be poured into water.   The water acquires

---

the odour of the wine because the wine is distributed through its whole volume ; and, in the same way, when ink is poured into wine the whole of the wine is thereby blackened.

So, too, in the human body, the vital moisture immediately diffuses whatever is received, and more quickly than in the examples we have cited.

But under what form the substance received becomes transmuted depends entirely on the nature of the members that receive it, just as if bread be conveyed into a man it becomes the flesh of a man, if into a fish the flesh of a fish, and so on. In the same way, it must be understood that the substances received are transmuted by the natural power of the members, and are appropriated according to the nature of the parts which take them up. A like judgment must be passed upon medicines, namely, that they are transmuted into members according to the properties of those members. For the limbs gain their own force and virtue from the substance of medicines peculiar to themselves, according to the good or bad dispersion of them, and according as the medicine itself was subtle or gross. This is the case with the quintessence ; its transmutation will be stronger and more effectual. But if it be thick it remains the same, just as a picture acquires its tint, its beauty, or its deformity from its colours, and if these be more vivid it will be the same. Wherefore, in order that we may have experience of like matters to fall back upon in those things which happen to us, and that we may lay them up in our memory, so as to have them ready in case of need, we will write these nine books, keeping the tenth shut up in our own brain on account of the thankless idiots. Nevertheless, to our own disciples, these things shall be made sufficiently clear.

And let no one wonder at the school of our learning. Though it be contrary to the courses and methods of the ancients, still it is firmly based on experience, which is mistress of all things, and by which all arts should be proved.

THE END OF THE PROLOGUE AND OF THE FIRST BOOK OF THE ARCHIDOXIES ON THE MICROCOSM, OUT OF THE THEOPHRASTIA

THE SECOND BOOK OF THE ARCHIDOXIES

FROM THE THEOPHRASTIA OF PARACELSUS THEOPHRASTUS

CONCERNING THE SEPARATIONS OF THE ELEMENTS *

BEFORE we approach a description of the separations of the elements, we shall explain this separation, for the greater and clearer understanding thereof, seeing that certain matters written about the generation of things are not altogether consonant with the separations of the elements. For every matter is more readily brought to its appropriate end, where mature and intellectual consideration is given beforehand as to what its end may be. Thence the practice becomes clearer. We say, then, that the four elements exist together in all things, and out of them arises to every one of these things its predestined condition. In this way you may understand how it comes about that these four elements, differing so widely from one another, are able to agree and coexist without mutual destruction. Whereas the mixture of the elements is united and strengthened by predestination, it results that no weight is taken account of in them, but the power of one is greater than of another; by which, indeed, it is understood that in the digest and ferment of the predestination, that which is strongest will preponderate and conquer and subdue the other elements. In this way, the remaining three elements cannot attain their perfection, but stand related to that perfect element as

---

* The light of Nature teaches that God has separated and divided everything, so that it can exist by itself. Thus are separated light and darkness without any mutual damage, as day and night prove. Moreover, He has also separated the metals, each into its property. Thus, gold has its vein, iron has its minera, silver has its brilliancy. Lastly, every metal has its proper domicile. Moreover, He has separated from one another the marcasites and the genera of salts ; in the same manner also summer and winter, elements, herbs, fruits, and every growing thing, so that we hence see how God has created various species from a single Iliaster, while the species of his workshop surpass all the number of the sands. He has so adorned heaven and earth that what they contain can never be sufficiently recognised and considered.—*De Balneis Piperinis*, c. 1. The philosophy of separation in preparing specifics is thus developed in treating of the cure of ulcers. You must also understand concerning separation that there is nothing so noxious but that it has its peculiar use, like the spider, which in addition to its venom possesses a marvellous power for curing all chronic fevers. Good and bad are equally necessary in the constitution of our arcanum, for both the venom and the antidote, the sweet and the bitter, are in the body. Further, it has been proved that all colours and all savours exist in the body. Inasmuch as there exist three colours in Saturn—yellow, white, and red; in Mars also three—purple, red, and black; notwithstanding neither Saturn nor Mars are a colour. Thus we must judge it to be the case with the colours of Salts by means of separation ; and just as the redness of Saturn tinges with a red, and the yellow with a yellow colour, by virtue of separation, similarly, alumen works as alumen and salt as salt in ulcers. Concerning the wonders of separation it should be remembered what happens concerning vinegar. Who would judge by his senses that vinegar was present in uncorrupted wine ? Nevertheless, no one can doubt that it is there. Similarly, in Venus there is vitriol. Indeed, Venus herself is vitriol, and by separation can be reduced into vitriol. Yet no one would rightly say that Venus was vitriol. Wherefore we must judge concerning separation that it is of the form but not of the species, as in the conversion of Venus into a salt.—*Chirurgia Magna*, Part III., Lib. IV.

the light matter in wood. Wherefore they are not to be called elements, since they are not all perfect, but only one. When, therefore, we speak of the four elements, which finally exist in all things, we are not to understand definitely so that there are four perfect elements therein, but that there is only one such finished element, the rest remaining imperfect through the potency of that excelling element. Hence it happens that they are able to meet and coexist, because in three of them there is no perfection. On which account, too, no corruption can prevail by reason of their contrarieties. Moreover, that an element is predominant in one kind arises from the fact that it is hereto predestinated. Hence no corruption or confusion can accrue, as we lay it down in our treatise on Generations.*

Since, then, there is only one element specially present in everything, it avails not to seek the four elements in things, seeing that three of these elements are not in a state of perfection. In a word, we must understand that the four elements are in all things, but not actually four complications. The matter stands thus: A substance contains water, and then it is nenufar. Besides this element there exists in it no earth, air, or fire. There is no appearance of heat or dryness in it. It has no peculiar operation; but the predestination thereof is water; and the element of water is the only one under which is no dryness nor heat, according to its congenital nature. Though matters be thus, however, yet notwithstanding the three elements are involved in it, still the things have not their origin in those three elements which are not produced in a perfect state, nor have they beginning or aid from them, but from that predestinated element which is united to and impressed upon that particular kind.

Now, although this is at variance with the vulgar philosophy, namely, that one predestinated element has of its own nature the other three elements cohering with it, still it is credible that the element and the substance differ the one from the other. It may be understood thus: the substance is not from that element which gives thereto its special tinge and elementary form; nor, again, are these elements from the substance, but they agree at the same time uniformly, as the body and soul agree. But now every body, as for example of some growing thing, has its own conformation; so has the element. Although the element itself is not visible in the body of the growing thing, or tangible, or demonstrable, because the element is stronger, by reason of its subtlety, and subdues the other elements of the growing thing. They are all, however, in the body, but imperceptibly, just as when water is mixed with vinegar the water becomes like the vinegar; and although the vinegar shall have changed the whole essence of the water, still the complexion of the water remains unchanged, nor does it on that account become

---

* Both from the text and the notes of this translation the reader will see that there are several treatises by Paracelsus on Generations of various kinds, not excepting some fragments upon the generation of fools, who appear, in the days of Paracelsus, to have abounded in the high places of science and religion. The reference above may be either to the *Generation of the Elements* or to the first book *Concerning the Nature of Things.*

vinegar, but remains water as it was before. And although it does not display the properties of water, yet it does not follow that it does not still possess those properties.

In these propositions we wish to make it clear in what way the separations of the elements are to be brought about, concerning which two methods of practice need to be understood. One is that with which agrees the separation of the predestined element, and this we shall elucidate in our treatise on the Quintessence. The other is that to which belong the four substantial elements in growing things. By this it is understood that the Quintessence exists as a predestined element, and that it cannot be separated from itself, but only from the three elements, as follows in the treatise on the Quintessence. But when we speak of the separations of the four elements, we understand those four which are essentially in the body. Hence have arisen various errors, because the four elements have been sought in the predestined element, and also in addition the Quintessence, which cannot in any way come to pass.

Moreover, it must be known when the elements of bodies are to be separated, that one exists as fire, another as water, a third is like air, and a fourth like earth, according to their complications, because elements sometimes appear with their own forms, and at other times with complications, for example, water as water, air as air, earth as earth, and fire as fire. These matters must be subtly understood, and this can be done by means of a similitude, if they are taken for the union of the elements not visibly or in act, or according to the element of fire, but as the environment is warm and dry like fire. In this way its own nature, essence, and condition is assigned to each element, without any breach of propriety. For it is not supposed, because any particular herb is especially warm, as the nettle, that on this account it contains within itself more fire, but rather it is supposed that its own Quintessence is warmer than the Quintessence of the chamomile, which, indeed, has less heat. But the elements of a body receive less or more from their own substance, as, for instance, wood contains within itself more fire than herbs do ; and, in like manner, stones have in them more dryness and earth than resins have. Note, also, in like manner, that the bulk and quantity of degree in a Quintessence arises from the predestined element. And the intensity of degree in bodily elements springs out of the appearance of the substance which is unlike.

But we must come to the practice of separating bodily elements from all other things, and this is twofold. One, indeed, teaches us to draw out the three elements from the pure elements as from burning fire, from invisible air, from true earth, and in like manner from natural water, which have not an origin similar to the preceding ones. Another method of practice is with those things out of which these four elements exist, as we have said above, with such difference, however, that this exhibits more of the element of fire, of water, of earth, or of air, with a likeness to the form of the essential elements. When they have been separated in this way they can never be further dissolved, for instance, it is impossible that they should be corrupted beyond their complexions.

It must also be considered that the elements are found by separation to be formally like the essential elements. For air appears like air; and the same cannot in any way be enclosed, as some falsely think, for this reason, because at once, in the moment of separation, it levitates itself, and sometimes bursts forth as wind, and ascends sometimes with the water, sometimes with the earth, and at other times with the fire. And, indeed, this levitation or elevation in the air is very wonderful. Just as if the air were to be separated from the essential element of water, it would be done by boiling. When this begins to take place the air is soon separated from the water, carries off with itself the very light substance of the water, and in proportion as the water is diminished, so the air itself decreases according to its proportion and quantity.

And it must be remarked here that no one of the elements can be conceived or had without air, though of the rest one can be had without another. We do not, therefore, undertake the task of separating the air, since it is in the other three elements, just as life is in the body. For when it is separated from the body all things perish, as we clearly shew in the following practical treatise on separations. Four methods must be considered at this point; one, indeed, in watery bodies, that is, in herbs, which have more water than any of the other elements. A second is in fiery bodies, such as woods, resins, oils, roots, and the like, which contain within themselves more of the fiery substance than of others. A third is to be understood of earthy bodies, which are stones, clays, and earths. The fourth is airy, and this is in all the other three, as we have mentioned. In like manner, also, concerning the pure elements, there are just so many ways to be considered, in the same manner as has been said above concerning the four preceding.

Hence it is easy to learn what the elements are, and how they are to be separated. And among these the separations of the metals first meet us, wherein there are peculiar predestined virtues which are wanting in the other elements. For, although all the elements are alike in form, in heat, in cold, in moisture, and in dryness, still, the dryness or damp, or heat or cold, is not the same in one as in another. In some it is appropriative, but in others specific; and this in various ways, as in each kind they are produced peculiarly and essentially, since no kind of the elements is precisely like another in its properties.

So, also, it must be laid down with regard to the separation of marchasites, which differ from other substances both in the practice and in their elementary nature. For every kind is disposed in a particular separation, and must be dealt with in a special way. Stones and gems must demonstrate their elements afterwards, since they appear in no way similar to the others.

Then, too, salts exist in a peculiar and most excellent nature, with more abundant properties than appear in other substances. There is also a different essence in herbs, which in no way agree with minerals, nor can they be alike so far as relates to their nature. Moreover, the property of woods, of fruits, of barks, and the like, is peculiar ; so, too, of flesh, of drinks, and all comestibles, and of things not good or pure, but bad and impure, which have to be separated into their elements.

Of that separation concerning which we think, two methods are found. One consists of the separation of any element confined by itself in a peculiar vessel, without the corruption of its own forces, the air excepted.

There is another method of the separation of the pure from the impure from among the four elements, namely, in the following way. After the elements have been separated, that is to say, one from the other, they have still a dense substance ; and, on this account, there follows another similar separation of the already separated elements. Now, we purpose to make clear the practical method in all these cases. For in the first place it must be known that the quintessence of things is to be separated and extracted in this way, because, indeed, the elements of bodies in the nature of a quintessence are not subdued but are left with them. So it is able to tinge these elements more strongly or more lightly. Hence it comes to be understood that the forces in the elements do not perish when the predestined element, that is, the quintessence, is extracted ; for this itself is elemental and separable so far as relates to its elementary form, but not as to diverse natures, as is clear from the treatise on the quintessence.

By separations of this kind all elemental infirmities can be cured by one simple method, namely, if the one set of predestinations oppose the other, as we have laid it down in the treatise on predestinations. In these words we have sufficiently unfolded the initial stage of separations. Wherefore we can now speed on to the practice of them ; and here there is a tenfold variety : one of metals ; a second of marchasites ; a third of stones ; a fourth of oleaginous matters ; a fifth of resins ; a sixth of herbs ; a seventh of flesh ; an eighth of juices ; a ninth of vitreous substances ; and a tenth of fixed things. For these separations of the elements three methods are adopted : one by distillations ; a second by calcinations ; and the third by sublimations. In these are comprised all the exercises, as the application of the hands to the fire, the labour, and other necessary things which will be specified in the following pages.

THE END OF THE SECOND BOOK OF THE ARCHIDOXIES AND OF THE
FIRST PART CONCERNING THE SEPARATION OF THE ELEMENTS

THE THIRD BOOK OF THE ARCHIDOXIES

ON THE SEPARATIONS OF THE ELEMENTS FROM METALS

FOR the separation of the elements from metals there is need of the best instruments, of labour, of diligence, together with experience of the art and adaptation of the hands to this work.

Take salt nitre, vitriol, and alum, in equal parts, which you will distil into aqua fortis. Pour this water again on its fæces, and distil it again in glass. In this aqua fortis clarify silver, and afterwards dissolve in it sal ammoniac. Having done this, take a metal reduced into thin plates in the same way, that is, in the same water. Afterwards separate it by the balneum Mariæ, pour it on again, and repeat this until there be found at the bottom an oil, namely, from the Sun, or gold, of a light red colour ; of the Moon, a light blue ; of Mars, red and very dark ; of Mercury, white ; of Saturn, livid and leaden ; of Venus, bright green ; and of Jupiter, yellow.

All metals are not thus reduced to an oil, except those which have been previously prepared. For instance, Mercury must be sublimated ; Saturn calcined ; Venus florified ; Iron must be reduced to a crocus ; Jupiter must be reverberated ; but the Sun and Moon easily yield themselves.

After that the metals have been in this way reduced to a liquid substance, and have disposed themselves to a disunion of their elements—which cannot be done in a metallic nature, seeing that everything must be previously prepared for the use to which it is adapted—afterwards add to one part of this oil two parts of fresh aqua fortis, and when it is enclosed in glass of the best quality, set it in horse-dung for a month. After that, distil it entirely with a slow fire, that the matter may be condensed at the bottom. And if the aqua fortis which ascends be distilled by a bath in this manner, you will find two elements together. But the same elements will not be left by all metals alike. For from gold there remain in the bath earth and water ; but air is in all the other three, and the element of fire remains at the bottom, because the substance and the tangibility of gold have been coagulated by the fire ; therefore, the substance will agree in its substantiality. From the Moon there will remain at the bottom the element of water, and in the bath the elements of earth and fire.   For from the cold and the moisture is produced the substance and corporality of the Moon, which is, indeed, of a fixed nature, and cannot be elevated.   From Mercury there remains fire at the bottom, and earth and water are elevated upwards.   From Venus there also remains fire, and both, that is to say, earth and water, remain in the bath.   From Saturn there remains the element of earth at the bottom, while fire and water are held in the bath.   From Jupiter air remains at the bottom, while fire, water, and earth are elevated therefrom.

So it must be noticed that in the case of Jupiter the air supplies a body, and in the case of no other metal.   And of this, although some part ascends together with it and remains mixed inseparably with the other three elements, still it is not corporeal air, but adheres to, and concurs with, the others, and is inseparable from them.

And now, it must be remarked that the residuum, that is, the corporeal element which remained at the bottom, must be reduced into an oil by means of the bath with fresh aqua fortis.   So, this element will be perfected, and you will keep it for one part.   The rest you will separate by means of a bath in this way.   Place them in sand, and press them gently.   Then, first of all, the water will be elevated, and will escape ; afterwards the fire, for it is known by the colour when these two remain.   But, if the elements of earth and water should have remained, the water will ascend first, and afterwards the earth. But if it should be earth and fire, the earth is elevated first and the fire after- wards.   If water, fire, and earth be together, the water will first ascend, then afterwards the fire, and last of all the earth.   These elements can be so kept in their respective glasses, each according to its own nature ; as, for example, from the Sun, the warm and the dry, without any other property ; in like manner the cold and the moist, and the cold and the dry.   So, also, must it be understood of the others.   It must not be forgotten that the corrosive nature of the aquafortis must be extracted as we have handed down in our book on the Quintessence.

Concerning the Separations of the Elements out of Marchasites

Having previously set down the separations of the elements out of metals, it remains that we come to those which can be produced out of marchasites, and that we shew what they are.

Take of marchasite, in whatever form you please, whether bismuth or talc, or cobalt, granite, or any other, one pound ; of salt nitre the same. Beat them very small, and, burning them together, distil by means of an alembic without a cucurbite, and keep whatever liquid ascends.   But that which remains at the bottom, let this, when ground down, be resolved into a water with aqua fortis.   Hereupon pour the water previously collected, and distil it into an oil, as before directed in the case of metals.   By the same process, too, you shall separate the elements.   So, the golden marchasite is to be understood as gold, the silver as the Moon, bismuth as lead, zinc as copper, talc as Jupiter, cobalt as iron.   Let these directions suffice for the separation of marchasites in every kind.

Concerning the Separation of the Elements from Stones

The separation of the elements of stones or gems comes to be understood in the following way.   Take a stone well ground, to which add twice the quantity of live sulphur, and, when all is well mixed, put it in a luted pot into an Athanor for four hours, so that the sulphur shall be entirely consumed.   Afterwards, let what remains be washed from the dregs and the sulphur, and dried.   Let the stony calx be also put into aqua fortis and proceeded with as we have already laid down concerning the metals.   Stones, too, are compared with metals ; as, for instance, clear gems which are not white or tawny are compared with gold.   White, cœrulean, or grey, with silver or the Moon ; and afterwards the commoner stones with the other metals, as alabaster with Saturn, marble with iron, flint with Jupiter ; but dulech with Mercury.

Concerning the Separation of the Elements from Oleaginous Substances

All oils, woods, roots, seeds, fruits, and similar things, which have a combustible nature, and one fit for burning, are considered oleaginous ; and the separation of them is twofold, namely, that of the oleaginous substances and that of the pure oils.

The Separation of the Oleaginous Substances is as follows :

Take such a body, pounded, ground, or reduced to fragments in whatever way you can, wrap it up in linen, fasten it, and place it in horse-dung until it shall be entirely putrefied, which happens sooner in one case than in another. When it is putrefied let it be placed in a cucurbite, and on it let there be poured so much common hot water as may exceed four fingers broad ; then let there be distilled in sand all that can ascend.   For all the elements ascend except the earth itself, which you will know by the colours ; nevertheless, let the hot water first ascend, afterwards the air, next the water, lastly the fire, and the earth will remain at the bottom.   Of the pure oils, however, it must be understood that these do not require putrefactions, but they must be distilled alone and without additions.   Afterwards their elements, as it has been said above of others, must be separated, and these are discerned by their own colours. None otherwise with resins of liquid substance must it be done, such as pitch, resin, turpentine, gum, and the like.   But the corporeal resins which exist, such as sulphur, must be prepared in the following manner.

Concerning the Separation of the Elements in Corporeal Resins

Take sulphur very minutely ground ; let this be cooked to hepatic sulphur in a double quantity of linseed oil ; let it be shut up in a vessel and place it to putrefy in horse-dung for a space of four weeks.   Afterwards let it be distilled in an alembic slowly over a naked fire.   The air and the water first ascend, with different and pale colours.   Then, the heat being increased, the fire ascends and the earth remains at the bottom.   The colours appear pure ; the air yellow ; the water like thick milk, so much so that it can scarcely be distinguished from milk ; the fire like a burning ruby, with transparency and with all the fiery signs ; but the earth is altogether black and burnt.   The four elements having been thus separated, every one is perfect in its own elemental complexion, and without any admixture, as has been said above.

Concerning the Separation of the Elements from Herbs

So, too, in herbs, the element of water is principally contained when they are cold ; but if they are airy, then that element predominates.   In like manner must it be understood of fire.   The separation of those elements is as follows :

Take sage, and bruise its leaves.   After this place it for putrefaction, as aforesaid.   Then you will distil it by means of a *venter equinus*, and the element of fire will ascend first so long as the colours are unchanged and the thickness of the water.   Afterwards the earth will succeed, and some part of it will remain at the bottom, which part, indeed, is fixed.   Distil this water in the sun six days, and afterwards place it in a bath.   Then the element of the water will ascend first.   It is very minute, and is distinguished by the taste.   After the colour changes the element of fire ascends, until the taste, too, is altered.   Then at last a part of the earth is elevated, yet it is a very small portion, which, being mixed with air, is found at the bottom.   In the same way it must be understood of airy and watery herbs, of which the air ascends first, afterwards the water, and lastly follows the fire, according to the process laid down concerning sage.

Concerning the Separation of the Elements from Fleshly Substances

The separation of the elements from fleshly bodies, and from those which live with blood, comes to be understood thus, because the predominating element in them is more copious and is generally found last of all ; as, for instance, water in fishes, fire in worms, and air in edible flesh are the principal elements, as we describe in our treatise on the generation of animals.*

The Separation of the Elements from Fishes is as follows :

Putrefy the fishes perfectly.   Then distil by means of the *venter equinus*, and a good deal of water ascends.   You will renew this putrefaction and distillation, and increase it until no more water rises.   Afterwards distil what remains in sand ; then at length the fire ascends in the form of oil, but the

---

* This treatise must also be included in the long catalogue of the missing works of Paracelsus. But many refer-ences to the generations of animals will be found in the present translation.

earth remains at the bottom. Thus the whole substance of the fishes is separated into its elements. One need take no account of the fats and the marrows, but must suppose that everything is separated by the putrefaction, and divided into its elements. In the same way is it to be understood of worms, except that there comes forth from them not only water but more fire, unless they are aquatic worms, as serpents; in the distillation of which more wonderful things occur than it is possible to say. Of edible animals, too, it must be understood in the same way, of such, that is, as respectively disclose their elements by separation.

Concerning the Separation of the Elements from Watery Substances

For the separation into their elements of juicy and watery bodies, and of those which have the form of wateriness, as urine, dung, water, and the like, note the following process :—

Take urine and thoroughly distil it. Water, air, and earth will ascend together, but the fire remains at the bottom. Afterwards mix all together and distil again four times after this manner; and at the fourth distillation the water will ascend first, then the air and the fire, but the earth remains at the bottom. Then take the air and the fire in a separate vessel, which put in a cold place, and there will be congealed certain icicles, which are the element of fire. Although this congelation will take place in the course of distillation, still it will do so more readily in the cold.

Concerning the Separation of the Elements out of Water

Make the water boil by means of a dung-heap, and the earth itself sinks to the bottom. Putrefy at the proper time that which ascends, and let it afterwards be distilled by a bath. Then the water will ascend first, and afterwards the fire. Dung, vitriol, tartar, and similar juices, as alum, salts, and other substances of that kind, are to be distilled in ashes, with such an amount of heat and for so long a time, until they cease to rise, and the water and air have ascended, while the earth has remained at the bottom. Afterwards, by means of the heat, the fire will ascend. And in this place it is to be remarked that although the four elements have been separated, there remain still in the earth four occult elements, as though fixed. From vitriol remains a *caput mortuum*, which sublimate with sal ammoniac, and there will issue forth an oil, in which are water and fire, and the earth itself remains in substance. Separate those things which have ascended, and again there will ascend water, while the fire will remain at the bottom. So also must it be understood of tartar and of salts. And although there are in existence additional separations of liquids, yet we shall discuss them more amply when speaking of Transmutations. It must, however, be remembered that there are more elements in a corrosive earth than in ashes. Therefore, the separation must be made by sublimation, as we shall shew.

Concerning the Separation of the Elements from Glasses and those Substances which are of the Nature of Glass

As we lay down above concerning the resolutions of marchasites, so is it to be in like manner understood in this place concerning the glasses : namely, the principal consideration is that they are calcined with sulphur as stones are, then washed away with saltpetre and aqua fortis, and furthermore, as we have before made clear. Their elements also are recognised by the colours in the distillations, not as they shew themselves to the eye ; and this is the sum and substance of what we have to say concerning them.

Concerning the Separation of the Elements in Fixed Substances

The separation of the elements in fixed substances is brought about by sublimation, as we teach concerning salts and liquids ; with this difference, however, that these are to be calcined with salt nitre, and afterwards to be sublimated. And although there are many other things which are not set down in this place, nevertheless it is to be understood that the separations of all substances should be made in the ten ways already described. Furthermore, concerning the separation of the four elements, it is to be remarked that each of them can be again separated ; for example, fire as fire, air as air, water as water, earth as earth, as follows hereafter concerning the respective separations of them.

Concerning the Separation of Fire

It should be known that from the element of fire, the four elements may be separated in this way. When the fire is burning most violently, or ascends, take it in a receptacle or vessel perfectly closed, and place it in horse-dung for a month. Then you will find in that one element four elements, which, when you have opened the vessel, put into a receiver. Thus the vapour or air will mount into the vessel that receives it. Afterwards, distil that which remains by means of a bath, and the water will thus ascend. Next, by means of ashes, the fire will ascend and the earth will remain at the bottom. What is the force of these elements, and why are they described in this place, we will make more clear in other books.

Concerning the Separation of Air

Having received the element of air in a perfect glass vessel, and hermetically sealed it, you must expose and direct it to the sun for the whole of the summer. By circulating, the air is turned into moisture, which increases daily more and more. This quantity you will separate after the following manner. Let it putrefy in horse-dung for four weeks, and afterwards distil it by the bath, like fire. Concerning its potency more is said in another place.

Concerning the Separation of Water

Having filled a glass brimful, leaving no space empty, seal the vessel hermetically, and place it in a warm sun for a month, so that it may receive a daily and equal heat, and would boil, but it cannot on account of the vessel being full. When the time has elapsed, putrefy it for four weeks. Then open it, and distil it by means of an alembic with four necks. In this way the three elements are separated, and in the bottom will remain the earth of that water. The nature of this is said to possess much virtue in many cases.

Concerning the Separation of the Earth

The same process is to be observed with the earth as with the water, save only in the distillation. For this is like that which takes place with fire and is accomplished in the same way. This separation of the elements we have inserted at this point for several causes, because it is very useful, not only in philosophy, but also in medicine. Concerning the separations of the elements, we have thus far written with sufficient fulness. Though much more might be added, it does not appear to be by any means necessary.

Now, we will make clear the separation of the pure from the impure, according to the purpose of our design. This, indeed, is done in the same way as we teach with regard to Arcana and Aurum Potabile; so it need not be put forward here, though that process from its origin is not altogether identical with that which is laid down concerning Arcana and Magisteries. Nevertheless, I do finally assume the same way by the separation of the elements; since in this place those elements are separated after each one of them has been purged from the impurities existing therein, so that no deformity or impediment may arise from them, as might otherwise easily happen.

The End of the Third Book of the Archidoxies concerning the Separations of the Elements

THE FOURTH BOOK OF THE ARCHIDOXIES

From the Theophrastia of Paracelsus the Great

Concerning the Quintessence *

WE have before made mention of the quintessence which is in all things. Already, at the beginning of this treatise, it must be understood what this is. The quintessence, then, is a certain matter extracted from all things which Nature has produced, and from everything which has life corporeally in itself, a matter most subtly purged of all impurities and mortality, and separated from all the elements. From this it is evident that the quintessence is, so to say, a nature, a force, a virtue, and a medicine, once, indeed, shut up within things, but now free from any domicile and from all outward incorporation. The same is also the colour, the life, the properties of things. It is a spirit like the spirit of life, but with this difference, that the life-spirit of a thing is permanent, but that of man is mortal. Whence it may be inferred that the quintessence cannot be extracted from the flesh or the blood of man : for this reason, that the spirit of life, which is also the spirit of virtues, dies, and life exists in the soul, not in the material substance.

---

* *The Correction of the Quintessence.* The books which have been written by so many previous authors concerning the Quintessence, such as those of Arnoldus de Villa Nova and of Johannes de Rupescissa, whence afterwards, under a pretentious title, was composed the *Cœlum Philosophorum* [not to be identified with the work of Paracelsus which occupies the first place in this translation], contain nothing of any value. The mere fact that these writings embody a singular and new praxis abundantly demonstrates that their authors have misunderstood the essential nature of diseases, seeing that they have, as it were, devised one form of all diseases, regarding which they have, moreover, invented many marvellous things, adorning their conceptions with monstrous titles, all mere boasting, wherein there is no mention of philosophy, medicine, or astronomy. They are all a mere deluge of absurdities and lies. There can be no doubt that originally most admirable discoveries have been transferred from the chemical art to that of medicine, but the same have been since adulterated by sophistry. For certain people, when they have investigated a chemical preparation, wish to vary it immediately in hundreds of different ways, and thus the truth is foolishly confounded with lies. Now, it should be observed that the severest rebuke which can be given to such impostures is that of paying no attention to promises and to proud titles, and of believing only to that extent which is warranted by good sense and experience. Remedies which require more knowledge of a practical kind than has ever been possessed by monastic pseudo-chemists may, in careless hands, give rise to the most malignant diseases. It may be observed, for example, that the preparations of Mercury which are used against Luis Gallica and, indeed, all remedies adopted in the cure of this disease, cannot be properly prepared without great skill in chemistry. Pretenders in pharmacy will vainly vaunt such decoctions unless they can compose vitriolated salts, alums, and things similar, by purely chemical artifice, seeing that there are recondite secrets in the remedies which conduce to the cure of Luis Gallica. Yet there is no need to write a new correction ; it is enough to adhere to the legitimate mode of preparation, giving no faith to the hollow pretences of alchemists. These men certainly promise more than they are ever able to perform ; but this is common to alchemy and not a few other callings, namely, that their professors boast of the harvest before they have finished the sowing.—*Chirurgia Magna, De Imposturis in Morbo Gallico*, Lib. II., c. 13.

For the same reason, animals, too, because they lose their life-spirit, and on that account are altogether mortal, also exhibit no quintessence. For the quintessence is the spirit of a thing, which, indeed, cannot be extracted from things endowed with sensation as it can from those not so endowed. Balm has in itself the spirit of life, which exists as its virtue, as a force, and as a medicine ; and although it be separated from its root, nevertheless, the life and virtues are still in it, and for this cause, that its predestination was fixed. Wherefore the quintessence can be extracted from this, and can be preserved, with its life, without corruption, as being something eternal according to its predestination. If we could in this manner extract the life of our heart without corruption, we should be able to live without doubt, and without the perception of death and diseases. But this cannot be the case ; so from this circumstance death must be looked for by us.

When, therefore, the quintessence of things exists as a virtue, the first thing we have to say is in what form this virtue and medicine are in things, after the following manner. Wine contains in itself a great quintessence, by which it has wonderful effects, as is clear. Gall infused into water renders the whole bitter, though the gall is exceeded a hundredfold in quantity by the water. So the very smallest quantity of saffron tinges a vast body of water, and yet the whole of it is not saffron. Thus, in like manner, must it be laid down with regard to the quintessence, that its quantity is small in wood, in herbs, in stones, and other similar things, lurking there like a guest. The rest is pure natural body, of which we have written in our book on the Separations of the Elements. Nor must it be supposed that the quintessence exists as a fifth element beyond the other four, itself being an element. It is possible that someone may think this essence would be temperate, not cold, not warm, not moist or dry, but this is not the nature of its existence. For there is nothing which exists in this temperature by which it is alienated altogether from the other elements ; but all quintessences have a nature corresponding with the elements. The quintessence of gold corresponds with fire ; that of the Moon with water ; that of Saturn with earth ; and that of Mercury with air.

Now the fact that the quintessence cures all diseases does not arise from temperature, but from an innate property, namely, its great cleanliness and purity, by which, after a wonderful manner, it alters the body into its own purity, and entirely changes it. For as a spot or film, by which it was formerly blinded, is removed from the eye, so also the quintessence purifies the life for man. All natures are not necessarily of one and the same essence one with the other. Nor do those which are fiery on that account manifest the same operations by reason of their complexion ; as, for instance, if anyone should think that the quintessence of Anacardus should have a like and identical operation with the quintessence of gold, because both are fiery, he would be greatly misled, since the predestination and the disposition make a difference of properties. For as every animal contains within itself the life-spirit, yet the same virtue does not exist in each, simply because they all consist of flesh and blood, but one differs from another, as in taste or in virtue, so is it with the quintessence, which does not acquire its virtue from the elements by a simple intellectual process, but from a property existing in the elements, as we lay down in our book on the Generation of Things. Thence it happens that some quintessences are styptic, others narcotic, others attractive, others again somniferous, bitter, sweet, sharp, stupefactive, and some able to renew the body to youth, others to preserve it in health, some purgative, others causing constipation, and so on. Their virtues are innumerable, and although they are not exhausted here, yet they ought to be thoroughly known by physicians.

When, therefore, the quintessence is separated from that which is not the quintessence, as the soul from its body, and itself is taken into the body, what infirmity is able to withstand this so noble, pure, and powerful nature, or to take away our life, save death, which being predestined separates our soul and body, as we teach in our treatise on Life and Death? In this place it should be equally remarked that each disease requires its own special quintessence, though we tell of some which are adapted to all diseases. How this comes about shall be explained in its proper place.

Furthermore, we bear witness that the quintessence of gold exists in very small quantity, and what remains is a leprous body wherein is no sweetness or sourness, and no virtue or power remains save a mixture of the four elements. And this secret ought by no means to escape us, namely, that the elements of themselves, without the quintessence, cannot resist any disease, but are able to effect only this and nothing more, that is, to produce heat or cold without any force : so that if a disease be hot, it is expelled by cold, but not by that kind of cold which is destitute of force, or things made frigid with snow ; since, though, these are sufficiently cold, yet there is no quintessence contained in them by the power whereof the disease might be expelled. Wherefore the body of gold is powerless of itself ; but its quintessence alone, existing in that body, and also in its elements, supplies the forces hidden therein. So, likewise, in all other things, it is their quintessence alone that cures, heals, and tinges the whole body, just as salt is the best seasoner of any food. The quintessence, therefore, is that which gives colour, whatever it be, and virtue ; and gold when it has lost its colour, at the same time lacks its quintessence. None otherwise must it be understood of metals from which when the colour is taken, they are deprived of their special nature.

In like manner is it with stones and gems ; as, for instance, the quintessence of corals is a certain fatness with a red tint, while the body of them is white. The quintessence of the emerald, too, is a green juice, and the body thereof is also white. None otherwise must we judge of all other stones, namely, that they lose their nature, essence, and qualities when they lose their colours, as we particularly shew in our book on the extractions of them.

The same, also, should be understood of herbs, plants, and other growing products. So, too, of flesh and blood, from which no quintessence can be extracted, for reasons already laid down. But, nevertheless, a certain resemblance to a quintessence can be extracted from them by us in the following manner. A morsel of flesh still retains life in itself, because the flesh is yet supplied with all its nature and force. Wherefore there is life in it, which, however, is not true life, but still the life is preserved until putrefaction sets in. So this difference must be noted, whereby dried herbs and the like are to be looked upon in the same way as flesh. That green spirit which is their life has gone from them. Dead things, therefore, can be taken for a dead quintessence, as flesh is able to put forth all its powers from itself, though specially separated as to one part from the body. So, too, with blood and with dried herbs. These, indeed, though they are not living quintessences, none the less demonstrate how a dead quintessence displays some virtues. But metals and stones have in them a perpetual life and essence, and they do not die, but so long as they are metals or stones, so long does their life last. Therefore, they also exhibit perfect quintessences which, in like manner, can be extracted from them.

And now we must see by what method the quintessence is to be extracted. There are many ways indeed: some by additions, as the spirit of wine; others by balsamites; some by separations of the elements, and many other processes which we do not here particularise.

But by whatsoever method it takes place, the quintessence should not be extracted by the mixture or the addition of incongruous matters; but the element of the quintessence must be extracted from a separated body, and, in like manner, by that separated body which is extracted. Different methods are found by which the quintessence may be extracted, for instance, by sublimation, by calcination, by strong waters, by corrosives, by sweet things, by sour, and so on, in whatever way it may be possible. And here this is to be taken care of, that everything which shall have been mixed with the quintessence by the necessity of extraction, must again be drawn off from it, so that the quintessence may remain alone, unpolluted, and unmixed with any other things. For it cannot be that the quintessence shall be extracted from metals, more particularly from gold, which cannot be overcome by itself, without exhibiting some appropriate corrosive, which can afterwards again be separated from it. In this way, salt, which was water, is again extracted from water, so that this water is free from salt. And here this consideration comes in, that it is not every corrosive which is adapted for this purpose, because they cannot all of them be separated; for if vitriol or alum be mixed with the water, neither of these can be separated from it afterwards without loss or corruption, so that they leave behind them a sharp residuum, for this reason, because each of them is watery, and thus two similar things meet, which ought not to occur in this process. Care must therefore be taken that a watery body be not taken for a watery, or an oily body for an oily, or a resinous body for a resinous one; but a contrary ought in every case to separate the quintessence and to extract it, as waters extract the quintessence of oily bodies (which is explained in the case of the metals), and oily substances the quintessence of watery bodies, as we may learn concerning the quintessences of herbs. The corrosives, therefore, after the separation and extraction of the quintessence, must be again separated, and this will be easily done. For oil and water are readily separated, but not so oil from oil; nor, in like manner, can water be separated from water without admixture; and if this is left it may cause great damage to the quintessence. For the quintessence ought to be clear and spotless, and collected without the admixture of anything, that it may possess a uniform substance by means of which it can penetrate the whole body. In truth its subtlety and force cannot be probed to the foundation, no more than its origin whence it first proceeded can be fully known. It has many grades: one against fevers, as in the case of opiates; another against dropsy, as the essence of tartar; one against apoplexy, as that of gold; one against epilepsy, as that of vitriol. The number of these is infinite, and incapable of being proved by experiment. Wherefore the greatest care and diligence should be shewn that to every disease its true enemy may be assigned. In this way Nature will afford help beyond belief, as will be made more clear in what follows. We cannot speak of the grades in the same way as grades are applied to simples in medicine, for this reason that there is no possible comparison between the grades of a quintessence and the grades of simples, nor ought the comparison to be attempted; but when such gradation is made it is found that the excellence and virtue of one is greater than of another, but not the complexion. For it must not be set down that the quintessence of Anthos is hotter than the quintessence of lavender, or the quintessence of Venus is drier than that of the Moon; but the grade of anything should be determined by its great and more excellent virtues, namely, after this manner. The quintessence of antimony cures leprosy, and the quintessence of corals cures spasms and contortions. In order to learn, then, which of these occupies the better and higher grade, there can be no other judgment than this, namely, that the quintessence of antimony is higher and more potent, inasmuch as leprosy is a more severe disease than colic and its belongings. According to the property, therefore, which it has against different diseases its grades are considered. This is, moreover, the case in one and the same disease; for one essence is more powerful than another for curing leprosy. The quintessence of juniper expels it, and the quintessence of amber, the quintessence of antimony, and the quintessence of gold. Now, although there are these four essences in all which cure leprosy, they do it with a difference, since with regard to the cure thereof they do not occupy the same grade. For the essence of juniper drives away this disease by the extreme purgation and purification which it introduces into the blood, and so consumes the poison that it is not so perceptible. Hence it is reckoned in the first grade of that cure. The quintessence of amber also takes away the poison and more. It purifies the lungs, the heart, and the members subject to leprosy; wherefore, the second grade is assigned to it. The quintessence of antimony, beyond both the virtues already spoken of, also clears the skin, and sharpens and renovates the whole body in a wonderful manner; so, then, it holds the third rank. But the quintessence of the Sun by itself fulfils each of these tasks, and then takes away from the roots all the symptoms of leprosy, and renovates the body as honey and wax are purged and purified by their honeycomb. For this cause it occupies the fourth rank.

In this way the grades of the quintessence can be learnt, and the one distinguished from the other, that is to say, which of them is higher or more excellent than another. Even simples should be known by their properties. For whatever be the property in a simple form, such is the property of their quintessence, not more sluggish, but much stronger and more excellent.

But now, moreover, let us learn the differences of quintessences, for some are of great service to the liver by resisting all its diseases, some to the head, others to the reins, some to the lungs, and some to the spleen, and so on. So, too, some operate only on the blood, others on the phlegm, others in melancholy only, others only in cholera, while some others have effect only on the humours, some on the life-spirit, some on the nutritive spirit; some operate on the bones, some others on the flesh, some on the marrow, others on the cartilages, some on the arteries; and there are others which have effect only against certain diseases and against no others, as against paralysis, the falling sickness, contractions, fluxes, dropsy, and so on. Some, also, are found to be narcotics, others anodynes, some soporific, some attractive, purgative, cleansing, flesh-making, strengthening, regenerative, and some stupefying, and the rest.

Some, too, are found which renovate and restore, that is, they transmute the body, the blood, and the flesh. Some are for preserving the continuance of life, some for retaining and preserving youth; some by means of transmutation, others by quickening. Moreover, this must be understood, that some have a specific form, others an appropriated, others an influential, and others a natural form. In a word, more of their virtues exist than we are able to describe, and their effects in medicine are most wonderful and inscrutable. In different ways it happens that some quintessences render a man who is a hundred years old like one in his twentieth year, and this by their own strength and potency. But what man is able to trace the origin of so great a mystery, or to ascertain from whence the first materials naturally take their rise? It belongs to our Supreme Creator to make these things so, or to forbear. For who shall teach us, that we may know by what powers the quintessence of antimony throws off the old hair and makes new hair grow; why the quintessence of balm destroys the teeth, eradicates the nails of the hands and the feet, and restores new ones; the quintessence of rebis strips off and renews the skin; and the quintessence of celandine changes the body and renovates it for the better, as colours renovate a picture?

There are far more matters than these, which here we omit, and reserve for making clear in their proper chapters.

How, then, at length could it be that we should relinquish that noble philosophy and medicine, when Nature affords us such wonderful experiments in and from them? Of these all the other faculties are destitute, and so take their position nakedly in mere cavil. Why should not this fact delight us that the quintessence of the carline thistle takes force away from one and affords it to another who uses it; also that the quintessence of gold turns inside out the whole leper, washes him as an intestine is washed in the shambles, and likewise polishes the skin and the scabs and makes a new skin, loosens the organs of the voice, takes away the whole leprous complexion, and makes him as if he had recently been born of his mother.

Wherefore we will turn our mind to the making of such quintessences, pointing out the way for the extraction or composition of them--one for the metals, another for marchasites, another for salts, another for stones and gems, another for burning things, another for growing things, another for spices, another for eatable and drinkable things, all of which, with their belongings, we will endeavour to make clear in a proper series. But it must be noted, in the practice with quintessences, a good knowledge of theory and of natural science is required; theoretically, that is, of the properties of things with regard to natural diseases. One must not be ignorant that there is a difference between a quintessence, aurum potabile, arcana, magisteries, and other things of that kind; such as this, for example, that a quintessence cannot be again reduced to its own body, but aurum potabile is easily transmuted again to its metallic body; so that far more noble virtues exist in a quintessence than in the other things.*

---

* The distinction which subsists between the quintessence of gold and potable gold is illustrated by the following citation which, though it occurs in an independent treatise, offers only some small variations from a passage in the *Chirurgia Magna* which has been already given in the note on page 76 of the first volume :—We have already said that it is scarcely possible for contractions to be cured in any other way than by medicines existing in the supreme grades, such as is potable gold and the like, concerning which we have treated in the larger grades, whence great care must be taken concerning them, as is stated in the Book about the Quintessence. It is called potable gold so often as it is reduced, together with other spirits and liquors, into a substance which can be drunk. The oil of gold is a golden oil made out of the substance only, saving addition. It is called quintessence of gold when a reddish tincture is extracted therefrom and separated from its body. A virtue, or at least an active force, exists in the tincture. The dose of potable gold is Э i., when needed. The dose of golden oil should not exceed ten grains of barley in weight. That of the quintessence should not exceed three similar grains in good water of life, or some other water of equal subtlety. It should be taken morning, noon, and evening, according to the requirements of the medicine, and without the addition of corrosives or corruptives, which can neither alter its nature nor be mixed therewith.—*Description of Potable Gold.* Take of potable gold, pulverized and dissolved in salt, ℥ i., with a sufficient quantity of distilled vinegar. Perform successive separations upon the whole by means of distillation till nothing of the acquired savour shall remain. Then take water of life ℥ v., pour them into a pelican, and digest together for a month, when you will have perfect potable gold, the practice of which you must learn from our book on the quintessence. Though it has not been described in glowing colours there is no equal medicine found in this age.—*Description of the Water of Life.* Take 10lbs. of *vinum ardens;* of roses, balm, rosemary, anthos, cheirus, both species of hellebore, marjoram, ana m. j. ; of cinnamon, mace, nutmegs, garyophylli, grains of paradise, all peppers, cubebæ, ana ℥ ii. ; of the sap of chelidonia, tapsus, balm, ana lib. ss. ; of bean ashes ℥ v. Let all these be mixed and digested together in a pelican for twelve days ; then separate, and use for above process.—*Description of the Oil of the Sun.* After the sap has been separated from the gold in the way previously stated, let it boil for fifteen days in the digestive compound which follows. Let it afterwards be separated by the bath, when a thick oil will remain at the bottom, and this is unalloyed gold, which use as above.—*Formula of the Quintessence.* Take as much as you please of gold which has been repurged by royal cement or antimony. Remove its metallic quality, or malleation, by means of the water of salt. Wash away the residue with sweet water. Extract its tincture with spirit of wine. Lastly, elevate the spirit from it

While we thus speak of the quintessence, the difference of one from another should be learnt, and then also what it is in itself. And although we have already sufficiently explained it, nevertheless, practice calls us another way by which also the condition and nature of the quintessence can be found out. For although these do not appear in the form of the quintessence, nor are they produced in the same way, nor do they consist of one element only as the quintessence should, still, none the less, we should judge of the quintessence of these things, which is of more importance than that they should be called a quintessence. It should rather be spoken of as a certain secret and mystery, concerning which more should be written than we have written concerning the quintessence. But since we have made that clear in the books of the Paramirum, we pass it by in this place.* The number of the arcana and of the arts of the mysteries is infinite and inscrutable, and many methods in them are met with worthy the attention of the clearest human intellects. Among these arcana, nevertheless, we here put forward four. Of these arcana the first is the mercury of life, the second is the primal matter, the third is the Philosophers' Stone, and the fourth the tincture. But although these arcana are rather angelical than human to speak of, nevertheless, we shall not shrink from them, but rather we will endeavour to trace out the ways of Nature, and we will arrange that everything which proceeds from Nature

---

and the quintessence will remain at the bottom.—*Construction of the Water of Salt.* Take by itself the purest and whitest pounded salt, which is produced by nature without decoction, boiling, or any of those processes by which salt is usually made. Liquefy it several times, pound it very fine, mix it with the juice of raphanum roots, dissolve and distil it, and when a reddish green appears again distil it five times, combining it in equal weights. Dissolve laminated Sol in this liquor till it becomes powder. Let this powder be washed in most limpid water, and distil a sufficient number of times till the salt shall depart from it, which will take place soon, as it does not penetrate the interior nature. When the corrosive has been removed the gold will be found by itself.—*Extraction of the Spirit of Wine.* Take one measure of the best natural wine, red in preference to white, place in a capacious circulatory vessel for its better rotation, seal it, plunge it into the sea-bath, and let it boil for forty days. Afterwards pour it into a cucurbite, distil by the cold way till the spirit shall have gone and all signs of it shall cease. So finish. (The process varies from that of the *Chirurgia Magna.*) *After the sign has been given cease. What follows is the water of life, not spirit; either is efficacious. Then the sign is double, one of the spirit and one of the water of life.* Pour out this spirit of wine to the dregs, and in such a way that it floats over the surface to the height of six broad fingers. Close all openings completely with glass. Digest for thirty days, during which time the tincture enters the spirit. At the bottom there remains a white powder. Separate by art, and let the powder melt. Hence there is produced an aqueous metal or metallic water. Let the spirit evaporate, as we are taught by alchemy, and a sap resembling liquor remains at the bottom. Graduate this four times in a retort adapted to the quantity of the matter. This is performed by elevation, for it renders such like substances subtle, though it permits not the same to be developed beyond the fifth essence.—*De Contracturis,* Tract II., c. 2. Among other uses of potable gold, it is a tonic for the heart which is so efficacious that it is affirmed by Paracelsus to prevent all injury befalling that organ. Of like virtue is the liquor of gold and the substance of pearls reduced into the form of an oil and balsam. After these are enumerated the essence from the crocus of Mars and corals. The description of the potable gold in this connection is given as follows : Let the gold be calcined into yellowness by the royal cement of *Hell and Malch.* Then let it be separated from its impurity, and afterwards mixed with circulated water. Digest for twenty-four hours in a moderate fire, when the oil will flow forth and will float upon the surface of the water. Collect and drink it mixed with water of life. Proceed in the same way with pearls, only adding calx of chelidonia, and confine by means of distilled vinegar until they pass off into liquor. Proceed also in like manner with corals, dissolving them in vinum ardens mixed with *Hell.* Remove the vinum ardens from the putrefaction and you will have the liquor of coral. There are also other essences, as of crocus, chelidonia, mace, cesium, balm, etc., which are suitable for other complaints and affections of the heart, and some of which will fill the old, infirm, melancholy, and depressed with the greatest joy.—*De Viribus Membrorum,* Lib. II., c. 2.

* The reference here made is too obscure for verification. In the large literature of the Paramirum and its connected treatises, there are many observations upon arcana and mysteries, though few passages deal actually with the quintessence. Whatsoever is of importance has been embodied in this translation as notes to the text. That section of the present volume which is devoted to Alchemy as the fourth column of medicine may also be consulted in this connection, as it is a treatise derived from the Paramirum.

shall be capable of being naturally understood. Concerning the Mercury of Life, therefore, we profess that it is not a quintessence, but an arcanum, because there exist in it may virtues and forces which preserve, restore, and regenerate, as we write in our book on the Arcana.

In like manner, also, the Primal Matter, not only in living things, but also in dead bodies, operates more in the same manner than is thought to be naturally possible. So also the Philosophers' Stone acts, which, tinging the body, frees it from all diseases, so that even metals are purged from their impurities. In like manner the Tincture acts, which, as though it should change the Moon into the Sun, so also changes disease into health. The same things equally the other magisteries and elixirs do, and the aurum potabile, all of which are treated of severally in their own respective books.

Concerning the Extraction of the Quintessence from Metals

We will, then, briefly go through the extraction of the quintessences from metals. For, in our times, many persons have made numerous experiments with these, and copious results followed which obliged them to enter on other different ways. With regard to metals, then, it must be understood that they are divided into two parts, namely, into their quintessence and into their body. Both are liquid and potable, and do not mix, but the impure body ejects the quintessence to its surface, like the cream from the milk. In this way two fatnesses or viscous liquids are formed out of the metals, and these liquids have to be separated. The fatness of the body is always white in all metals ; but their quintessence is coloured, as we have before explained concerning the seven metals. All, moreover, have the same process, which is as follows :—

Let the metal be dissolved in water, and afterwards this solution distilled by a bath and drawn off. Let it be putrefied until it is reduced to an oil. Let this oil be distilled from small phials or cucurbites by means of an alembic, and one part of the metal will remain at the bottom. Let this be reduced to an oil as before, and be distilled, until all the metal shall have ascended. Afterwards let it be again putrefied for a month, and at length again distilled with a slow fire. Then the vapours will at first ascend and afterwards fall into the receiver. These vapours remove ; and there will ascend two obscure colours, one white, but the other according to the nature and condition of the metal. When they have ascended altogether they become separated in the receiver, so that the quintessence remains at the bottom and the white colour of the body floats at the top. Separate these two by means of a tritorium, and in another phial receive the quintessence, into which pour purified ardent wine and let it remain until the wine is completely acidulated. Afterwards let it be strained or separated from the quintessence, and let more fresh wine be poured on. Do this until you no longer perceive any sharpness. At last, pour on doubly distilled water, so that it may be washed and brought to its proper sweetness, and so keep it. In this way the quintessence of metals is prepared. But if you reduce the white portion, you will have therefrom a malleable, white, and metallic body of which it cannot be known under what species it is embraced. There are many other ways found out for extracting the quintessence. About these we keep silence, for the reason that they are not considered by us to be true extractions of quintessences, but only transmutations in which no extraction is produced or comes to be used.

Concerning the Extraction of the Quintessence from Marchasites

In marchasites also are found various methods for the extraction of their quintessence, and yet we scarcely judge these to be true quintessences. And although they are of greater virtue than their quintessence, as we teach of Arcana, Magisteries, and Elixirs, nevertheless, this our mode and manner of extracting the quintessence from all the metallic marchasites is like the true extractions of the metals. But the following is the reason why we before said that the quintessence is the supreme virtue of things, and now, on the contrary, do here say of Arcana that they are greater than the quintessences themselves : because all the Arcana contain in themselves the quintessences and, moreover, are reduced to such subtlety and acuity that they hence receive a far greater virtue than the quintessence does. The same is laid down, too, concerning their appropriate and specific quality. But the process for extracting the quintessence from marchasites is as follows :—

Take a pound of a marchasite very finely ground, of the eating water two pounds, mix them together in a pelican, let them remain in process of digestion for two or three months, and be reduced to a liquid. Distil this entirely with the fire and it will pass over into an oil which you will putrefy in the dungheap for a month. Afterwards distil it like metals, and in the same way two colours will ascend therefrom, one white, and the other the true colour of the quintessence. Leave the white, unless it be from bismuth or from a white marchasite, in which case you shall know one from the other by the density. Take the lower one, and reduce it to its sweetness as was said above concerning the metals. In this way you have extracted the quintessence from marchasites without any corruption of their powers or virtues.

Concerning the Extraction of the Quintessence from Salts

The method of extracting the quintessence from salts is brought about in a somewhat peculiar way, so that their force may not be diminished, and is as follows :—

Take salts, which you will calcine perfectly ; and if they be volatile you will burn them. Afterwards let them be resolved into tenuity and distilled into water. Put this water in putrefaction for a month, and distil it by means of a bath, when the sweet water will ascend, which throw away. That which has not ascended put again in digestion for another month, and distil as before, repeating the process until no sweetness is any longer perceptible. By this means you now have the quintessence of the salt at the bottom, scarcely two ounces in weight out of a pound of the calcined or burnt salt. Of this salt, thus extracted, though it shall be only common salt, an ounce and a half seasons food more than a pound and a half of the other. For only its quintessence is present, and the body is taken away from it by means of the liquid solution.

In this way is separated the quintessence of all salts; but it is extracted in another way from alum and from vitriol, as follows :—

These will not allow of their being calcined into flux like salts. Therefore, after their calcination it is necessary to burn them, and resolve them according to the usual method. After they have been resolved, pour on them again the waters which have proceeded from them, and go on according to the method prescribed in the case of salts. For much of the essence ascends with the moisture, which again subsides at the bottom in process of composition and putrefaction, and so they meet together in one.

Concerning the Extraction of the Quintessence from Stones, from Gems, and from Pearls

The method of extracting the quintessence out of stones, gems, and pearls, with all of which the process is the same, is of all others the most excellent, and in its operation is very subtle and ingenious. A very small quantity of this quintessence is to be obtained from gems, and the more subtle and pure the gem is, the more minute is the essence. It is scarcely worth the trouble to extract the essence from dense, great, and cheap stones, since little virtue exists in them, whence it happens that very little comes forth from them. The process is of the following kind :—

Take gems, margarites, or pearls, pound them into somewhat large fragments, not into powder, put them into a glass, and pour on them so much radicated vinegar as will exceed the breadth of four or five fingers. Let them be digested for an entire month in a dung-heap, and when this is over the whole substance will appear as a liquid. This you will lighten with other radicated vinegar, and by shaking mix them together. The vinegar then acquires the colour of the stone. Pour the coloured liquid into another glass, again pour on vinegar as before, until the whole has no longer any colour. In that colour is contained the quintessence; the residuum is the body. Then take the colours, suffer them to be cooked to dryness, and afterwards wash often with distilled water until all becomes sweet as above. At length let this dust be dissolved on marble. In this manner you will have the quintessence of the gems and pearls. It must be remarked, however, in the colour of pearls, that they themselves are resolved into the colour of thick milk, and their body is sandy and viscous. In like manner is it to be laid down about the crystal. Its quintessence comes to the top, a certain viscous body remaining, by which the success of this kind of extractions may be known.

Concerning the Extraction of the Quintessence from Burning Things

Those things are called burning which are used neither for food nor for drink, and which of their own nature burn and keep alive the fire in the bodily substance. The method of extracting the quintessence from these is as follows:—Take such a body, very finely pounded, place it in a glazed pot, until it be full, seal it with the seal of wisdom so that it shall not breathe forth, and burn it in a circulatory fire for twenty-four hours, so that it shall remain at an even temperature, while the pot glows like the coals. Then take it out of the fire, let it putrefy in dung for four weeks, and afterwards distil it. Whatever ascends, let this be placed in a *venter equinus*, in order to distil all the moisture from it, and again set it to putrefy until no moisture any longer issues from it ; and then at length the quintessence of the body which thou hadst taken remains at the bottom. In this manner the quintessence is extracted from all things which contain oil in them, or resin, or pitch, or anything of that kind, as out of turpentine, fir, juniper, cypress, and the rest ; and in like manner out of all seeds, fruits, and similar things.

But it must be remarked that many more methods of extracting the quintessence from these are handed down elsewhere, ways and modes by which it comes forth quite odoriferous, subtle, and clear ; but these methods are not extractions of the quintessence; they are rather certain magisteries of these things, by which certain portions of the quintessence mount up at the same time in the process of mixing. They are not, however, perfect quintessences. For the essence of woods is a certain fatness or resin and a thick substance, whence it is not extracted in the form of magisteries. And the cause is this. The quintessence of the turpentine tree heals wounds, but when, in the above-mentioned way, it has been extracted from other magisteries, it does not heal wounds, because it has not in itself the fundamental power of the quintessence. The magisteries, indeed, are separated from the quintessences on this principle, that they only concern the complexions and the four elements, which, however, is not the nature of quintessences. Moreover, these magisteries receive it spiritually and not materially in its proper essence, as is clear from the chapters.

Concerning the Extraction of the Quintessence out of Growing Things

Those things are called growing in this place which fall and grow again, as herbs, leaves, and the like. For extracting the quintessences of these, several methods also have been found out by the addition of other things. But they should be extracted without the admixture of anything, so that they may retain their taste, colour, and odour ; and that these properties may be increased in them, not diminished. Thus, if the quintessence be extracted from musk, amber, and civet, their body afterwards stinks, so that they are no longer of any account in taste, in odour, or in nature. So of other things in this class it must be understood, so far as relates to the extraction of their quintessences. But we do not discuss musk, amber, and civet in this chapter, because we write specially of them elsewhere, and at present we are only treating of growing things, as the lily, spike, leaves ; for the extraction of the quintessence from which the following is the process.

Take growing things, bruised as completely as possible. Put them in some fitting vessel, and set this in a *venter equinus* for four weeks. Afterwards distil them by means of a bath ; again let them be placed in horse-dung for eight days, and once more distilled by the bath of Mary.

Thus the quintessence will ascend by the alembic, but the body will remain at the bottom. If any of the quintessence shall have remained at the bottom, putrefy it still further, and proceed as before. Then take at length this distilled water, add it again to this growing thing, and so, by means of a pelican, let them be digested together six days. Thereupon the colour will be dense. Abstract this by means of the *balneum maris*, when the body will disappear, the quintessence remaining at the bottom. Separate this by a retortive process, that is, by pressure, from the dregs, and let this quintessence digest for four days. In this way you will have it perfect in odour, in juiciness, in taste, and in virtue, as well as consisting of a thick substance.

Concerning the Extraction of the Quintessence from Spices

We will now teach the method of extracting the quintessence from spices, as musk, civet, camphor, and the like. First the quintessence ought to be reduced to another form, and at length to be separated therefrom. In that same process of separation the quintessence is found, as follows :—

Take oil of almonds, with which let an aromatic body be mixed, and let them be digested together in a glass vessel in the sun, for the proper time, until they are reduced to a paste. Afterwards let them be pressed out from their dregs. In this way the body is separated from the quintessence, which is thus mixed with oil, from which it is separated in the following way :—

Take rectified ardent wine, into which let the aforesaid oil be poured, and let them be left in process of digestion for six days. Afterwards let them be distilled by ashes. The ardent wine will ascend, and the quintessence with it. The oil will remain at the bottom without any of the quintessence remaining. Afterwards let this wine be distilled by the *balneum maris*, and the quintessence will remain at the bottom in the form of an oil distinct from all similar ones.

Concerning the Extraction of the Quintessence from Eatables and Drinkables

The quintessence of food ought to be none otherwise than in a form similar to that wherewith we should be fed, namely, flesh. And although, as we said before, no quintessence can be extracted from flesh, nevertheless we are easily able to extract from it, so far as it is its own, that which is equivalent to the quintessence, as follows :

Take some eatable thing, cut up, put it into a vessel or jug, carefully luted, and suffer it to boil for three days. Then strain off what is in the pot, and distil it by means of a bath. Thus there will ascend first of all a kind of wateriness, and when this has entirely passed over, the quintessence will be left at the bottom. This is the chief nutriment beyond all others which we could put down or describe. And in respect of its nutritive power it is equal to a quintessence. The quintessence may also be extracted out of drinks in various ways ; but this is the true process which we consider chiefly useful and convenient for the work in hand :

Take anything drinkable, enclose it in a pelican, just as it is, with its whole substance. Let it digest for a month in horse-dung. It will be still better to let it stay a year or more, and you will find in the pelican a certain something digested. Separate this by means of the bath, afterwards by ashes, and lastly by fire. In this way you will have three quintessences, which, in like manner, are in all drinks, for many reasons, which we enumerate in our treatise on Generations.

These three quintessences having been extracted, and each put into its proper vessel, the two latter should be further digested and then placed in the *balneum Mariæ*, when there will ascend more of the previous quintessence. Keep on doing this until no more of the former ascends, and in this way they are separated as completely as possible.

What we have so briefly taught about the quintessence of all things, and the short method of its extraction, ought not to rouse wonder in any at the rapid course of our hand and pen. For they are all well and completely handed down, and not so succinctly is the quintessence written of by us but that the work and labour necessary for it are clearly demonstrated. What need is there of much writing, which shall only nauseate ourselves and our readers, if we do not take into consideration that exercise and experience teach everything?

But how wonderful are its virtues and powers if it be extracted in the aforesaid manner we have even now partly taught, and shall make more clear in the last books that which belongs to this part of the Archidoxies ; and thus we shall have fully described the quintessence of all things. And although many before us have, in various documents, written great things about quintessences, still we do not think of their writings as a quintessence, the cause whereof we have already sufficiently adduced.

We have even learnt from them that verdigris was accounted the quintessence of Venus, when it is not so, but the crocus of Venus is the quintessence of Venus, which is thus to be understood. The flower of brass is a transmutation with the substance, at once dense and subtle, and extracted from every complexion of Venus. Wherefore it can be no quintessence, but the crocus of Venus, as we have taught, is the true quintessence : for it is a potable thing divided from the body without corrosion or admixture, very subtle, even more so than one cares to write in this place, for the sake of avoiding prolixity. So also, the crocus of Mars and its rust have hitherto been considered a quintessence, but it is not so. But the crocus of Mars is the oil of Mars. Concerning these things we set down more about transmutations in our philosophy.

A quintessence, therefore, is thus to be understood, namely, that it is nothing else than a certain separation of virtues from the body, wherein exists the whole virtue and essence of medicine. But what are the flower of brass, and the flower of Mars, and many similar matters, is handed down in the treatise on Magisteries.

None otherwise must it be judged of vegetables and herbs and such things than concerning the metals. And although we have put forth lofty and numerous virtues of the quintessence, nevertheless, only the smallest part of their forces and qualities has been told. But we have principally made it clear how these things are to be understood ; yet least of all have we been able to tell what and how great are their powers and virtues. From this may be hinted how great is the power which we have in our hands, only let us know how to use it well. Hence, also, is made clear the cause why man was created, and all things on earth were made subject to him ; and also why it is that nothing good or ill happens without a cause, which we set forth more clearly in our book on the Nature of Things.* For that foundation brings with it a faith fixed on the Creator, and a hope of His love towards us, as of an excellent father for his children. So, then, we must not snatch at any shadowy and vain faith ; but it is right that we regard only God and Nature, and the Art of Nature. Wherefore, with good reason, we call only on Him in this life and for evermore, and believe that only which we see to be so, receiving or approving nothing that does not agree with Nature, or which is beyond Nature.

THE END OF THE FOURTH BOOK OF THE ARCHIDOXIES ON THE QUINTESSENCE

---

* In the first book of the treatise entitled *Philosophia Sagax*, man is considered as the quintessence of the macrocosm. This point is frequently a subject of consideration in the transcendental physics of Paracelsus, and some reference has been made to it already in the annotations of the first volume. Man is a certain extract of the whole machine of the world derived no otherwise than the physician extracts the strength and essence from a herb, the result of which process is called a quintessence. That which the despoiled body is to the extracted virtue, so is the world to man. By so much as the body is weaker the quintessence is more efficacious ; the more is extracted the less remains in the original matter. In the case of the macrocosm, however, there has not been so exhaustive an extract as is performed by medicine. That only has been taken which was needed for man, and what remains is sufficient for his nourishment. Now, concerning Adam it must be known that he was made man in the image of God ; his wife, Eve, was made, and not born, out of Adam. It was not the will of God to make a double extract from the world, but one only, which is the quintessence of the Microcosm. He extracted, therefore, the man, not the woman. . . . Man is generated from putrefaction. Putrefaction takes place through the operation of the macrocosm, through the elements, and through the stars, in the father and mother, as by instruments which Nature has bestowed. Not that the exterior world works here, but the microcosm through the quintessence. . . . From the father and mother no intellect, or sapience, is born, but only out of the firmament, by the operation of the quintessence and the microcosmic virtue. . . . The quintessence which is made in man is retained, being ordained for seed, whence children are born. That sperm which is the quintessence retains the nature, essence, and property of the mass and clay of the earth.

THE FIFTH BOOK OF THE ARCHIDOXIES

FROM THE THEOPHRASTIA OF PARACELSUS THE GREAT

CONCERNING ARCANA

HAVING treated of quintessences, we will now turn our attention in due course to write about Arcana, since we understand more about them than about the forces of quintessences; wherein experience teaches us that there is the greatest difference on account of the very powerful operations whereby are demonstrated to us by most evident signs what things are better, more useful, or inferior in their powers. In this way we are able to avail ourselves of one or the other, according to their usefulness in medicine. Indeed, the ancients often thought that the arcana were quintessences, since they saw that they were much more subtle than dense substances, and knew that they operated in a wonderful manner through their subtle nature. But this error of theirs arose, not from reason, but rather had its origin in a lack of practice; since there was among them no knowledge as to the determinate difference of the high degrees, but they esteemed every higher and highest degree as quintessences. This difference, however, ought to be known and defined, not by practice alone, but also, and rather, by the operations of medicine.

Before, then, we treat of arcana, we must see and know why they are so called, and what an arcanum is, since it has so excellent a name, and well deserves to have it, too. That is called an arcanum, then, which is incorporeal, immortal, of perpetual life, intelligible above all Nature and of knowledge more than human.* Compared, indeed, with our corporeal bodies, arcana are to be considered incorporeal and of an essence far more excellent than ours, the difference being as great as between black and white. They have the power of transmuting, altering, and restoring us, as the arcana of God, according to their own induction. And although there is not eternity in our arcana, or that harmony which is celestial, nevertheless, compared with us, they ought to be judged celestial, since they preserve our bodies more than can be done or found out by Nature, and operate upon them in a wonderful way by their virtues. None otherwise, therefore, should these

---

* The arcanum is, as it were, a potent heaven of medicine set within the hand and the will of the physician.—*De Aridura.*

natural arcana be compared to our bodies, so far as medicine is concerned,* than as the secrets of God are.   Nor shall we fear to write that these arcana are higher and greater than ourselves, and that they have the greatest power of inaugurating life in us, as witness those four which we shall set down. Neither shall we take any heed of the empty stories told by those slavish dwarf-divines, since we consider that they have no more understanding than a blind man has sight.   One arcanum, then, is of a single essence ; another is the arcanum of Nature herself : for the arcanum is the whole virtue of a thing, excelling a thousandfold the thing itself.   We are able, therefore, fearlessly to assert that the arcanum of a man is every gift and virtue of his which he retains to eternity, as we teach in another book of these Archidoxies.   An arcanum, then, comes to be understood in two ways : one is perpetual, the other is quasi-perpetual.   This quasi-perpetual arcanum we judge to be like that which is perpetual according to the esteem and predestination of it. Four arcana only have been known to us from our boyish years ; with these we will complete this book, and work out for ourselves a sufficiently praise-worthy memorial, so that we may never forget them, praying the Supreme God that of His mercy He will allow our human flesh to attain to many years, so that we may lay up a long and gentle repose for our age, to hope faithfully in Him, and in no way to doubt that He, since He deigned to take our human nature, will allow us to enjoy it, so that we may in no way be disappointed of our hope, as we confidently expect will be the case.

Relying thus upon this hope, we will begin by making clear what is the difference between these four arcana, as to the labour, the art, and lastly, the virtues.   For this purpose, a final and conclusive knowledge of the virtues of each is required.   Generically, first of all, they keep the body in health, ward off diseases from it and drive them away ; they enliven the depressed spirits, freeing them from all sorrow ; they protect from all sickness, and happily conduct the body even to its predestined death, which has no end save by a lessening of consumption, as we lay it down in our treatise on Life and Death.

And although we have already made clear their virtues in a general way, and their nature, nevertheless they differ very much in particulars one from the other, so that no one of them operates altogether in the same way as another, or fulfils its virtues, but they are different both in manner and form, each with its own proper and peculiar ways.   So, then, the *Prima Materia* is the first Arcanum ; the second is the Philosophers' Stone, the third is the *Mercurius Vitæ*, and the fourth is the Tincture, for even thus we will set down the

---

* Now, the difference between arcana and medicines is this, that arcana operate in their own nature, or essence, but medicine in contrary elements.   Yet arcana do not prefer themselves to medicine.   Medicines are those things wherein it is understood that cold is to be removed by heat and superfluity by purgation.   Thus, there are reckoned substances of the arcana which by their natures are directed against the property of their enemy, even as one pugilist is opposed to another.   Accordingly, the conflict of Nature is such that craft is circumvented by craft, and all things that we possess naturally in the earth, the same Nature also requires to be preserved in medicine.   This is, therefore, the part of the physician, namely, to act not otherwise than as if two enemies opposed one another, who were equally cold or equally glowing with heat, and are both armed with similar weapons.   Since the victory lies between these, so must you also understand concerning man, that there are two combatants, soliciting natural aid from one mother, namely, the virtue alone.   The arcana also operate with like virtue.—*Paragrinum*, Tract II., *De Astronomia*.

practice of them in order, after we have explained their modes of operation as follows :—

In the beginning, it must be remarked, concerning the Primal Matter, that it puts forth its predestination, to which it is foreordained, entire, and from its first origin to its final end well-defined and exemplified. For as the seed gives of itself the entire herb, with renewal of all its forces and consumption of the old essence, so that the former substance, nature, and essence have no further operation, so do we say of the primal matter, that we are born from one seed like something growing in the field according to its growing nature. According to the aforesaid example, the primal matter introduces new youth into a man, just as a new herb springs forth from a new seed in a new summer and a new year.

The Philosophers' Stone, which is the second arcanum, perfects its operations in another form, namely, this: As the extrinsic fire burning the spotted skin of the salamander, renders it pure as if it were newly born, so, also, the Philosophers' Stone purges the whole body of man, and cleanses it from all its impurities by the introduction of new and more youthful forces which it joins to the nature of man.

The Mercury of Life, which is the third arcanum, gives proofs of its operations like those which the kingfisher displays, which is renewed every year at its annual period and endued with new plumage. Even so this arcanum casts off the nails from the hands and feet of a man, the hair, skin, and everything that belongs to him, makes them grow afresh, and renews the whole body, as was said above concerning the kingfisher.

But the Tincture, which is set down by us as the fourth Arcanum, displays its operations like the Rebis, which makes gold, or the Sun, out of the Moon and other metals. None otherwise does the tincture affect the body of man, and take away from him his corruption and impediments, changing all into the greatest purity, nobility, and permanence.

How, then, can it be that we should withdraw from this noble art of medicine, or from philosophy itself, when we see so clearly their force and power, which alone confirm us, and deservedly so, that we should place the greatest confidence in them? For we have not applied our minds to believe, learn, or imitate things which cannot be proved and attested by most true and certain evidence conjoined with reason. When Christ hung upon the Cross, if the sun and moon had not been affected with a kind of sympathy, so that they were deprived of their light and darkened, and had not the earth been shaken with a fearful tremor, and had not other signs been manifested at His nativity, no one would now believe in Him. They naturally teach us to see and know this, that Jesus Christ is God, and took human nature upon Himself. We may say the same of the Arcana, that they cause and even compel us to believe in them, so that we should not withdraw from them even up to the time of death, but rather strictly and constantly, among many hindrances, every day go on to give thanks to God. So neither the eclipse nor the moon will detach anything from us.   We will therefore put forth the practice and elaboration of these four Arcana, by which we are able to drive away the accidents and corruptions of our youth, and to rejoice in them as our eternal Arcanum rejoices in its eternal life.

Concerning the Arcanum of the Primal Matter *

Since we have sufficiently pointed out concerning the Primal Matter, whence it proceeds and what it is, we must understand that it is based not only on men, but on all bodily creatures, that is, on everything that is born from any seed.   Whence it may be inferred that if it has its operation in any created body and perfects it, as we have before declared, it is able also to preserve trees from corruption, herbs from being dried up, and also metals from rust, concerning which the same thing must be understood of men and of brute beasts.   So, then, a tree which is now almost consumed with age, and daily more and more verging on its own corruption, not from defect of root or of nutriment, but of its own proper virtue, can be renewed by its own primal matter just as we said about the skin of the salamander, and so may arrive at another age according to its predestination, nay, even at a third, a fourth, or more.   On this principle virtues are to contribute to it, namely, in order that its corruption and destruction may be now and again renewed in a long process of time.   No less is this to be understood of herbs, which last only for a single year, because their predestination is no longer.   For they, even when they begin to be dried up, are renewed by their primal matter, so that they remain green and fresh for another annual period, or for a third, a fourth, or more.   The same thing understand of brute beasts, as, for instance, old sheep and other animals.   They can be renewed for a fresh period of life, having received their virtues, such as milk and wool, like young sheep. Equally, too, can man be led on from one age to another, as we have said before.   From this it ought be known what the primal matter is according to its essence.   In created things, such as have no sensation, it is their seed ; but in created beings endowed with sensation, it is their sperm.   For it must be known that the primal matter is not to be taken from the thing out of which this created body is produced, but out of the produced and generated material. For the primal matter has in it such virtues that it will not allow the body which is born of it to go into consumption, but abundantly affords whatever is necessary for the supply of every requirement.   Indeed, death only arises from the destruction or infection of the living spirit.   Now, that spirit grows out

---

* According to one of his treatises on turpentine Paracelsus held that the ultimate matter is contained in that seed wherein God has digested each thing, but those more especially which are subject to natural growth, created by Himself.   The primal matter is gross seed.   It is the ultimate and not the primal matter, which, on the authority of this treatise, is useful to man.   According to the tenth chapter of the *Labyrinthus Medicorum*, every growing thing whatsoever is in its first matter without form, or unformed.   For example, the fir, the beech, the oak, are all in the beginning seed, wherein there is nothing which ought to be, that is, nothing which we should expect to find, having regard to what they become subsequently.   But if such a seed be put into the ground, it is needful that it should first putrefy.   Otherwise nothing is produced from it.   When it putrefies, it dissolves altogether, and from being a seed it becomes nothing, but, at the same time, the putrefaction is the first matter from which the tree develops.

of the sperm, or out of the seed, and is altogether a spermatic substance ; therefore it can be helped by its like. For wherever the like is given as a help, there is introduced a new period of life, for many causes which we do not detail in this place, but make clear in our " Philosophy."* Furthermore, although we did not propose to write in this book anything about the nourishment and renovation of trees, since we undertook to treat only of the medicine of the human body, still, let those facts about the trees and other transmutations of this kind be set down, in order that, parabolically, and by those examples, we may make our meaning easier to be understood. The quintessence of the seed of the nettle (otherwise the lavender) if it be poured on to any root of its own herb, so that this herb may receive its tincture and be affected by it, it remains another year as in the former year, not putrefying until that second year shall have been completed.

In like manner, if the quintessence of the seed of quinces be poured on the root of a quince tree, the tree remains green to the end of another year, and also produces flowers and fruits. In the same way, the quintessence of cherries causes trees to put forth their fruits twice in a year, as if in two summers ; one is the middle summer of the seasonable cherry trees, and the other summer is made like the former.

Not only, therefore, is it fit that we should speak of the quintessence of the sperm, but also concerning the arcanum of the sperm, from which proceed far more wonderful things than we have already pointed out.

So then, let us first make clear the process of this practice. In the first place, it is alike among men and animals. Secondly, it is made out of primal matter only in the following way :

Take primal matter, let it digest in a flaccum in a resolutive digestion for a month. To this let there be joined the addition of a monarchy of equal weight. Let them be suffered to digest again from one to two months. Then distil this matter [? by the cloth] and what ascends will be the arcanum of the primal matter, concerning which we are here writing. And let no one wonder at the brevity of this method or process, for complexity is apt to involve much error.

CONCERNING THE ARCANUM OF THE PHILOSOPHERS' STONE

I am neither the author nor the executor of that Philosophers' Stone, which is differently described by others ; still less am I a searcher into it, so that I should speak of it by hearsay, or from having read about it. Therefore, since I have no certainty thereof, I will leave that process and pursue my own, as being that which has been found out by me through use and practical experiment. And I call it the Philosophers' Stone, because it affects the

---

* Paracelsus has bequeathed to his followers a Philosophy of the Four Elements, a Philosophy addressed to the Athenians, an Occult Philosophy, Five Philosophical Tracts, and a vast system under the title of Philosophia Sagax. The connection of the living spirit with the sperm is discussed very largely in most or all of these, as, indeed, elsewhere in his writings. Under these circumstances there is scarcely any need, as, in fact, there would be great difficulty to distinguish the special section to which reference is here made.

bodies of men just as their's does, that is, just as they write of their own. Mine, however, is not prepared according to their process; for that is not what we mean in this place, nor do we even understand it. We do not set down in this our practical treatise the process of the operation, since we have before mentioned it in the beginning of this book when we were writing of its force and its effects, which it has by means of separation.

Concerning the entrance of this penetration, you shall also further note, by which entrance it penetrates the body and all that therein is. For by that penetration it restores and renews it, not that it removes the body altogether, and introduces a new body in its place, or that, like the primal matter, it infuses its spermatic arcanum thereinto, but that it so purges the old whole body as the skin of the salamander is purged, without any injury or defect, and the old skin none the less remains in its essence and form. In like manner, this Philosophers' Stone purifies the heart and all the principal members, as well as the intestines, the marrow, and whatever else is contained in the body. It does not allow any disease to germinate in the body; but the gout, the dropsy, the jaundice, the colic, fly from it, and it expels all the illnesses which proceed from the four humours; at the same time,

purges bodies and renders them just as though they were newly born. It banishes everything that has a tendency to destroy nature, none otherwise than as fire does with worms. Even so, all weaknesses fly before this renovation.

This Philosophers' Stone has forces of this kind, whereby it expels so many and such wonderful diseases, not by its complexion, or its specific form, or its property, or by any accidental quality, but by the powers of a subtle practice, wherewith it is endued by the preparations, the reverberations, the sublimations, the digestions, the distillations, and afterwards by various reductions and resolutions, all which operations of this kind bring the stone to such subtlety and such a point of power as is wonderful. Not that it had those powers originally, but that they are subsequently assigned to it. Something like this is to be understood in the case of honey, which, by its elevations becomes far sharper than any aquafortis or any corrosive, and more penetrating than any sublimate. Such a property of sharpness it has not by nature, but this proceeds solely from the elevation, which changes all the honey into a corrosive. In these effectual Arcana, too, it must be considered that those who use them, as well as the children sprung from them, live afterwards endued with such health that no sickness or ailment, or anything like a flaw, afterwards happens to their bodies, but they are adorned altogether with such a subtle and pure complexion of Nature, that it is impossible a more noble state of the complexion can be induced. For that most choice and excellent medicine effectually renews and purifies, and introduces an incorruptible life, which cannot be contaminated by any kind of life. It suffers nothing to become enfeebled, but secures that men shall live in the highest nobility of Nature, while it advances their offspring to the tenth generation.

This Philosophers' Stone not only transmutes one weight, but this transmutes another, and this again another, and so on, in so far that these mutations might be extended almost endlessly, just as one light kindles a second, and this second a third. So it should be understood of the Philosophers' Stone in relation to health, as out of a good tree good seed and good shoots are born, out of which again good trees are produced. The power and potency of the Philosophic Stone is exalted to so wonderful an extent that it is impossible to trace how it can be naturally brought about; and unless most evident signs lay open to our eyes, it would be incredible that men could perfect and accomplish such wonderful things; since the virtue of that operation passes from generation to generation without any break. On the other hand, by the mercy of God, it exists in one body, and at length, according to their deserts, it is either denied to others or conceded as a special act of grace.

Now let us set down the process of our Philosophic Stone in the following manner.

In the name of the Lord, take Mercury, otherwise the element of mercury, and separate the pure from the impure. Afterwards let it be reverberated even to whiteness, and then sublimate this by sal ammoniac until it is resolved. Let it be calcined and again dissolved, and digested in a pelican for a month. Then let it be coagulated into a body. This body no longer burns nor is consumed in any way, but remains in the same state. Those bodies which it penetrates are permanent in the cineritia, and cannot be reduced to nothing or altered; but the stone takes away every superfluous quality from sensible and insensible things, as we have before related. And although we have set down a very short way, nevertheless it requires a prolix labour, difficult in its adjuncts, and requires an operator who is affected by no weariness, but is in the highest degree active and expert.

CONCERNING THE MERCURIUS VITÆ

Next in order we wish to write concerning the *Mercurius Vitæ*, the virtue of which far excels the virtues of the two preceding Arcana; for that virtue consists, not in the art or in the operation, but in the *Mercurius Vitæ* itself, like which we have never known any simple anywhere existent; forasmuch as that nature and property is innate therein, not from the virtues of the quintessence or of the elements, but from the specific quality of its predestination. Neither has it only the virtues of transmuting persons and other essential things, but also of renewing every growing thing, and their likes, out of the old quality into a new, after the following manner. The *Mercurius Vitæ* reduces Mars into its primal matter, and again transmutes it into its perfect matter, so that out of it iron is again made. It also renovates gold in the same way, because it reduces it into proper mercury and tincture, and again digests it into gold, so that it becomes a metal as before.

Nor, in truth, does it operate in metals only, but similarly in other bodies, as in herbs. When their roots are suffused with it they will bring forth a second crop of flowers and fruits. If the first seed has fallen off, and they are at that time suffused with the same, they will produce second flowers and fruits, irrespectively of the season.

And none otherwise must it be understood of animals than of men and other things. When this Mercury is applied it renovates all their old and consumed members, and restores the defective and lost powers into a youthful body or abode, so that in old women the menses and the blood flow naturally as in young ones. It also brings back aged women to the same perfection of nature as the younger ones.

Concerning the arcanum of life it is further to be remarked that its forces exist so potentially in its specific form that it separates the old from the new, or age from youth ; that it augments the latter, and so renovates the period of life. Whence it may be inferred that youth and its powers do not fail on account of old age, but that these exist equally in the old as in the young. The corruption, however, which grows up with youth is so strengthened that it takes away the powers, whence old age is recognised.

As soon, therefore, as this corruption is separated from youth, that youth again manifests itself without let or hindrance. Now this must be understood thus : When any body or corpse putrefies, the quintessence thereof is not putrefied, but remains fresh and unconsumed, and is separated from the corpse into air, or sometimes it is scattered into the earth or into the water, according to the place whither it goes. For there can no destruction of the quintessence occur, a fact which must be clearly noted and regarded with admiration, as we teach in our treatise on Corruption and Generation.* So, also, a rose putrefied in dung retains its quintessence in itself, even whilst in the dung. Though everything becomes fœtid and putrid, still, in the separation of the pure from the impure, the quintessence lives without spot or blemish, though the bodies are noisome corpses. Thus, therefore, we say of the *Mercurius Vitæ* that it separates corruption, even as in wood it separates that which is decayed. So powerful, also, is it in man, that, after the corruption shall have been separated from him, the quintessence is again excited, and lives as in youth. And this is to be understood thus : Not that the *Mercurius Vitæ* stirs up a new essence, as some persons may malignantly interpret our opinion and experience, but that the essence and the youthful spirit whence proceed the forces of youth remain unconsumed, none the less, though, being oppressed, they are beheld as dead. The *Mercurius Vitæ* removes the impurity, whence it happens that the aged life recovers most effectually its powers as they were before. As we said above when speaking of the king-fisher, it is renewed after death for this reason, that the quintessence does not

---

* The reference does not correspond to the treatise on the Nature of Things, and there is no other extant under this exact title.

withdraw from its abode. But if that dwelling-place be dissolved by putrefaction, then the quintessence is received into that upon which it lies. Whence
it happens that there are often found wonderful conditions of Nature in
growing things, not naturally existing in them, but by some accident of this
kind, as we set it down in our book on Generations. In this way, then, the
matter is to be understood. In dung there meet and are accumulated many
corruptions, as from herbs, roots, fruits, waters, and other like things.
Whence it happens that fields are rendered fat and fertile, not on account of
the corruption, but on account of the quintessence existing in the dung, which,
betaking itself to the roots, exhibits power in the growing things though the
body, that is, the dung vanishes and is reduced into nothingness, its substance
being consumed. Wherefore human dung has great virtue in it, because it
contains in itself noble essences, as of the food and the drink, concerning
which wonderful things might be written. For the body receives from it
nothing save nutriment, but not the essence, as we write in our treatise on
Nutriments.*

We will come, then, more strictly to the practical view of *Mercurius
Vitæ*, which, as we said before, perfects its operations in a wonderful manner,
as, for instance, by shedding the nails of the hands and the feet, and
plucking out grey hairs by the roots, it strengthens youth, so that corruption
cannot demonstrate old age by those signs unless a second old age be
again attained.

Now, approaching the practice, let us treat with alchemists in a few
words; for there is no need to write much or to preach with prolixity to
them. Pretenders and fools we will altogether exclude. Let this, then, be
the method of practice :

Take essentialised mercury ; separate it from all superfluity, that is, the
pure from the impure. Then sublimate it with antimony, so that both may
ascend and become one. Then let them be resolved upon marble and coagulated four times. This being done, you will have the *Mercurius Vitæ*, about
which we have before spoken, and with which, as with an arcanum, we will
console our own old age.

Concerning the Arcanum of the Tincture

None otherwise is the arcanum of the tincture to be understood, namely,
that it takes away all the inconvenience from old age, all diseases, and whatever corrupts the health or has an influence contrary to it. This arcanum is a
certain tincture of such properties and conditions that it operates and induces
health, not in the same way as the preceding three arcana, but according to
its own name. The tincture *tinges* the good and the evil, the dense and the
subtle. None, otherwise, does this perfect its operations on the body so as to
transmute corrupt and ill-disposed complexions into sound ones, just like that

---

* This is another treatise which apparently is wanting.

tincture which makes Luna out of Mercury ; it does not separate from it what is evil, but it tinges both the good and the evil, so that in the end they both turn out to be excellent.   So, too, this tincture tinges the dropsical and jaundiced body and makes it sound, not because the dropsy is taken away, the origin driven out or separated from the good, but it is transmuted into good, as it should be, and is settled in its highest and best grade, in like manner as the corrupted dung may, by the subtle corruption of art, be converted into an elixir, which drives away all corruption ; that corruption not being separated, but the entire substance being transmuted into another quality and nature.

The same is to be thought concerning this tincture, that it tinges the body apart from all separation of the evil from the good, or expulsion of man's original essence, but by his renovation.

Yet it should be known that the tinged body no longer lives in its old form, but, like a metal, is transmuted into another, as into copper or some other.   Saturn has not in itself its old quality, but the quality of the tincture itself.   None otherwise must it be thought of tinged bodies which have received the tinging of the tincture, that they exist no more in the former life from which they were transmuted by the tincture, but far nobler, better, and more healthful is the condition of the body and the form than its native origin was ; and it is like natural gold made out of iron by the tincture, as we have also written concerning Transmutations.   If, therefore, this tincture is a transmuter of bodies into better ones, as in the case of metals which so few know or have tried, there will be as many and various corporal as there are metallic tinctures.   And as of these one is always better than another, so with the corporal tinctures.   It must be considered that some tinge and are tinctures naturally, as the crocus, flower, sulphur ; some are made by art, as a stone, realgar, and others.   It is highly necessary that these things should be understood, because the beginning and the entrance to these tinctures which they exhibit is not unimportant.   Moreover, it is to be remarked that these tinctures ought to be made for the seven principal members, and that to every one of them its own property ought to be assigned and given, as to the heart those that serve the heart, to the brain those which belong to it ; and such tinctures should be prepared from metals, herbs, and like things which are proper.   So it will happen that by them the whole body will be tinged.   Nor will it suffice that it be tinged by one tincture only, but as one tincture tinges only one metal, so must it be judged of these.   The practical preparation of the Tincture should be as follows :—

Take the essence of the members, from which you will separate the elements.   Afterwards put the fire of them in digestion, and leave it so long until there remains nothing more at the bottom and nothing of the matter appears substantially.   Then take the matter, and the glass luted in this way with the lute of Hermes ; put it in a moist, cold place, where again it will be resolved into visible matter.   That visible matter is the one concerning which we have written.   With these few words we conclude : for if we were to write more concerning it, that would be a handle for the derision of the stoics.* From this we would fain be exempt, and speak only to alchemists.

---

* In another place Paracelsus affirms that arcana are not old things but new, not ancient but recent productions. The ancient productions are substances and forms as they exist in the world. And as the form of these things is no good to us, but must be resolved and renovated in order that it may be of use, so there is needed in addition the removal of all properties, whether of heat or cold. That is, unless the solatrum lose its cold it does not become a medicine. Similarly, unless anacardi lose their heat they will in no wise prove a remedy. In short, unless all the old natures pass away and are removed, and are brought over into a new birth, they will never be made into medicines. This removal is the beginning of the separation of the bad from the good. Thus, therefore, the latest medicine, that is, born recently, is left, without any complexion and the like, a pure and absolute arcanum.—*Paramirum*, Lib. II., *De Origine Morborum ex Tribus Primis Substantiis*, c. 2.

THE END OF THE FIFTH BOOK OF THE ARCHIDOXIES FROM THE THEOPHRASTIA, CONCERNING ARCANA

THE SIXTH BOOK OF THE ARCHIDOXIES

FROM THE THEOPHRASTIA OF PARACELSUS THE GREAT

————

CONCERNING MAGISTERIES

THE preceding books on most excellent medicines having been finished, we propose to add this one treating of the Magisteries: and first to declare what a magistery is. It is, then, that which, apart from separation or any preparation of the elements, can be extracted out of things; and yet by the addition of something the powers aud virtues of those things are attracted to that material and kept there.

Those virtues do not in any way proceed from Nature, so far as their operation is concerned, nor from a specific virtue, but from a mixture whereby virtues of the same kind are attracted. If vinegar be poured into wine it renders the whole vinegar. Now, this is a magistery. When wine is poured upon honey the wine is so transmuted; so there is no magistery in that case. Those things, therefore, must be considered which relate to a magistery, even as wines do to vinegar. For such as are perfect, or, as should be the case, ill adapted for this purpose, cannot produce a magistery. So, then, the natures of things come to be considered. In the same way, the difference of the extractions of magisteries must be noted, as, for instance, out of metals, marchasites, stones, herbs, and the like matters, by those things which are not metallic, but which become like metals, just as wine is made like vinegar in its powers, its virtues, and its taste. The cause that wine does not appear different from vinegar is, that a nature like vinegar exists in it, whence it happens that the appearance of their natures is the same. If, in like manner, the nature of metals be but pure, it equally appears in their magisteries, but yet it is not of the same property.

Moreover, mention must be made of additions. Those things which are taken for this purpose, though they are not of one and the same complexion, power, or action, nevertheless agree in preparation, since that which results from power of this kind is appropriated and not complexioned. By means of these substances the metals afford their magisteries, which, indeed, are no less to be accounted of than quintessences on account of their virtues. Gold in its magistery lays down all its quality and complexion in one essence. It must not for that reason be supposed that because the body is of no account the rest will therefore be infected with that failing ; for in this case its leprosy is in no way injurious, but the thing is entirely good. Sugar, again, is wholly sweet whilst as yet it remains in the body and is not separated, and it can be so prepared, while retaining its sweetness, that it shall turn out by far sweeter and more efficacious than it was before. But the quintessence when extracted is not sweeter than when in or with its own body. Wherefore, this body by no means brings loss to it. But still its virtue separated from the body is less than when prepared in or with it. The magistery amends it more than Nature can be supposed to do.

The same must be understood of stones, which enter into the number of magisteries, and also of their bodies, namely, that what is assumed for their use is not deficient in its virtue, but is a sufficiently powerful magistery. This you may understand thus, as when sulphur is lighted and burnt up ; that which burns is the very smallest part ; and so likewise stones ; for example, the crystal, when it is reduced to a magistery, brings all stones to their first matter, and grinds them in a wonderful manner, in the same way as its essence does, for this reason, that the nature is tinged by the quintessence, and may be held for a quintessence, as in the case of the vinegar and the wine, of which the one makes the other like itself without any defect. In like manner, it is not only in stones that tincture of this kind is made, but also in metals, as the quintessence of gold tinges all its body into a pure quintessence, which light we think great, and even too great, for it is the light of all the secrets in our Archidoxies : wherefore we kindle coals with a light mind, so that we may investigate the final end of these secrets of Nature. We derive our instruction from examples proved experimentally by ourselves, as in magisteries, and chiefly in that of gold, which, containing the body and the quintessence, is drawn therefrom just as the quintessence is. Wherefore our magisteries are known to be dowered with special virtues, and we write them down to our own praise even until death.

We speak also in like manner concerning the magisteries of herbs, which indeed are so efficacious that half an ounce of them operates more than a hundred ounces of their body, because scarcely the hundredth part is quintessence. So, then, the quantity thereof being so very small, a greater mass of it has to be used and administered, which is not required in the case of magisteries ; for in these the whole quantity of the herbs is reduced into a magistery, which is not then, on account of its artificial character, to be judged inferior to the true extracted quintessence itself. One part of this being exhibited is of more avail than a hundred parts of a similar body, for this reason, because the magisteries are prepared and rendered acute to the highest degree, and are brought to a quality equal to a quintessence, in which magisteries all the virtues and powers of the body are present, and from these its own helping power arises to it. For in them the penetrability and the power of the whole body exist, from the mixture that is made with it. For the body receives none of those things with any desire except such as are spiritual to it. Whence it happens that it attracts that magistery and mingles itself therewith none otherwise than as gold attracts to itself the *Mercurius Vitæ*, and is mingled therewith. Now, iron does not do this, because they are not, in their composition, so much in agreement with each other. So, then, both the body of it and this magistery are amalgamated, and become one ; and hereof many examples may be found which need not be brought forward in this place.

Some of the marchasites, too, perform the office of a medicine in a like way, although with a difference : that is to say, they leave their body, and the best part of them, as the juice, is extracted, and none the less exists as a magistery, though the body be separated from it. This, however, must be understood, that it is not the body of the marchasite, but rather of the earth or the mineral in which the marchasite exists. Its virtue, therefore, is not efficacious of itself, that the earth or the mineral should be separated from it, but it abides therein as in a marchasite, which, indeed, it is thought to be. In order, therefore, to make it clear concerning a marchasite, what it is, whence produced, and with what virtues endowed, we wish to relate its practice in the following treatises, and also to describe the art of its preparation. The process, however, cannot be comprised in one general explanation, but the matter must be treated of particularly, concerning each special case, as of metals by themselves, marchasites, too, in particular, and also stones and herbs.

There is, likewise, reckoned a special magistery of the blood, which is taught in a peculiar form and manner. In it is considered what virtues and forces of man exist, and what its nature contains in itself, in what there is any defect produced, and so on ; but still, without diminution of the natural creation, but that it may be considered as a complete and perfect work in all its parts, as a bird with all its feathers.

Concerning the Extraction of the Magistery from Metals

First of all, we will set down the magistery from metals, and make that clear which openly shews itself to be of wonderful virtues, which are to be known according to the tenour of its essenee. The process of them must be carried on without any corrosives and all things which are complexioned contrary to the metals. For, from the conjunction or mixture of contrary things, the essences are corrupted, so that, on account of this error, they put forth no virtues, since the one contrary or the other predominates and prevails.

Seeing, therefore, that great account is to be taken of agreement, only the temperate will agree therewith. Consequently you must know what is a temperate thing. A temperate body is one of a certain complexion ; this receives another into itself, and is incorporated with that which shall be joined to it, so that it no longer displays its old and special complexion, but only the virtues of that which has been added to it. So, forsooth, *vinum ardens* has in itself a full and perfect complexion ; but yet, whatever is put therein, it, so to say, complexionates more abundantly, fulfilling its operation thus, according to the power of that which was put into it.

Since, then, it attracts the virtues of other things and subdues its own, it is on this account said to be temperate, and is deservedly so called. Here some differences call for notice ; we understand the elements only, so that we are able to say of oil that it draws to itself foreign natures and hides its own element. From anything like these a magistery ought to be made, so that the virtues of the metals may pass into the same temperate thing, may be cleansed and purified therewith, and may be distilled to their own end. A magistery like this, after it has been purified, shall be called potable, because it can be taken in drink : whence it comes that the magistery of gold is called *aurum potabile*, that of the moon *argentum potabile*, so also iron, lead, quicksilver, etc., will be capable of being made potable, and so called and described according to their complexions, in relation to which processes they have far greater operation than it becomes us to describe. After the following manner, with one temperate thing, and by one process, as also by one practical method, all the magisteries of the seven metals can be made. The practical method is as follows :—

Take circulatum thoroughly purified, even to its highest essence, in which place very thin metallic plates or filings of any metal you please, perfectly and subtly wrought and purified. When these two have been put together in a sufficient weight, let them be circulated for four weeks, and by means of this temperate medium, the leaves will be reduced into an oil, and into the form of a fattiness swimming at the top, coloured according to the condition of the particular metal which you shall separate by means of a silver attractorium from the circulatum. This same is the potable gold or silver. The like is extracted with the other metals, and may be taken as drink, or with food, without any harm.

The Extraction of the Magistery out of Pearls, Corals, and Gems

The Magistery from precious stones is to be understood in the same way as from metals, according to the virtues which each stone has in harmony with its nature. It must be remarked, however, that for stones there is no need of a temperate medium, or of any addition whatsoever, because the solution of them is not the same as the resolution of metals, but the magistery of them is extracted in another way ; in the practice of which three processes are understood, one for gems, another for marchasites, and a third for corals, according to which stony growing things may be brought each into its own magistery. In the process of stones, the colours need not be considered, nor their brightness be taken into account, since all their magisteries have a white colour. So also pearls, with the exception of corals, which retain their colours in a remarkable way above others in the magistery. In them, therefore, the colour is chiefly to be noted ; for the magistery of them, together with the body and the element and the whole essence, is extracted by means of additions, without any corruption. None the less, however, they may again be restored to their perfection.   Wherefore, in respect of their generation and nature, they cannot be compared to stones, and yet they have a stony condition.   This, too, may be done : the whole colour may be extracted out of the body of corals into another medium, and afterwards from their body a form may be made as if from clay.   After this formation the colour can be again infused, so that it becomes a coral as before.   This cannot be done with pearls or gems, which are unable to return to their perfection by a similar process, but they remain in their magistery, their essence not having been corrupted.   We have even seen this penetrate glass and instruments, and form them according to its nature.   The magistery of the magnet has drawn things towards itself, in the form and after the manner of the material magnet, and afterwards it has fixed itself in the glass and tinged that, so that this also has attracted needles and straws.   None otherwise must it be understood of the rest.   Therefore they must be kept in gold only.   More of these things have occurred to us than is credible, whereof we put together this record, so that from this inducement other persons also may investigate the arts and the magisteries.   Since, therefore, they exhibit a demonstration apart from other extractions, we will now teach the practice of them in the following manner.   First we will speak of gems.

Take gems, having first bruised them and calcined them according to the reverberatory condition, together with common salt-nitre, of an equal weight, that is to say, one pound.   Let these be burnt together into the form of lime, and afterwards washed in *vinum ardens* until no surplus matter can be found. After the washing, calcine this substance again, and cause the whole of it together to pass through the *vinum ardens*.   Let this evaporate by boiling ; and thus you will have an alkali, which dissolve in water and keep.   There is no reason why you should fear to make use of this administration, since, however sharp and calcined it may be, with all its sharpness it only affects that which resists it ; and it is so subtle, that one single drop alone tinges the whole body so as to produce a remarkable virtue.

Pearls, too, are to be reduced to a water in the following way, namely : Take corrected vinegar, in which place bruised pearls, and let them digest therein for a month ; so that they will be reduced to a water.   Then distil it, and separate the *acetum* from it by a bath, having done which, you shall find the pearls in the bottom resolved into water.   This is the magistery of pearls or unions.   And although the process of this practice is a short and easy way, nevertheless, give credence to one who has tried it ; for the operation of these same things is indeed wonderful.   This action of their virtues, however, is not produced by art, but is placed in the nature of them, which is hidden in a dense substance, on account of which it cannot operate, just as is the case with a dead body ; but by making a resolution its body will be vitalised.* Enough of this. Corals, however, must be ground and calcined from the commencement with salt nitre ; then, afterwards, they must be prepared as gems are, and also resolved. So you have the magistery of corals, the virtues whereof I very much and specially wonder at, which God has bestowed on this growing thing, which also do operate powerfully and wonderfully, even as they grow.

CONCERNING THE EXTRACTION OF THE MAGISTERY OUT OF MARCHASITES

Concerning the magistery of marchasites thus much must be known ; that they are only minerals. So, then, the mineral is not brought to its magistery, but only the true marchasite, as is also ascertained with regard to metals, since these do not pass into their magistery unless they shall have been first separated from their mineral. And although marchasites cannot well be separated from these, still this may none the less be done in the magisteries. There are, indeed, different kinds of marchasites, such as the golden, the silver, the golden talc, the white, the purple, the tin or bismuth, antimony, granite, and others of a like nature, for all of which there is only one appropriate extraction. In like manner, the virtues and powers of them manifest themselves in medicine according to the conditions of metallic operations. And although they do not exist as metals, none the less they have the properties of them. Wherefore, we set down only a few details about them, because we treat them more at length in our book on Extrinsicals.† There is a difference to be noted, according as they come together and agree one with another, as gold and the marchasite, antimony and lead, which, indeed, in their fabrication and constellation are compared one with the other, but are none the less separated in their virtue. For in some marchasites there are more virtues than in the assimilated metals. This we see to be the case with lead and antimony, of which the one cures morphia, alopecia, and the like, and all scabs and scars, the leonine, the elephantine, the Tyrian, etc., which the magistery of lead or Saturn does not affect. So, then, the properties of this kind are to be noted, which are sometimes latent not only in great things but also, and that more abundantly, in smaller ones. Let us, therefore, go

---

* Among other virtues which Paracelsus attributes to the pearl is the increase of milk in women when the supply is deficient. Let the pearls be resolved into a liquor, otherwise an elixir, but in the body a ferment. Whatsoever other things are required for this ferment, these the body supplies. The quantity administered must be regulated by the experience of the physician.—*De Aridura.* Another method for the extraction of a medicament from pearls is as follows : - The matter in the form of a potable sap is extracted from pearls, and this is so powerful that it can scarcely find its equal among sperms. *Formula for compounding the Sap of Pearls.*—Take one pound of purest distilled vinegar, half a pound of circulated water of life, and four ounces of cleansed pearls. Reduce to a fine powder, prepare, digest with others for a month in a circulatorium. The matter of the pearls will then have sunk to the bottom like a thick liquor. Then separate the waters by effusion from the sap. Of this sap administer doses of six grains. There are many other processes for the extraction of this kind of sap, but the above is the most useful of all.—*De Contractiuris*, c. 3.

† The book concerning Extrinsicals, and that also concerning the Generation of Wines, mentioned a little further on, are unknown. A similar remark will apply to the treatise on the Properties of Things, which is cited in the seventh book. There is probably little doubt that many of these apparently missing works are merely the mis-stated titles of others which have come down to us. For example, the treatise on Physics which is quoted in Book VII. may be extant under a different name, or again, the reference may be to a projected treatise which was never accomplished.

on to investigate this reason, why antimony has more virtues than its metal. Its body is not fixed, and not sufficiently digested into its own perfection, as is the case with Saturn, whence it assumes a volatile property. But the very material out of which it is produced is privative, and of a cleansing character, from its own natural property, as we set down in our book concerning Generations. Hence it happens that it purifies gold and silver more than does the fire or any other element. Thence, too, it proceeds that it cleanses and purges the body, even as gold and silver are freed thereby from all their impurities. The magistery of antimony drives out leprosy in a more than credible manner. And so, too, must it be understood of the others.

Let us now approach the practice by which we teach the preparations of the magisteries from all these substances, in the following manner :

Take a marchasite, ground very small, and add to it so much dissolving water as will cover the breadth of six fingers. Let it be dissolved and subsequently putrefied for a month. Afterwards let it be distilled and separated, as we teach concerning the metals, which having done, you will have the magistery of whatever marchasite you selected.

Concerning the Extraction of the Magistery from Fatty Substances

None otherwise must it be thought concerning the extraction of the magistery from fatty substances, as the fat of amber, of resins, of oils, and of other things, even as they appear in like material substances. Concerning these there are three methods of extracting their magisteries. One special method is that of amber, another of resins, and the other of fatty substances, such as oils, tallows, butter, and the like. For amber does not in any way admit of the same process as fatty substances, since its virtues would be lost. Resinous substances, again, do not allow the same practice as amber does, for it would be destructive to them. We will therefore teach the preparations of those magisteries in three ways, since such excellent virtues appear in fatty substances, and these in many forms. In cases where essences produce no effect these fatty substances render aid ; for they have that property on account of their specific and appropriate virtue, which is not found to be so perfect in other substances, neither, indeed, is it so. None otherwise than as the difference is between corals and gems when compared must it be understood of these things, since the practice is as follows ; and first concerning amber :

Take of amber well ground as much as seems good, and of circulatum a sufficient quantity. Digest them in a flat vessel in ashes for six days. Afterwards let the circulatum be distilled off from thence, and again poured on. Let this be repeated until an oil is found at the bottom ; which oil is the magistery of amber. This has revealed to us its wonderful virtues. May it continue so to do !

Resinous bodies are reduced to their magistery by the following method :

Take of turpentine, gum, or resin, as much as you will. Place this in a luted glass vessel and let it be digested by itself for a month in a warm digestion. Afterwards, having mixed it with dissolving water, let it be boiled in *vinum ardens* for half an hour only. Then distil it in a blind alembic and let all stand for one day. Thus you will find certain oils distinct from one another, which separate. Each one of these is a magistery in its own nature.

But the magisteries of oils are made without the addition of dissolving waters ; and these in like manner have virtues corresponding with the virtues of their matter.

The Extraction of the Magistery from Growing Things

If we speak of growing things, we understand as growing things those which are green and afterwards wither, and again become green in their season, even as they were before. The magisteries of these are made and extracted from them in different ways, as, for instance, in one way from trees, in another from herbs ; the difference whereof consists in this, that the former is wood, the latter putrefiable matter. Therefore the leaves and flowers must be prepared as herbs are. Accordingly, we will set down these Magisteries separately. The preparation of the magistery of woods is as follows :—

Take wood, cut sufficiently small, which put into a glazed pot able to stand the fire, and closely shut up. Let it be burnt in a coal fire for four hours. Then take it out and putrefy it in glass for a month. Afterwards distil it in ashes even to its last spirits, and when these are perceived, presently cease, so that the magistery may not acquire any evil odour from the fire. In this way you will have the magistery of the wood which you undertook to prepare.

So, too, may seeds, roots, barks, and the like, which contain an oil in their material substance, be extracted. There lies hid in these extractions a greater art than is spoken of or understood, though the process is here shewn in its entirety.

Herbs, however, and other things of that kind, must be mixed with *vinum ardens*, and putrefied with it for a month. Then they must be distilled by the *balneum maris*, and that which is distilled must be again poured on. This must be repeated until the whole quantity of *vinum ardens* shall be four times less than the juices of the herbs. Distil this in a pelican with new additions for a month, and then separate it. When you have done this, you will have the magistery of that matter or that herb which you selected.

We now wish to make clear the magistery of wine, which, indeed, appears to be endowed with countless virtues, since it receives a nature of the same kind from many virtues lying hid in the earth, as we set down in our treatise on The Generation of Wine. Now, it should be known that the magisteries of wine are produced in two ways, one of which we pass over in silence, because the practice thereof is common, such as we use in many extractions of wines.   Some practical methods for this magistery of wine are here set down.   Certain persons seek to extract it while the wine is still new and boiling, during the time of purifying.   Some bury it, and leave it so a 100 days (or years).   Most people separate it without fire.   However, this may be done, I do not wish to write of it here.   I will only hand down that method which I have proved experimentally.   First of all, it should be known that wine is a very subtle spirit, though small in quantity, and contained in much phlegm.   And although this be a quintessence, nevertheless, a magistery can be made from it, but by using a superior practice and process.

It should be known, too, that there are more and greater virtues in that wine which has not yet deposited its tartar, since there are sometimes more virtues in the tartar than in the wine itself.   The age of wine, too, is more commendable than its youth or newness, for its spirit is more digested by the lapse of a longer than of a shorter time.   Moreover, it must be borne in mind that the wine for this purpose has to be buried in the cold earth, and the vessel containing it must be closely shut below and above without any vent.   In this way it is preserved for many hundreds of years without tartar.   We are unwilling to speak here of any prolonged time, which would be tedious ; but, nevertheless, let this be committed to memory.   So, too, that is not a magistery of wine which is extracted out of must, or new wine, but a magistery of must. Nor is it a sign of art to distil it with its own fæces, or its own phlegm, as with *vinum ardens*, because in this way the spirits of the wine, which are in its essence, are destroyed.   This should on no account be done.   So, the oftener the best vinegar is distilled, the farther does it withdraw from the spirit of wine.   Wherefore, the utmost care must be used so as to see that the quintessence shall not be in any way corrupted in the magisteries, but that it shall rather be increased and fortified in its virtues.   Again, when it is separated without fire, it cannot in any way be a magistery, since the substantiality is lacking to it.   Also, it should be known that the spirit of the wine must be kept along with its substance and not with its phlegm.   For in wine two substances are found : the one is vinous, in which is the spirit of the wine, and from which it cannot be separated ; the other is phlegmatic, which is mixed with the dregs, and with the sweet water.   These should be separated from the true substance, as a metal is from its mineral, or from the earth.

It must be known concerning wine that the dregs and the phlegm are, as it were, the mineral, and that the substance of the wine is the body in which the essence is preserved, even as the essence of gold is latent in gold.   According to which we put the practice on record, that so we may not forget it, as follows :—

Take very old wine, the best you can get as to colour and taste, and of the same as much as you please.   Pour this into a glass vessel, so that a third part thereof may be full.   Close it hermetically, and keep it in horse dung for four months at a continuous heat, which heat do not allow to slacken.

Having done this, then, in the winter season, when the frost and cold are excessive, let it be exposed to them for a month, that it may be frozen. In this way, the cold thrusts the spirit of the wine, together with its substance, into the centre of the wine and separates it from the phlegm. Throw away that which is frozen, but that which is not frozen you must consider to be the spirit with the substance. Having placed this in a pelican with a digestion of sand, not too hot, let it remain there for some time. Afterwards take out the magistery of the wine, concerning which we have spoken. As to those additional processes which are in existence, and are put in practice, we will speak clearly when we shall treat of elixirs. Coming to an end here, we will ignore some of those other processes concerning wine which have little attraction for us.

Concerning the Extraction of the Magistery out of Blood

We wish also to unfold our opinion concerning the blood, in which there are very many and wonderful virtues, exceeding belief, but still sufficiently evident, chiefly on this account, that the blood exists out of the best root and most potent fountain of the heart, as we make clear in the treatise on The Composition of Man.* In this exists, and can exist, no defect, for it has its conditions according to the nature of the heart, and is a costly treasure of the whole nature, with all that is therein.

Here someone may urge that the blood, when it has flowed out of the veins, will be soon deprived, by the nature of the case, of those virtues which renew and sustain that blood. But this is not so; for it can be preserved in the essence, as we shall point out below. Let us, therefore, consider the small number of men who live with healthy body and blood. Wherefore, care must be taken that men of this kind be brought to a renewed quality and essence by arcana and the quintessence, as we have before mentioned, so that the blood may flow from them incorrupt and healthy. Nor do we speak of man's blood only, but also of the blood of the sperm, which we describe in our secrets, in which blood there is no disease or alteration, but the most wonderful blood of the human sperm, which we intend to take in this place, and this for many causes not here set forth. We speak also of the blood of bread, which is to be taken for the same use and in the same way. For there are in it such virtues as we are scarcely able to scrutinise thoroughly, nor do we undertake the task of doing so to the full. In the same way it may be understood of other nutriments and comestibles, in all of which blood is present, although we do not see it in them any more than in bread; yet, nevertheless, by putrefaction, as in the stomach and the liver, it becomes blood. So also every condiment which is taken therewith is changed after the same fashion as in the body. We are unwilling to write of this blood more largely or at

---

* The *Liber de Generatione Hominis*, to which allusion is apparently made, exists only in the form of a fragment, and is concerned chiefly with the seminal philosophy of Paracelsus.

greater length, especially for the reason that we cannot make the subject agreeable to any but ourselves. We determine, therefore, to take rest in sleep, and then, waking from our pleasant slumber, we will go on to speak further about this blood. Let each thing prevail so far as it can by its own virtues, and according to what it has in it : for out of a good thing much good proceeds, and this always stands forth for our consideration. Nor shall we speak only of the blood of edible things, but also of potable things, which exhibit simply blood to our body. There can also be extracted from blood quintessences as well as arcana ; but of these former here we make no mention, having set before us to treat only of magisteries and to comprise each in this one :

Take blood, shut it up in a pelican, and suffer it to rise up so long in a *venter equinus* until the third part of the pelican shall be filled. For all blood in its rectification is dilated according to the quantity, and not according to the weight. When this time is fulfilled you must rectify it by a bath. In this way the phlegms recede, and the magistery remains at the bottom. Having shut this up in a retort, and hermetically sealed it, distil it nine times, as we have taught in the book, "Concerning Preparations." In this way you will arrive at the magistery of blood.*

---

* The treatise Concerning Preparations makes no reference whatever to the extraction of a magistery from blood.

THE END OF THE SIXTH BOOK OF THE ARCHIDOXIES FROM THE
THEOPHRASTIA CONCERNING MAGISTERIES

THE SEVENTH BOOK OF THE ARCHIDOXIES

FROM THE THEOPHRASTIA OF PARACELSUS THE GREAT

CONCERNING SPECIFICS

IT remains, now, that we speak also concerning specifics, in which reside many marvellous and great virtues, which do not derive their origin from Nature, so far as they are warm or cold ; but besides these qualities they have one nature and essence, as we have mentioned in several places.   Such a specific takes its nature from externals, as when a piece of wood is thrown into the fire and is burnt; that burning is not part of its peculiar nature, but the being wood is.   Specifics, therefore, are produced from a conjunction, as when mastick and colophonia are blended, an attractive substance is generated, which is neither of these ; or, when terebinth is coagulated, a stone is produced therefrom which attracts iron to itself like the magnet.   There are kindred substances which have such virtues, but from their compositions and from without. Ellebore, too, is compounded from the liquid of stone and earth. From the composition of these two a specific proceeds. So the oil of cherries and acetum after their digestion produce a laxative, though neither of them in its own nature has a laxative property. Wherefore, such specifics are produced out of their own nature by composition of elements and of the proper nature, just as tincture or colour, which is not derived from cold and heat, but from composition, as galls with vitriol produce ink, though neither the one nor the other is black by nature.   So, also, sal ammoniac and urine produce a black colour, when both of them are white. It is after the same manner with specifics. They have their origin from externals ; but some things which assume such virtues from without may exist in any herb, not in any one kind only, which must be understood in the following way.   Wherever the magnet has grown, there a certain attractive power exists, just as colocynth is purgative and the poppy is anodyne.   This arises from the composition which exists in them. Hence it is that every magnet is attractive and all colocynth is purgative, But such is not the case with specifics from without, of which the condition is as follows : if one flint has the virtues of a magnet, and another like it has not such virtues, this latter would be an external specific.

But it seldom happens that a peculiar condition is found in one herb and not likewise in another which is similar to it.   Then, again, though many similar specifics arise from influences, nevertheless, we will not here discuss at length whether they arise from the same influences or not.   We prefer to reserve this discussion for its proper place in our Physics, and to pass it over here.   Moreover, many specifics are found—the odoriferous ones, to wit— which derive their origin from composition and digestion, as water of vitriol with sal ammoniac possesses the odour of musk, though neither of the constituents has this by itself.   There are many such, which become odoriferous when they were not so before, and acquire a notable fragrance, like a rose or a lily, though they had no odour before, yet by labour, by digestion, and separation, this fragrance is eventually aroused.   In like manner, cow-dung is a fœtid excrement, but if it be elevated, it acquires the odour of ambergris, while the residuum which is left at the bottom smells worse than human fæces.

There are some diaphoretic specifics, producing sweat, which acquire a similar virtue by composition, as a burning coal placed on fat earth emits a vapour.   So ginger placed on a body burns, and is extinguished like lime when water is poured on it.

This heat accrues to the ginger from the sharpness or asperity which it contains within it, and with a warm element it is coagulated like a stone, which, when placed in the fire, is brought to the same degree of heat.   Indeed, every diaphoretic is a calx of the liquid of the earth, as we state in our treatise on Generations.   In like manner, purgatives themselves come from composition, as in the case of rhubarb, which is also the calx of a liquid, with a certain difference intervening : for as a Tartarus draught it is resolved into water, and has with it some liquid in itself when it is put in a damp place ; so also rhubarb and other purgatives have a manifold origin, as being a calx of the earth.   Some remove the cholera, as rhubarb which is like calcined tartar. Some take away phlegm, as Turbith— so written because the word was wanting in the autograph.   With these realgar is dissolved, but nothing else. Others cure melancholy, like senna, which is to be understood after the manner of nitre, which resolves stones as nothing else does.   Some act on the blood, as manna, which, like arsenic, resolves sublimates.   Thus we must judge concerning the difference of these substances, seeing they are divided one from the other as we have mentioned.   Some which are strengthening are derived from composition, as sperm lacking strength, from which, nevertheless, a glandule is produced according to its predestination.   So then the strengthening specific is a certain predestinated body by the predestination of its composition.   But the carline, which is not produced in this way, attracts to itself the virtues of other herbs, and takes away their strength, which it then possesses solely, just as the sun attracts to itself the moisture from wood, as we declare more at length in the treatise on Generations.

In like manner, also, some mundificatives acquire that virtue by composition, as when the calx of the earth is transmuted into another form by liquid, as resin, honey, gums, pitch, etc.   Similar alterations exist in the flowers of Venus which at first are a purgative, as should be the case with a calx. Afterwards they are reduced by fire into a styptic, so that they lose their purgative properties and have a mundificative effect. It is the same with corrosives which are salts, and sometimes calcined in earth, sometimes substantiated (if one may so say) into one matter, as we set it down in the book on the Different Species of Salts. Many and various are the properties of this kind in things, which we handle in our treatise on The Properties of Things, and about which we have said sufficient here. Also why some are hard and others, on the contrary, soft, this we are unwilling to disclose here. Our Archidoxies do not treat of the whole principle, but only the special subject of compounding specifics and bringing them to the highest grade of Nature. With these we will now deal in due order, putting forward our own experience of them, and leaving behind those crooked haranguers who talk about God and understand nothing but hypocritical rites and similar fables, who also are the enemies of such as practise these arts and arcana. They are plunged in worldly glory, romancing and cavilling through their capacity for much speaking. At it they go with both shoulders (as folks say); they flatter and pride themselves upon being wise, when they are only stupid and fatuous, and for filthy lucre's sake they befool their fellow men. Now, then, let us go on to treat concerning specifics according to our usual custom, and leave those University doctors, who only read and think—be their success what it may— to gnash their teeth against us if they will.

Concerning the Odoriferous Specific

Now, then, let us speak of odoriferous specifics, in what way and in what form they are produced. And, first, as concerns their powers. An odoriferous specific is a matter which takes away diseases from patients, just as civet destroys ordure, with its odour. This specific mingles itself with the foul smell of the ordure; and the smell of the filth can no longer hurt or remain. The stench is tinged with fragrance, so that the good odour is as strong as the bad one was before. Since there is nothing which can take away the good odour from civet or musk, it is transmuted, as we prove in many passages. Hence it happens that occasionally some of the excrement is mingled with the musk, because this penetrates more readily than any lily with all its operations. It is well known that more of bad odour than of good is met with; for as tyrus is taken in theriaca in order to penetrate all the limbs at once with another influence, so, in our opinion, does the odoriferous specific act.

By means of the odoriferous specifics diseases are cured in persons who cannot take medicines, as in cases of apoplexy and of epilepsy. Many odours exist which relieve the epileptic; and many, too, which aid the apoplectic. They may not cure them, but they pave the way for a cure. A force of this kind, brought to bear on the body, immediately stirs up the blood, and by forcing this to the heart, revives it to an indescribable degree. We will therefore prescribe an odoriferous specific which shall serve as a foundation for compounding such specifics against all diseases. The process is as follows :—

Take of white lilies, anthos, basilicon (carbon), cardamum, and roses, one handful respectively, with two handfuls of spike. Pound into a coarse paste. Add two quarts of the juice of oranges. Let all these be digested in a pelican for the space of one month. Afterwards separate them from their dregs with the hand, or, what is still better, with a press. Put this again in a pelican, and add one ounce each of mace, caryophylli, and cinnamon ; half an ounce of ambergris, two ounces of musk, and one ounce of civet. Having ground these very minutely, put them in the aforesaid pelican, and digest them in dung with the other ingredients. Then add half an ounce of gum arabic in solution, also one ounce of similarly dissolved tragacanth. Place these in a closed glass vessel and indurate them with the clear part from white of eggs. You will presently see the mixture assume the form of glass. Break the vessel, take out the stone, and you will have the odoriferous specific, concerning which enough has now been written. It would, however, be well to add to it aurum potabile.

CONCERNING THE ANODYNE SPECIFIC

Many causes combine to induce us to write about this anodyne specific. There are some diseases in which all arcana fail us, with the exception of this anodyne specific, which works wonders. Nor need we be surprised at this, when we see water extinguish fire. Just in the same way the anodyne specific extinguishes diseases. There are many reasons why it does this, but we pass them over in silence.

That which rests or sleeps does not, in the course of nature, cause any discomfort. A paroxysm, if it sleeps, is not felt ; if it does not sleep, its operation is accomplished. We console ourselves with the reflection that many anxious cares and cases of melancholia are removed by simply sleeping.

And note this : Sleep does not necessarily apply to the sufferer only, but may be predicated of the disease itself. We, therefore, compound a specific which operates on the disease alone, not on the entire patient, as is the case in those applied to fevers. What we put forward would be deadly to the whole man, but is salutary in its application to the disease. We attack the disease itself, and so elaborate our remedy that the ailment shall have, or can have, positively no effect on the body. The following is the formula : Take of Thebaic opium, one ounce ; of orange and lemon juice, six ounces ; of cinnamon and caryophylli, each half an ounce. Pound all these ingredients carefully together, mix them well, and place them in a glass vessel with its blind covering. Let them be digested in the sun or in dung for a month, and then afterwards pressed out and placed again in the vessel with the following : Half a scruple of musk, four scruples of ambergris, half an ounce of crocus, and one and a half scruple each of the juice of corals and of the magistery of pearls. Mix these, and, after digesting all for a month, add a scruple and a half of the quintessence of gold. When this is mixed with the rest it will be an anodyne specific, capable of removing any diseases, internal or external, so that no member of the body shall be further affected.

Concerning the Diaphoretic Specific

Let us now speak of the diaphoretic specific, whereby every disease is cured which can or should, from its nature and properties, be treated by sudorifics. Such a disease is removed barely better than by any medicament. By means of the diaphoretic, a cold disease grows warm, and is removed by that heat. It has often happened to us that a cassatum of twenty years' standing has been cured by the diaphoretic; and other diseases which they call intercutaneous, or which have their seat in the marrow. On these the quintessence has no effect, and still less the strengthening arcanum; principally for this reason, that there is not sufficient strength in the heart to drive away the cassatum unless this be done by the diaphoretic alone. As the sun warms a frozen stone, and liquefies the hardest ice, just in the same way the diaphoretic exerts its powers on a disease which can be overcome by no other power, however excellent and good. Now, although the flammula be a very warm herb, still its heat cannot be compared in the faintest degree with that of the sun; and so these same warm diaphoretics differ one from the other. Wherefore we put forward at this point a specific which, as it were, summarises the diaphoretic properties. Take one pound of ginger, half an ounce each of long pepper and of black pepper, three drams of cardamum, and one ounce of grains of paradise. Let these be ground to powder, and placed in a glass vessel with an ounce and a half of best powdered camphor and two ounces of dissolving water. Let this mixture remain in a sealed glass vessel in sand so long until its digestion is completed. Then separate therefrom the dissolving water, let it putrefy for one month, and circulate it for one week. Afterwards press and keep it. This is the best and most powerful diaphoretic, acting more vehemently than could be believed both in cassatum and in other diseases. Enough on this point.

Concerning the Purgative Specific

We must also describe the purgative specific. And although we shall have to consider its complexions and the like, still we base our remarks on a more solid foundation, and accept only those specifics which exhibit not one disease or another, but all diseases. Hence it is gathered that whatever in the case of cholera is superfluous, and of no moment, is removed by this specific. Similarly also in cases where the symptoms show the phlegm or the blood at fault, or in melancholia, as also in abscesses and other affections which cannot be cured by mere complexions, nor be resolved by purgatives, as is the case with many diseases.

Our endeavour is solely to remove the peccant matter, whether it be corrupt or not, whether in the form of an abscess, or in complexion, or arising from any admixture. Paying no heed to the prolix and useless discourses of the University doctors, we give our thought simply to sanitation, and, with this end in view, we shall build up our medicaments. Of such a nature, for example, is tartar, which, by its nature and properties, takes away all putrefactions and is not inclined to cholera, to melancholy, to phlegma, nor to the blood, but expels all that is useless from the body, or that can do harm to it. In like manner, vitriol purges every complication from which diseases arise. Coloquinth does not drive out all cholera or purge what arises therefrom ; nor does rhubarb effect this. Neither does turbith arrest or evacuate all that proceeds from phlegma. Neither does lapis lazuli remove the impediments produced by melancholia, or manna drive out all bad blood ; but these two, besides other of our purgative specifics which we do not mention here, accomplish this, principally in all those putrefactions and superfluous fæces of the body, whencesoever they arise, just as water washes linen rags, and soap clears them from all defilement and impediments of any kind. So also do these act singly on single diseases. Let us then lay down a specific drug which, to our thinking, acts thus in the way just described. Take the magistery of tartar and the magistery of vitriol, mix them together, and afterwards add equal parts of the quintessence of crocus. Put them in a pelican, let them be digested for a month in sand, and let the drug of which I have spoken be carefully preserved. Concerning the other things necessary for this, greater intelligence exists among the erudite than among the ignorant.

And not only human beings may in this way be purged of their diseases and superfluities, but also trees and herbs. For in growing things, just as in persons, infirmities prevail, and their remedies, too, have grown, as we remark elsewhere. The magistery of vitriol heals the defects of anthos, when it cannot vegetate perfectly, and causes it to grow excellently, as we state in our work on Plants.* With these few words, therefore, we now conclude this subject.

Concerning the Attractive Specific

Before we begin to speak of the attractive specific, it should be known that this attractive specific draws to itself everything that is superfluous in the body, and leads out whatever evil may adhere to it. This frequently happens ; and the result is proved to demonstration.

Some attractive specifics are so adapted to the flesh that they attract to themselves a hundred pounds of flesh, just as a magnet attracts iron. It has happened in our time that an attractive of this kind has drawn a man's lungs into his mouth and so suffocated him. It has also occurred that, in another case, the pupil of the eye was drawn from its position right down to the nose and could never be removed thence. Attractives are found to act not only on

---

* The *Herbarius Theophrasti* contains no reference to the subject, and selections have already been made from that chapter in *De Naturalibus Rebus* which is devoted to vitriol.

iron, but also on wood, herbs, flesh, and water. We have seen a plaster which attracted so much water that a vessel could be filled from it, and the water ran down from that plaster just as from the roof of a house. In the same way, lead, tin, copper, silver, and gold can be drawn by the composition of attractives. By these same attractives, too, the branches may be torn from trees, a cow lifted into the air, and many other effects produced, which we have detailed in our secret writings as a Thesaurus; so that we may admire and venerate that man alone who has brought it about by his art that so many discoveries should be made; demonstrating as we do these incredible operations which so far excel Nature as it is constituted in itself.

We will, then, lay down certain attractive virtues for the body, whereby what is evil and corrupt may be extracted from it and separated from what is good, so that the attractive may be applied on an emunctory to the particular spot where any defect has shewn itself, as on an ulcer which may be taken on an emunctory, or if a glandular swelling has arisen this must first be opened like an emunctory. We know from experience that by means of an attractive of this kind the pestilence has been extracted in a way that it would not become us to describe here. When this medicament has been used, no sick person has ever perished, however severe the disease with which he was affected. The following is the recipe for the attractive specific: Take of the quintessence of all gums of every kind one quart, a pint of the magistery of the magnet, one pound of the element of *carabis igneus*, of the element of *mastix igneus* and of myrrh, each three pints, and ten ounces of the element of scammony. Of these make an ointment with wax, gum, tragacanth, and turpentine, using it as directed above.

CONCERNING THE STYPTIC SPECIFIC

We now have to speak of the styptic specific, the virtues whereof are very great, more so even than the others. When people see with their own eyes so many wonderful works of Nature, they urge us, in their delight, not to desist from them, but to tax our memory so as to impart what is offered by these arts of ours. And if, perchance, something was discovered by the old physicians and philosophers (which we deem by no means certain), that does not disturb us in any way, since all those things they wrote about were mere blind gropings. We delight in that great Nature which presents itself to our hands, and we rightly pass by the works of these ancients whom we esteem to have been blind, as we mention in many places. Is it not subject for wonder that styptics are so strong and possess a power in their quintessences, arcana, and mysteries, whereby two fragments of iron are so fastened together that afterwards they can only be separated by fire? What is still more remarkable is, that one piece of copper can be so joined to another by similar styptics that they can no more be disjoined either by fire or by water. By a like attractive styptic a heap of stones can be conglutinated into a mass like a rock. In like manner, sand and lime, by means of such styptics, are welded together in perpetual compaction harder than marble.   So far we speak of hard material. Let us go on to other kinds.   By the same method we have seen leaves joined together so that they were taken for one concrete natural growth ; for instance, the leaf of the lily with the leaf of the rose in one combination.   Blacksmiths, acting on my advice, have made their weldings as compact as though they had been originally one solid mass.

We have seen, too, by means of these styptic specifics, the lips of a person so firmly drawn together on a single washing, that afterwards they could only be separated forcibly by instruments, the blood flowing freely in the process.   Some persons, too, for a joke, closed up another's anus with these styptics so firmly, that when it was necessary to go to stool it had to be opened with a gimlet !   So, too, in the case of wounds and in rupture of the bladder, we have seen the compaction made so effectually that no opening occurred in our time or during the life of the patients.   Whatever limb this styptic material touched, it so contracted the orifice or the bare flesh that it could only be made smooth with a file.   No water softens the force of such styptics, which is really much greater than we have described.   So far, however, as relates to medicine, we will detail the styptic specific as follows :—Take quintessence of bolus, that is to say, of iron, and the quintessence of carabis, otherwise called cathebes, one pound of each.   Let these be digested in ashes for a month, and afterwards put in a pound and a half of dried tartar.   Keep this body until required.   This and similar styptics are inscrutable in their own bodies, but when employed on separate substances, they attract in an incredible manner, through the nature and condition of their great dryness.   So, then, these styptic specifics may be called styptic beyond all others.

Concerning the Corrosive Specifics

We would now describe in addition the corrosive specific, wherein marvellous powers are naturally present.   It has this wonderful property, as compared with the old corrosives, that it completely annihilates metals, so that no body is any longer found in them, just as wood disappears in the fire.   It is true that the consumption of metals is brought about by aqua fortis, but still some part of their bulk remains undiminished and unchanged in its essential nature, so that it can be once more brought back by means of fire to its original body and matter.   But this is not the case after consumption brought about by the corrosive specific, and for this reason : because no matter is any longer found in this case which can by any means be brought back to a metallic nature, any more than ashes can be again turned to wood.   It should be known, too, that this corrosive acts upon flesh in a way to which nothing else is comparable. Its velocity in one moment penetrates the hand like an awl.

Now, we recount this in the interest of medicine, with the following end in view.   In the body a good deal of putrid superfluous flesh grows up, in the way of ulcers, like fistulas, scrofulous cancers, etc., all of which can be cured by corrosives of this kind. In this specific a styptic virtue exists of very great power, by which it acquires special curative faculties. It might really be called rather a fire than a medicament, seeing that it consumes iron chains and bars in a way beyond belief or power of description. We give the recipe for it as briefly as possible : Take one pound of aqua fortis rectified from its *caput mortuum*, half a quart of sublimated mercury, two ounces of sal ammoniac. Mix all these and let them be consumed ; then add an equal weight of mercurial water. No adamant can resist this corrosive. Although the same is understood concerning the quintessence and the arcanum, namely, that the skin may be taken away by either mode of cautery, and new skin superinduced in its place, as in leprosy, morphea, serpigo, lentigo, pannus, etc., against all these this corrosive specific comes into use ; but we omit it on account of its violent action. We choose in preference this mixture, with which the skin should be washed, when it falls off and is laid bare. Afterwards it can be consolidated in due course. The mixture should be as follows :—Take one pound of the juice of the flammula, four and a half of cantharides, and ℥ ij. of the Ignis Gehennæ aforesaid. Mix and use as above.

CONCERNING THE SPECIFIC OF THE MATRIX

One would speak of the specific of the matrix for several reasons, specially on account of the affections to which it is liable. But in this place we hesitate to treat of those substances which warm or refrigerate it, since these results can be brought about by magisteries and arcana. Here, however, we set down two specifics, one for suffocation of the matrix, the other for promoting or restraining the menses. Suffocation can only be remedied by the specific, though not elementated or prepared for this object, but in its common form or essence, adapted to the said purpose even as the skin of the fig is. As soon as ever its fume enters by the vulva the disease is expelled. And this is chiefly remarkable on account of its being such a cheap simple ; and though it be prepared, the essence of its fume, wherein lies its sole virtue, is not lost. For promoting the menstrual discharge the method is the spleen of a castrated ox reduced to a magistery or quintessence. This is an admirable provocative not only in young, but in old, women.

As to restriction, the best method is to use the quintessence of coral, the oil of iron, or *ferrum potabile*, which restricts more than anything else.* There is no need to describe the properties at greater length, since this would be too prolix for our Archidoxies. It should be remarked that, under the above-mentioned compositions, that is to say, the incarnatives, the conglutinates, and the specifics, are comprised laxatives, mundificatives, and the like, under purgations themselves ; and so with others, as the deoppilatives under

---

* By this oil of iron every genus of humid, fluid, and flaccid ulcers, those also which have swollen and ruddy lips, are completely cured, and the older and more deeply rooted the ulcer is, so is the remedy more ready and easy.—*De Tumoribus, etc., Morbi Gallici*, Lib. X.

purgatives and attractives.   Here we would conclude our book on Specifics, as an aid to memory, lest we forget them.   In like manner the confortatives are also recounted in single chapters.

THE END OF THE SEVENTH BOOK OF THE ARCHIDOXIES FROM THE THEOPHRASTIA OF PARACELSUS THE GREAT CONCERNING SPECIFICS

THE EIGHTH BOOK OF THE ARCHIDOXIES

From the Theophrastia of Paracelsus the Great

Concerning Elixirs

AFTER many of the most secret mysteries of Nature, we desire to treat compendiously of Elixirs; and that not in vain, since we see that there is latent in them the chief conservation, which compels us to bend our mind thereto without any rest.* For every elixir is an inward preservative in its essence of that body which shall have taken it, even as the extrinsical balsam is an external preservative of all bodies from putrefaction and corruption, a fact which is evident enough in balsam, that is to say, it preserves bodies so that they abide many hundreds and thousands of years without corruption or change.†

When, therefore, we see a gift like this in balsam, which preserves dead bodies and keeps them in incorruptibility, it is to be understood equally that by this same gift, or mystery, the sound and living body can be far better, more usefully, and more conveniently preserved. But we have not this naturally that these mysteries of Nature and those constituted above Nature, by which we may preserve the body inwardly and outwardly from all contrariety, should be known to us; but among them many things meet us which are most occult.

Concerning elixirs, then, it should be known that these have not their operations from Nature, nor from their complexion, but that they are mysteries rather than specifics, leading us to the very highest admiration of the Creator by many demonstrations. Yet they are implanted in Nature herself, so that they are in her, as is seen in the case of balsam. If, therefore, it be possible to preserve dead bodies, still more so living bodies themselves can be preserved. And there is no reason why we should heed the words or arguments of our adversaries, but rather it is well that we should disclose our own, by which means we desire to guide towards the true foundation of the internal balsam, not heeding their triflings and their useless words who talk of a limit

---

* The *Scholia in Libros Paragraphorum* define the elixir as a ferment, a quintessence, the pure substance separated from the impure.

† Of balsam in general, Paracelsus says that it is a temperate thing, neither sweet nor bitter, nor pontic; it is a liquor of salt and a salt of liquor. Therefore it most powerfully preserves from putrefaction. - *De Tartaro*, Tract. IV., c. 2.

of death and of its predestination, and close it in with determinate points. For God our Father gives us life, and along with it medicaments by which we are able to defend and sustain it. If, therefore, the limit of death were laid down at a precise point, it would of necessity follow that this other theory should be false, which is not the case ; but as long as we have power and knowledge, we possess the capacity of sustaining our life. For Adam attained to such an advanced period of life not from the nature or condition of his own properties, but simply from this reason, that he was so learned and wise a physician, who knew all things in Nature herself with which he sustained himself during so long a period. And there were many others, in like manner, who used similar remedies.

There were in the days of Adam many who died without reaching his age, and some did not even attain our limit, as we are now constituted since the Deluge ; and these died just as we die, because they were unskilled in those arts which Adam and others understood ; and hence it happened that they were deprived of life before their due time, nor did their foods or drinks help them.

Since, therefore, we are enabled from such examples as these to discover naturally that a protracted life proceeds from Nature, we desire to investigate what it is in which Nature and the gift of God consist. Some things preserve a dead body from putrefaction for one year only, as the *oleum laterinum*. Others preserve it for ten years, as the corrected oil of the philosophers. Some preserve it for twenty, as the water of honey. Some for fifty, as the distilled preservative. But some preserve it for ever without end, as balsam. Some preserve it only for eight days, as salt ; others for a night, as distilled water. Some preserve it longer, as *vinum ardens*. There are also some others which preserve the body from corruption in a new and strong essence of Nature when a man is confirmed by them, as aloes, citrine, and myrrh. Some preserve bodies from corruption only by their great tincture, which is so powerful that it admits no evil, nor suffers it to grow or to enter secretly, as gold, the sapphire, pearls, arcana, magisteries, and the like, as has been before written about these.

We purpose, therefore, to describe a preservative against all corruptions of the living and the dead body. But it is considered that a preservative of the living body must be taken at the mouth, so that there may be no member of the body which does not receive that same preservation, or which may not be informed by attracting to itself the benefit thereof. Moreover, it must be remarked that the spirits of the excrements existing in the bowels are so strong that they fight against a preservative, for this reason, that nothing which is putrefied can be embalmed or preserved, for it has not in itself any essence as recently dead flesh which is embalmed has. But this agrees with a preservative none otherwise than as worms do with the best herbs, and as a putrefied substance does with one that is incorruptible ; since whatever is putrefied cannot be further corrupted or be changed, since there is no movement in it. On the other hand, a preservative cannot in any way be putrefied, for it is like gold, which never rusts. They are mutually separated one from the other, so that each fulfils its own proper function. This, therefore, is said, because, in lapse of time, the excrements are able to overcome the preservatives; but this cannot occur in dead bodies, because they are disembowelled, or, if not yet disembowelled, they are coagulated by death, just as the blood withdrawing from the veins coagulates.

We call this preservative an elixir, as if it were yeast, with which bread is fermented and digested by the body. Its virtue is to preserve the body in that state wherein it finds it, and in that same vigour and essence. Since this is the nature of preservatives, namely, that they defend from corruption, not in any way by purifying, but simply by preserving. The fact that they also take away diseases is due to the subtlety which they possess. So, then, they do not only preserve, but they also conserve. They have a double labour and duty, that is to say, to prevent diseases and to keep the essence itself in its proper condition.

Nor do they do this in human bodies only, but in all bodies possessed of sensation. Thus, also, dead wood can be preserved from corruption, just as any body that is treated with balsam. Herbs, too, can be preserved in their essence none otherwise than a living body can: since those conservations which apply to herbs keep them in the same essence as that wherein they find them, so that they flourish and remain as fresh as in the fields and gardens, or elsewhere, to the fifth or sixth day. If they find them with flowers, they preserve these; if with fruits, these also.

Nor wonder at this, when it is possible for dead wood to live again, and for iron to be so fixed that it shall never rust again, and likewise for sulphur to be made incombustible, all which things are beyond the understanding of a simple man. The cause of all these things we set down more broadly and fundamentally in the treatise on Conservations. Nor must they be considered impossible, since many other things which are deemed impossible can be most easily accomplished. We wish, then, to speak of the conservation of balsam by the distinctions of ages in the following manner.

ON PRESERVATION AND CONSERVATION BY ELIXIRS

We purpose to write of the first elixir, which conserves the body in that essence where it finds it, and does not suffer it to putrefy or grow weak, but conserves it in the spirit of life, so that no accident can happen to it. It also brings it to a third period of life, or to more. As to its use, the operation on dead bodies differs from that on the living, since the dead must lie in the balsam nights and days, but the living and healthy bodies neither can nor need do this.

Wherefore, this elixir must be understood to be of use only for life, and for the heart and those parts especially in which the life chiefly flourishes, that is, the spirit of life dispersed throughout the entire body. And it preserves the spirit of life in that virtue wherein a dead body or corpse is kept from putre-faction; because, if a wound or an ulcer may be externally preserved from putrefaction and from every mishap, so, also the inner body is arranged so as to be defended from all adversity.   Wherefore, we arrange the elixir which is directed to the spirit of life none otherwise than as yeast operates in paste; and with reference to the body, as when a tree is tinged in its root, so that the same colour may never withdraw from it.   In this way the whole body is preserved, when the tincture is more or less scattered amongst all the members and penetrates them none otherwise than as the whole metal is tinged into gold, and becomes gold, or is preserved from rust.   So, also, is it in the conserved body; there is no member comprised in it which is not full of the elixir.

After that, being dispersed throughout the whole body, it shall have acquired its virtue, and already is exercising its operation by itself, no corruption can happen to it by contact, because the life of every member is full of the elixir as the body is tinged with the balsam.   But it should be understood that it is not necessary the whole body should be affected with the balsam by means of the elixir which has been taken; for where the spirit of life has only been surrounded therewith at its root, this suffices for the conservation of the body.

Now we must come to the practical method; and first of all treat concerning the elixir, which by the conditions of the powers of balsam preserves the whole body from decay.   Then concerning that which preserves the body by the potential power of salt.   Thirdly, concerning the elixir of sweetness, which supports the body in its conservation.   Fourthly, we will treat of the elixir which enters the human body by the powers of a quintessence.   Fifthly, another shall be appended which is truly noble by the force of its great subtlety: for it resists all the enemies of Nature, by which resistance it never suffers the body to fall into disease.   In place of a conclusion, we will add that elixir which, by the forces of its own proper nature, is endowed with similar conser-vations.

CONCERNING THE FIRST ELIXIR, THAT IS, OF BALSAM

Take of the true and very best balsam, well known to us, one pound. Let this be put into a glass covered with a blind alembic, and together with it pour in two ounces of the quintessence of gold and one ounce and a half of the essence of the greater circulatum.   Let all these be digested together with a slow fire, so that the vapours may ascend night and day.   Afterwards let the fire be increased so that some drops may adhere, and may fall down, drop by drop, for two months.   At length let them remain in horse dung for four months, so that they may have their digestion without intermission.   When this has been done the elixir is finished.   It must be understood that this balsam or elixir becomes a fermentation which is developed and mingled with the root of life, and has the power of ruling the life in a good essence, so that no nature can resist it.   None otherwise than as arsenic overcomes Nature for evil does this elixir, on the other hand, overcome it for good, defending the body. The dead body is kept safe by this odour so that it cannot putrefy when it is put in the tomb, and is covered up so that it may not evaporate. Much more do its own virtues remain in a living body, which we hope to have sufficiently explained in this place.

CONCERNING THE ELIXIR OF SALT, BY THE FORCE OF WHICH THE BODY IS CONSERVED

There is no less power and virtue in salt than in balsam, whereof we have spoken ; for this reason, that flesh is preserved from putrefaction by salt for days, years, and long periods ; and that in different ways, and one way more than another. On the same basis it would be possible to conserve and to preserve the body ; not because we mean to use salt in the precise way as with dead flesh, but from it should be compounded the elixir of salt, which penetrates materially the spirit of life, so that it lives by the salt as salted flesh does. For this elixir is so subtle that it can be brought to bear on the spirit of life. Thus these two meet closely in one conjunction, so that the one is tempered by the other to perfection, just in the same way as salt perfects certain food in point of taste, without which it could scarcely in any other way be brought to perfection in respect of unity. And it must be remarked that the elixir of salt is a fermentation in which exists a certain tincture whereby the whole body is penetrated. It is also an inconsumable thing, which is not consumed in the body by digestion along with natural things, but is fixed as glass in the fire, which does not at all perish in the process. This fixed elixir so fixes the body that it becomes permanent in life, none otherwise than as when a metal is fixed, in which case no damp afterwards, no corrosive, or anything of the kind, can injure it, or produce rust in it. So, then, it may be inferred from hence that the elixir is a fixed body, like gold, into which nothing impure can penetrate so as to hurt it. The practical preparation of this elixir of salt we will set down as follows :—

Take salt, prepared in the best possible way, very white and pure. Let this be placed in a pelican with so much dissolving water as may exceed six times its weight. Let them be digested together in horse dung for a month. Afterwards let the dissolving water be separated by distillation, and again poured on and separated as before. Let this be repeated until the salt is converted into an oil, to which let there be added an eighth part of the quintessence of gold. Let them be digested together in a pelican and in horse dung for four months, and afterwards circulated for a month. Let there be added another part of circulated wine ; and let them so remain in ascension for a month longer. When that time has elapsed you will have the elixir of salt, concerning which we have made for ourselves a memorial according to rule for the relief of our ancient days.

CONCERNING THE THIRD ELIXIR, NAMELY, OF SWEETNESS

We are certainly assured that bodies may be preserved from corruption by sweetnesses ; but by what forces this is done we set down in the books of the Generations of Honey,* Sugar, Manna,† Thronus, and the like ; and we are unwilling to repeat it in this place on account of the writings of the ancients. We are able to transmute sweets into an elixir, the preparation of which rather conserves the living body in its conserved essence than a languishing body. For it is the property of all specific sweetnesses, that they are neither corrupted nor do they allow this to be corrupted, unless by contrary things they become liable to corruption, as out of honey and bread worms and ants are produced, as also out of sugar and thickened milk. Out of manna and water is produced a corruption like dung. Many more similar compositions may be made, by which the sweets pass into corruption. That this may be obviated, the following is our intention and experience, namely, that in this composition such a thing should be taken as will not prevent the sweetness from remaining in its proper essence, and such as may be without the corruption of other things. In this way, it has the virtues of a balsam for preserving the dead fleshly bodies of corpses or other things. A like sweetness is the balsam of the earth, and some others of the dew, because it derives its origin from them. Now, therefore, we will set down the elixir of thronus, since no sweetness can be compared therewith, and it contains more mysteries than could be believed, as we state in our treatise on Generations. By the preparation of this may be gathered the methods of preparing other sweetnesses. Let that of thronus be as follows :

Take as much thronus as you will, which place in a pelican, and set in the sun for digestion during two months, or, better still, for the whole summer. Afterwards let there be added a fourth part of the quintessence of gold, and so let them be circulated together for two months. Keep this. Although this method is short, nevertheless the elixir made by it is wonderful in the case of very aged persons.

---

* In the description of the essence and property of honey, it must first of all be understood that the prime matter of honey is the sweetness of the earth which resides in naturally growing things, and is extracted out of the property of the same by magnets.   Hence you ought to gather that in every prime matter all that concerns growing things is col-lected, just as when several colours are combined one only appears though all are present.   Similarly, also, the seed is the wood, the leaves, and the branches, not actually and in the present, but for the future, if it be brought forth and grow.   So, also, in the prime matter of things there is a similar composite from which all growing things attract what they require. . . .   Now ye must know that honey in its first matter is a resin of the earth, but resin is a gum not of all flowers and growing things, but of some only.   Of some there is a resin produced, of others a sulphur and bitter-ness, and of yet others something else.   That which grows in flowers and locusts is conceived with the form and appear-ance of honey.   It does not yet dwell in its ultimate matter, being perfected by the Sun and the Moon.   These two planets, however, can bring nothing to its final perfection without all the assistance of a celestial operation, which in the case of plants, etc., is the summer star. . . .   Honey, therefore, is a terrene spirit at first, but if there be applied the influence of summer, it generates a corporal spirit, that is, the spirit which in the sun and moon was terrene becomes a corporal spirit.   Bees can remove the same and take it away to their hives.   This is the prime materialised matter, for honey and wax are one ; if they be separated like chalybs and iron, it is then called a separated corporal from a materialised matter.   For as the alchemists, in a circulatory or pelican, circulate the spirit of wine, so, also, the summer star in natural growing things circulates liquid.   So honey emerges from the earth into the materialised matter, which is the subject of the bees, and this materialised matter requires further perfecting. . . .   There is a threefold kind of honey in all growing things – the lowest, middle, and topmost.   It is the last which the bees seek, and they find it in the flowers, where it is purest, the gross being always relegated by Nature to the lowest place.—*De Melle.*

† Manna is the chiefest and most excellent nutriment and the marrow of locusts.   It is the highest natural preparation of the star.   The food of those bees that are fruitful in honey and wax is this manna and tereni-abin.—*Ibid.*

Concerning the Fourth Elixir, which is that of Quintessences

Similarly quintessences can be reduced to an elixir, which, like balsam, conserves living as well as dead bodies. In this place we make very slender mention thereof, because it has been previously demonstrated in the process of quintessences. So, then, advancing a little farther, we will set down such things as we are mindful of as making for preservation and conservation. Afterwards out of the elixirs of those processes we will teach the composition of one elixir equally profitable to the body as the three preceding. It must be understood that this elixir of quintessences has in it a secret virtue which daily tends to restoration, and endeavours to renovate and restore the whole body. So, then, it produces something more than a mere conservation; for it also renovates, not, however, so perfectly as we have described above concerning quintessences and arcana, but with inferior force, because the conservation and the restoration of these cannot co-exist; still, by this method renovation is disposed towards conservation, in the following way :—

Take the quintessence of chelidony and of balm, each two ounces; of quintessence of gold and quintessence of mercury, each half an ounce; of the quintessence of saffron and of all the mirobolanes, each one ounce. Let all be mixed together and remain in digestion of the sun, enclosed in a blind alembic, for two months. Afterwards add one ounce and a half each of the quintessence of wine and the magistery of the same, and let them be again digested as above for a month. Then keep it as a treasure, not only for preservation, but also for restoration.

The Fifth Elixir is called that of Subtlety

Now, we have thought that we ought to set down something concerning the elixir of purity or of subtlety, which conserves by the force of its great purity, as is the case with the corrected oil of the philosophers. This suffers nothing which has been anointed with it to putrefy. The same effect is produced by the corrected laterine oil, and many others, whereof the property, however, is not to preserve from putrefaction, but they acquire this, and take it as their property, from the preparation and the labour bestowed, as distilled or corrected wine does not allow putrefaction any more than digested wine does, and this, moreover, is not changed by the fire.

The water of honey by its preparation resists putrefaction, so far as concerns sensible bodies; though the crude substance thereof does not produce this effect, but is itself liable to putrefaction. Wherefore we set down an elixir of subtlety, since, just as mercury itself, which is volatile, is fixed and becomes permanent through its own water, so the human body also is fixed into consistency and permanence. Now, although this may be done by many other methods than that which we here describe, nevertheless, we mention only those which are experimentally known to ourselves. Not that on this account we wish to detract in any way from the others; only we say that all these things have not yet come to our knowledge and experience. The process of this elixir is as follows :—

Take olive oil, honey, and *vinum ardens*, one pound of each. Distil them all together after the manner of the alchemists, and do this thrice. Afterwards separate all the phlegm from the oils, which are distinguished by their many colours. Put all these oils into a pelican and add to them a third part of the quintessence of balm and of chelidony. Digest for a month. Afterwards keep it for use. No sensible or insensible body can resist it, on account of many causes and properties which we are unwilling to set down in this place.

THE SIXTH ELIXIR, WHICH IS THAT OF PROPRIETY

Equally from natural objects a perfect elixir can be extracted, as out of myrrh, saffron, and alöepatic. As to what forces it proceeds from, that we set down in our treatise on the Generations thereof. Here we only put forth the process, leaving the origin, which we often treat of elsewhere.

Take of myrrh, of alöepaticus, and of saffron, each a quarter of a pound. Put these all together into a pelican, set them in sand, and let them ascend for a month. Then separate the oil from the dregs by means of an alembic without burning. This oil suffer to digest for a month, together with circulatum of equal weight. Afterwards preserve it. In this elixir are all the virtues of the natural balsam, and, moreover, such a conservative virtue for old persons, more than it seems right to assign to it. For not only one period of life seems to proceed from it, but four, seven, or ten. It is scarcely possible to express its force and natural powers, but, so far as our judgment goes, it has been sufficiently elucidated, nor do we think it requires fuller explanation.

THE END OF THE EIGHTH BOOK OF THE ARCHIDOXIES FROM THE THEOPHRASTIA OF THEOPHRASTUS PARACELSUS CONCERNING ELIXIRS

THE NINTH BOOK OF THE ARCHIDOXIES

FROM THE THEOPHRASTIA OF PARACELSUS THE GREAT

———

IN the preceeding books we have treated of internal diseases. Now we have to write of those which arise without, and to set down remedies for the same. Although we insert nothing in these books as to the origin of diseases, internal or external, we will lay down the origins of the medicaments against them, and afterwards the composition of similar remedies for external ailments. Some refer only to wounds, and by these remedies a wound can be healed in twenty-four hours. This is to be understood in the following manner :—A wound that has been inflicted requires nothing else save that it be again connected or conjoined, just as two planks are connected with glue. You should on no account allow wounds to lie open, but should endeavour to refill them with flesh. This is a matter for the rustic rather than for the physician. Consider that when the lips of the wound are joined, as the planks by glue, they are more than half healed. What has to be done by some kind of medicament is to bring together each side as well and as closely as possible. Thence it follows that when the lips touch, and the compression of the medicament aids Nature, the cure is complete. So that no wound, in which no fracture of a limb is involved, can be so bad as not to be cured in twenty-four hours. Bones cannot be connected in the same way as flesh can ; so in this place we do not speak of them. You can understand the matter by an example. When some limb is altogether cut through, before the veins are dead, and while they are still warm and fresh, let them, as soon as possible, be moistened with the medicament, the wound joined together and its two sides connected exactly like two sticks fastened together with glue. Thus they are healed and grow together. This the medicament effects, which Nature takes care of by its power of resiccation, and heals through that power whereof we have spoken above. But it should be understood that the medicament for wounds should not be incarnative, or mundificative, or attractive, because such medicaments draw out the putrid fluxes and cause much matter to be formed. Moreover, the opening or cavity of the wound has to be filled with flesh ; and this is done very slowly and, in consequence, with a good deal of peril, and without a magistery. This is also the case with old-standing ulcers, which in course of time have become loaded with discharges, whence it happens that they can only be cured with many accidents and much difficulty, indeed, sometimes can never be cured at all. Wherefore, there is urgent need of some medicament for them, such as we have mentioned, which also with a certain force compresses in like manner the skin and the cavities.

In like manner, it must be remembered in the cure of ulcers that the formation of flesh is necessary, and this cannot be brought about by mere compression. As we said in the case of wounds, so in that of ulcers, fistulas, and the like, all these ailments have to be cured by the force of medicaments. We lay down, therefore, two fundamental rules for this flesh-forming—one being incarnative and the other exsiccative.

And now to speak of other malformations of the skin, such as cicatrices, morphea, serpigines, pannus, spots, leprosy, etc., including all diseases arising from the skin. For these we prescribe the following method of treatment. First we ordain that all the skin shall be stripped off, just as the flesh is stripped from a calf. Afterwards a new skin must be induced by the proper medicament. It will be inferred that the skin must be removed by some medicament, and a new, pure, immaculate skin generated by some other in colour, like that which follows, so that not much of the flesh and the humour may be attracted thither; and thus, as we have said, any spots are removed. The origin of such removal we have not from the beginning mentioned here, because it has been treated of elsewhere, and it does neither good nor harm to our intention and to our present teaching. There are other diseases also, such as cancers, buboes, and the like, which require their special medicament to remove their origin, and to purge away all their impurities. This is best effected by the attractive specific. Then there is need of consolidation, as we show in our treatise on fistulas and the like.

Ruptures of the bones and the like can only be consolidated with an attractive styptic; and this we need not discuss afresh, since we have spoken of it elsewhere. In like manner, many superfluous growths are found, as strumæ, glandular swellings, etc., which ought first to be emptied and afterwards cured.

We will, therefore, divide surgery into three parts or methods of cure: one referring to wounds, a second to ulcers, and a third to spots. Cancer we shall cure only with an attractive specific, and afterwards treat with those medicaments which we shall describe below.

A REMEDY FOR WOUNDS

If it be necessary to have such a medicament as by its special nature shall connect the lips of wounds, as glue joins two boards, this must be accomplished by its very great dryness and its styptic qualities, which act on the flesh only in the following manner.

Take Samech which has been well burnt and calcined to whiteness. To this add a smaller quantity circulated. Afterwards distil, in order that a very dry *caput mortuum* may remain at the bottom, and that the whole glass may glow. Then pour in fresh matter again, as before, and do this until the circulated substance comes out quite sweet, as it is in itself. Then allow it to be resolved by itself.

That which is so resolved becomes a remedy for a wound ; in fact, it might be called " A Balsam for a Wound," because in our common German speech balsam is the same as Baldtzusammen, that is, *mox conjunctum* (soon joined), and the term is not derived from the Latin idiom. We are unwilling to speak in detail as to the virtues of this medicament, but generally, we assert, it has such an effect on all wounds that with one single washing we have cured many hundreds of these in a manner which is not credible, judging by natural methods.

A Remedy for Ulcer

Ulcers, it should be understood, must, in like manner, be compressed by a medicament with the addition of a generative virtue. The writings of the ancients are not worthy of our imitation ; they are malicious and wicked. What we have to consider may be expressed thus : " Compel them to come in ;" and we do it in the following way :

Take of the aforesaid balsam for wounds, and of balsam similarly made from rust—of Samech, for instance—one pound each. Mix these together, and add a pound and a half of oil of iron. When all these ingredients are mixed together, let them be placed on the ulcers, which must be washed daily, as shall seem expedient. Let a consolidative plaster be applied such as we prescribe for ulcers. Follow up your ligatures thus to the end, until a cure is effected. It must be noticed that the members have to be compressed with these ligatures, as we have pointed out at sufficient length elsewhere. Let this suffice on the subject of ulcers.

A Remedy against Spots

We have sufficiently explained the removal of the skin by a corrosive specific, and together therewith the cautery—how it is to be produced and adapted for use. After the skin has been removed, and with it the spot, the cure is as follows :—

Take the above-mentioned balsam for ulcers, and add to this washed turpentine, oil of worms, and oil of eggs, equal parts of each. With this wash all the flesh when it has been stripped of its skin. After this treatment nothing more is required. The property of such a medicament is that it induces a new colour along with the new skin, and a natural hardness, so that it can be no longer defaced by the previous spots.

Although it is true that such spots can be removed by many waters, as for instance, the water of bean flowers, of sigillum Mariæ and the like, as well as by human dung, still these do not come within our scope, since their purpose is not uniformly compassed, and the spots are much more effectually removed by the method we have pointed out.

Nor let any be surprised that we set down so few and such concise remedies for the whole range of surgery, and do not follow those surgical methods which the ancients have described and the moderns have adopted, like them, for their own uses. For by that method of medicine, so long as we followed it, we could never find or perceive anything well founded and certain. But we have used our own remedies according to our experience, and in this way we have found out the best medicaments in the whole of surgery, and have comprised them here in three paragraphs only.

And although, no doubt, other diseases can be found besides those which we have mentioned here, such as bullæ, alopecia, etc., they are, nevertheless, comprised under spots and scars, and must be cured according to the treatment prescribed for them, for many causes which are not here brought forward, but left to our own experience; we had not forgotten them. We have had hundreds and thousands of these wounds passing through our hands, and when we saw them cured so quickly and so effectually by these remedies, why should we imitate the tedious and empty processes of the ancients, forgetting those nearest to ourselves? Why, then, we use mundificatives, lotions, sutures, ligatures, corrosives, and the like—which are all effectual against wounds, and destroy them most thoroughly—we have given the cause of all this at sufficient length in our book on wounds. Why do we use different plasters, ointments, unguents, and the like, even in the cure of ulcers? Why ligations, anointings, and the like? Well, to go through all this would be tedious since it only leads us to enter on a prolix, intricate, and foolish course, where we seek mere accidents without finding anything. It is a mere superstition to pin one's faith to antiquity. For in surgery to debate as to the nature of fistula, cancer, or ulcer, or the like, and then to assign to each its special remedy is mere vain talk and waste of writing, which will not repay the outlay in paper; when all can be thoroughly cured and removed by one single remedy—as, for example, external leprosy, alopecia, serpigo, spots, and the like, pustules, the itch, and scars, which can all be thoroughly removed by one medicament and one method of practice, as can also artetic wounds by spears, weapons and bullets. With these few words we would close our treatise on surgery and bring it to an end.

THE END OF THE NINTH BOOK OF THE ARCHIDOXIES FROM THE THEOPHRASTIA
CONCERNING EXTERNAL WOUNDS AND THEIR REMEDIES

THE KEY OF THEOPHRASTUS PARACELSUS
BOMBAST VON HOHENHEIM

OR

THE TENTH BOOK OF THE ARCHIDOXIES *

From a German manuscript codex of great antiquity

WE had decided to write our Archidoxies, as also other books concerning Medicine, with especial clearness and lucidity, inasmuch as all the highest medical arcana cannot be prepared without true chemical encheiries (undertakings), nor yet can they be speedily exalted in grade, and it is notorious that almost the whole world, through its devotion to riches and earthly wealth, zealously pursues tinctures only and transmutations of metals in order to amass the greatest possible amount of gold and silver, to obtain which they stand in the greatest need of chemical preparations, which also they would like to find in concise form and easily in our Archidoxies. Notwithstanding, it was in consideration of the very great evil which might thence arise, and at the same time to oppose their malice, that we have concealed our doctrine, according to ancient philosophic method and cabalistic practice. I shew this, my doctrine, clearly to the upright and the perfect, yet leave it none the less dark to contemptuous and impious men.

Our writing according to the method of cabalistic philosophy is not due only to the lachrymistæ who gape after gold, but also to the majority of the

---

* The following editorial preface introduces the tenth book of the Archidoxies in the Geneva folio :—Behold, gentle reader, the tenth book of the Archidoxies, which has been long demanded by the wishes of all lovers of chemistry, but till now has not been included among the works of Paracelsus, because, being in the possession of few, It has been held among secret things, as a most precious treasure replete with great arcana. Once, indeed, it was printed in the vernacular tongue of the author at Mentz, but the envy of certain persons, who gazed with a fierce eye at the revelation of the mysteries therein revealed, suppressed the majority of the copies, and compelled very many persons, studious in chemistry and in the writings of Theophrastus, to make transcripts for themselves. We, in order to consult the public good, have thought proper to add to the nine books previously edited, this tenth book, rendered from German into Latin, by one who is skilful in both languages, so that those who are studious in this science may be saved from transcribing, and the number of the Archidoxies may be complete. The author has denied in several places of this work that he would write the said book, lest he cast pearls before swine, and lest these arcana should come to the knowledge of the unworthy and the impious, but he changed his mind at the persuasion and prayers of his friends, who impartially weighed the disadvantages of giving their mysteries to the public with the advantages which could be derived to the human race by their communication. The latter seemed to preponderate by far ; consequently, they at last obtained this concession, that he would entrust the work to certain of his familar acquaintances. It appears from the preface, which he prefixed, with how great an oath he bound each and every person who should obtain an exemplar that he should guard it with the greatest secrecy as a treasure of Nature and Art, and should have nothing to do with avaricious chemists, or with the ambitious and envious followers of Galen. Nor did the author forbid the publication without a reason ; he rightly rejoiced in the title of a key, whereby the doors of the preceding books are opened, and the bars removed, so that

followers of Galen and of Avicenna. The latter would very gladly avail themselves of our medicaments and arcana to repel chronic diseases, which otherwise, using the method of Galen, would be incurable—they would be glad, I say, to use them if they were able to find a short, sure, and easy method of preparing and administering the same, without renouncing the error of the heathen and false Christians, and providing also they could ascribe the honour, fame, and riches which they thence would obtain to the writings of Galen only and to themselves, out of envy ignoring my name and glorious achievements, but claiming the same for their own writings, keeping secret the fact that the whole art proceeds from me. For they themselves, being old doctors, do not wish to appear as if, at their advanced age, they were reduced to be disciples of an unpolished Swiss and younger apostle, and to make public profession of this circumstance, seeing that all detest him for subverting their principles. On account of this, their laziness, ambition, envy, and hatred, as also ingratitude, I have taught and philosophised in my Archidoxies, and in my other books, in the aforesaid manner, as has pleased me, which I will justify before God and my conscience at the last day, in order that those who desire to arrive at the fundamental principle of my Archidoxies, may publicly call themselves Theophrastics, acknowledge me as their monarch, follow me in their labours, frequent my school, and, *vice versa*, discard their old fathers. But although they may chance to secretly obtain some process from a miserable and simple rustic, or from elsewhere, they will not, however, understand the arcana of administering my medicaments, and will thence derive more shame than honour. Wherefore, although it has been shewn to them by Anicula, that the young of swallows, their cranium, and glue of the oak, are a sure remedy for the falling sickness, which is the case, yet even by this you will not effect a cure. Whence is this, and what is the cause? It is for this reason, namely, that you do not understand the mode of administering them and the great Ilech, nor will you learn from Galen unless you shall have frequented my school and learned philosophy according to Christ and not after Mammon.

---

an entrance for all is left open to the sacred and more secret penetralia of the divine chemical art. It is, however, by no means to be thought that anyone who has only slightly occupied himself with the school of Vulcan, who has seen nothing of importance, save obscurely, in this most noble art, or has tasted of its fruits for the first time, should fancy that the things herein contained are for his use, or should deceive himself with a false opinion, persuading himself with excessive credulity that they are intended for him. Let him not apply himself to putting in practice the prescribed formulas of the recipes, unless he wishes to be wise too late, like the Phrygians. No one, indeed, should suppose that food is here set before him which is made all ready for his palate; he who would obtain the kernel must first crack the nut, and then at length he will taste the sweetest fruits of chemistry. Not a raw recruit but a veteran soldier, not one slightly tinged but he who is completely permeated with chemical undertakings, not soiled by scholastic dust but by the smoke of coals and cinders, not accustomed to arguments which savour more of vain subtlety than of real usefulness, but acquainted with true and sound philosophy—those, in a word, who are most skilled in chemical practice—let such gird themselves to the execution of these matters. But let the profane crowd retire as far as possible. If any one doubt that this is the genuine production of the author, the peculiar style and mode of writing affected by Theophrastus should remove his difficulty, while the promises made in the preceding books and here fulfilled should place the matter beyond question. Nor let anyone be surprised that the author declared he would keep the tenth book locked up in his heart, or, to use his own expression, he would retain it in his occiput, whereas the prayers of his friends persuaded him to change his mind. Do you, therefore, gentle reader, being assured by this lawful witness of Theophrastus, enjoy the labour and work of so great a man, and accept with grateful mind our zeal in acquiring and publishing the treatises of our great doctor,—F. B.

Since, therefore, the glue of the oak does not fulfil your expectations, you imagine that it is too weak by itself, wherefore you correct it with other herbs, and make a great composition of sixty or more ingredients, which you digest and purge of dross. You do not, however, expel the disease even by this means, since you do not understand either simples or compounds, nor yet the method of administering them.

Had such persons, indeed, accepted my doctrine in a grateful manner, and had they cast out of doors the red bonnet, or fool's cap, received from Galen, and at the same time submitted themselves to my discipline, I would have put on them a better cap, that of Fortunatus himself, wherein is concealed more art than in all other writings. So, in the presence of none would they need to doff it, but, just as Fortunatus cured the king's daughter, might they cure chronic diseases.

But they are unworthy of better things, and are to be blamed for their mischief, since they are completely ignorant of the great secrets or the mysteries of the sanctuary of Nature, and as much concerning the celestial treasure, which, in these last days of grace, has been freely revealed to me from on high, which, indeed, make a true Adam and paradoxic physician, according to the days of Enoch, in the intellects of a new generation. But these ignorant persons boastfully refuse it. Wherefore, I pity them not, but leave them in their own ignorance.

There is no doubt, that in that very great multitude of men, mentioned in the fourth book of Esdras, the Lord God will reserve for Himself a small number of certain elect persons, who will desire faithfully to pursue my Theophrastic doctrine, to love the truth, and help their neighbours in their destitution and diseases, out of a true and unfeigned Christian love, not for the sake of filthy lucre or ambition, but for pure love of God ; and also that out of Nature's light the marvels of God may manifestly appear. At the same time, all are not born under such a constellation as to be able to perceive the sense of our books, however diligently they study, without divine aid. It is on account, therefore, of their sincere intention and love, and that they may understand our excellent writings and arcana of medicines, and may arrive at a blessed end ; and also, lest the most precious secret of Nature, divinely revealed to me should altogether be buried with me ; that we shall write this book, therein shewing in full light the principle of our Archidoxies and universals, and shall teach the preparation of singular arcana, the quintessence, prime entities, and magisteries.

But, lest this clear light itself should reach the ungrateful and unworthy, I exhort all who have a supply of this book, I bind you by the very great sacrament and oath which you have given to God in baptism, that you shall hide all these things secretly and as the noblest treasure of Nature, lest you admit any unworthy person. Do you rather venerate that treasure as a most blessed talent, and help your neighbour in adversity.

May God grant benediction and favour, that whosoever partakes thereof may rightly use it !

THE TENTH BOOK OF THE ARCHIDOXIES

OF THEOPHRASTUS

Comprised in Ten Separate Chapters.

CHAPTER I

CONCERNING THE SEPARATIONS OF THE ELEMENTS

IN all things four elements are commingled one with the other, but in each thing one of these four is perfect and fixed. That is the predestinated element in which dwells the quintessence, the virtue, and the quality; but the other elements are imperfect, and each an element only, in which is no more virtue than in any other single element. These are all, as it were, the abode of the true, fixed, and perfect element; whence, also, they are called qualified things. Now, the fact is that some persons think the body to be a true element and quality, and that it in some way displays the virtue of a true element. This is because the body, like the three imperfect elements, is tinged and qualified, each according to its own nature, by the fixed, perfect, and predestinated element, as by its indwelling inhabitant.

For example, in some bodies the element of water predominates, in others that of fire excels, in some earth, and in others, again, air. If, then, the predestinated element has to be separated, it is necessary that the house be broken up; and this breaking up or dissolution of the house is brought about in divers ways, as is clearly said in my Metamorphosis concerning the death of things. If the house is dissolved by strong waters, by calcinations, and the like, care must be taken that what is dissolved from that which is fixed must be separated by common distillations. For then the body of the quintessence passes over like phlegm, but the fixed element remains at the bottom. But since we are little concerned about the house or the dwelling, it is necessary to find it in a fixed, predestinated element, and thence to extract it after the mode of a quintessence, so that that fixed element may be dissolved by other stronger artifices than calcinations, sublimations, and so on, and the pure separated from the impure. The pure is the quintessence; but the impure is the superfluous tartar, which is mixed up with every generation, concerning which see the book on Tartareous Diseases.

But since my theory is given at length in other books of Archidoxies, of Metamorphosis, and the Generations of the Paramirum, I am on that account unwilling to cause any weariness, but will briefly point out the practical method. Prepare a metal according to the process in the book on the Death of Things ; reduce it to a liquid substance according to the method which we have taught in the book on the Separation of the Elements ; separate by continual cohobations and putrefactions the three imperfect elements ; and then the fixed element, of whatever kind it may be, remains at the bottom, and in this way these four elements are correctly separated.

CHAPTER II

CONCERNING THE QUINTESSENCE

Abstract the volatile portion, which passes over in the separation of the elements, several times from that which is fixed, so that the quintessence, which partly was raised with the phlegm, may be again conjoined. Take the fixed element that remained after the separation of the three imperfect elements, of whatsoever sort it may be, then dissolve it in its proper water, each accord- to its nature, as we have said in the Archidoxies concerning the Quintessence. Keep it in the highest state of putrefaction, distil it by cohobation, and the rest by descent. Putrefy still a little, distil, and join all. Then distil it in a *Balneum Mariæ*, even to oiliness. Then break it up with the subtle spirit of wine by boiling ; the impure will sink to the bottom and the pure will float on the surface. Separate this by means of a tritorium, and, in order that it may at the same time lose the nature of the aqua fortis, pour on a greater quantity of the spirit of wine, which frequently abstract until the quintessence turns out sweet. Lastly, wash it in common cold water. In the same way, it must be understood of marchasites, stones, resins, herbs, flesh, watery and fixed substances, that first of all, according to the teaching of the book on Separations, the three imperfect elements shall be separated ; and that afterwards measures shall be taken with the fixed element according to the instruction given in the book on the Quintessence.

By eating or corroding water, understand *acetum* mixed with spirit of wine, and that which being frequently abstracted from the spirit of salt nitre becomes *acetum*. In this the fixed elements of marchasites should be dissolved, purified, and elevated by means of an alembic ; then, lastly, corrupted by spirit of wine, so that the impure may sink to the bottom, and separate itself from the pure.

With regard to the essence of gems, by radicated *acetum*, understand that you have a sharp *acetum* corrected with bricks, a sufficient number of times from the tartarised matrix of *acetum*. Dissolve therein the gems, which have been first calcined by sulphur, putrefy, and then separate the pure from the impure by breaking them up with the spirit of wine.

From fruits, herbs, and roots the essence is easily perfected, so that you dissolve the imperfect elements by the highest secret putrefaction of extreme heat. Then putrefy in dung, drive out by descent all that can go out, and from thence abstract the injurious imperfect body of the moisture by distillation in a bath. Thereupon there will remain at the bottom the predestinated element. Separate this from the impure residuum by corruption with its proper spirit, or with spirit of wine. Abstract this, and you will have the pure quintessence.

The extraction of the quintessence from salts (for example, vitriol, common salt, salt nitre, antimony, etc.) is accomplished in this way : cohobate them frequently with their own proper liquid, or with water, putrefy with phlegm, and abstract the body from thence after the manner of phlegm, even to the fixed spirit. This dissolve in water, or its own proper liquid, and separate in heat with spirit of wine, the pure from the impure.

CHAPTER III

CONCERNING MAGISTERIES

Magisteries are deservedly to be called mysteries, on account of the great tinctures which they display in an appropriate body, such as *acetum* or wine ; and as we mention elsewhere, so here also we teach that the one thing to be considered is with reference to the agreements which are adapted to the extraction of magisteries. For if you take distilled *acetum* you must not tinge water but wine into *acetum* if, indeed, the tincture or the *acetum* was made from wine. If you have well and rightly understood the magistery of *acetum*, you will also sufficiently understand the book of Magisteries. In the magistery of *acetum* it is to be understood that from corrupted wine, by a fermentation which is naturally adapted to it—for example, by tartar—you make, first of all, the tincture, that is, the *acetum*. Then with a small quantity of this same *acetum*, you shall thoroughly tinge a large body of wine, previously corrupted and putrefied, in a short time, into the best *acetum*. If, therefore, you purpose to convert metals into a magistery, and altogether to tinge the whole body into an essence, you must take a principal and an open metal to which all the others in Nature are cognate. That you must corrupt in its own matrix which has been placed in water, and is called the Mother of all Metals, purge it from its superfluous elements and reduce it to its primal liquid entity, that is, the sharpest metallic *acetum*. As often as all the metals are digested therein, they are of necessity transmuted by it into *acetum*, that is to say, into a quintessence. But as wine must be already in some way previously corrupted, if, indeed, good *acetum* is to be quickly prepared from it, so, in like manner, metallic bodies, too, must be previously corrupted, or putrefied and mortified, as is said in the Metamorphosis concerning the Death of Things ; and then they are truly called potable.

In this manner, also, the magisteries of marchasites are to be prepared, in which almost more virtue is found than in metals, just as the other magisteries are prepared ; and by our dissolving water is to be understood the water of salt.

But the magistery from gems is that you first of all calcine them with sulphur for four hours, then reverberate them, and afterwards burn them with nitre. Then boil them with simple water eight hours, filter, coagulate, and extract them with spirit of wine.

The magistery from gums and resins, for example, from turpentine and amber, is made after the following manner: First boil in spirit of wine, then corrupt in fresh spirit of wine mixed with a dissolving water, of salt, for instance, then distil from it.

The magistery of herbs, in like manner, as also of all spices and fruits, is thus accomplished. First of all, let them be fermented like must. Then extract the spirit, as from the dregs of wine. In that spirit digest the putrefied herb, frequently renewing it with fresh herbs until the spirit shall have become quadrupled in quantity. But since frequent mention is made in our Archidoxies of First Entities, and since the chief foundation is hidden in them, we will here briefly add the preparation of our water of circulated salt, which is here required, but was omitted.

THE PREPARATION OF CIRCULATED SALT

In our other books we have sufficiently shewn and made clear that the true element is water, or the sea, as if the true mother of all metals, and from its primal essence it received the sperm of the three principles, of which none before me has made any mention, only they built up their principles from sulphur and mercury, neglecting all mention of the third principle, that is to say, of salt which lies in the sea. But, having been taught by experience, I have in my other books touched upon the fact that the first entity or quintessence of the element of water is the centre of metals and minerals; and I have elsewhere added that every fruit must die in that wherein is its life, so that it may afterwards acquire a new and better life, and thus by laying aside the old body be brought back to the first entity. Wherefore we will add here the extraction of the centre of the water, in which metals ought to lose their body.

Take first the true element of water, or, in place of it, some other salt not yet boiled to dryness, or even purified salt of a gem. Pour two parts of water mixed with a little radish juice; putrefy in an accurate digestion, the longer the better. Afterwards let it congeal; putrefy again for a month, and then distil by a retort. Urge the residuum with a strong fire, so that it may melt. Reverberate it in a retort over a continuous fire. Dissolve it on marble. Pour upon it the water that flows from it, and putrefy it again. Distil it once more even to oiliness; mix it with spirit of wine, and the impure will fall to the bottom, which separate, but the pure will be crystallised in the cold. Pour the distilled matter on it again, and cohobate until a fixed oil remains in the bottom, and nothing sweet afterwards passes over. Digest for a month; and then distil until the arcanum of the salt passes over through the alembic. Do not grudge this protracted labour, for this is the third part of all the arcana which are hidden in metals and minerals, and without it nothing fruitful, nothing perfect, can be brought about.

But although there are several ways for extracting the first entity of salt, this is the most useful and expeditious; and after this is that other way which we have mentioned as the elixir of salt, namely, that fresh salt being mixed with dissolving water, which is the distilled spirit of salt, should be putrefied and distilled until the whole substance of the salt shall be dissolved and reduced to a perpetual oiliness, the body being removed thence as phlegm. In this way it is taught that the arcanum or magistery of vitriol and tartar, and of all other salts, is to be prepared.

CHAPTER IV

Concerning First Entities; and primarily concerning the Extraction of the Quintessence or First Entity of Common Mercury

If the common mercury is to be reduced into its first liquid entity, then it must be previously mortified and brought out from its own proper form. That is done by various sublimations with vitriol and common salt, so that at last it becomes like fixed crystal. Then dissolve it in its own matrix, that is to say, in the first entity of salt. Putrefy for a month; corrupt with fresh arcanum of salt, that the impure part of it may be precipitated to the bottom, but the pure turned into crystals. Sublimate the stones in a closed reverberatory; when it is sublimated, always invert it until it grows to redness. Extract this sublimate with spirit of wine rectified to the highest point. Separate the spirit of wine, dissolve what remains upon marble, and digest it for a month. Pour on fresh spirit of wine, digest for a time, and distil it. Then the arcanum of the first entity of mercury will pass over in a liquid substance, which is called by the philosophers a very sharp metallic *acetum*; and in our Archidoxies the Greater Circulatum. And the same is to be understood concerning antimony, gems, and herbs.

CHAPTER V

Concerning Arcana

What we say concerning arcana is to be thus received: that they are nothing else than a graduated quintessence or first entity. And under the first arcanum of the primal matter, we wish to be understood the first material or first essence of the limbus of man. Also we understand the first matter of the mercury of salt, for that is most closely conformed. Wherefore, according to the process of the first entity, you will reduce all to a liquid substance, then join it again with a monarchy, as if with the living unreduced body of that thing, and so promote it for distillation.

What we think concerning the arcanum of the Stone shall be made clear in the succeeding practice. But by the arcanum of the *Mercurius Vitæ* we intend a living fire, so that the mercury of common life shall be essentialised with the quintessence of salt and be vitalised with the first ens of antimony, as if by a celestial life. But the arcanum of the tincture unfolds itself, wherefore we omit it here.

CHAPTER VI

Concerning the Arcanum of the Stone, or of the Heaven of the Metals

What we have in one place and another theoretically advanced concerning the arcanum of the stone we here pass over; and I say that this arcanum must not be sought in rust, which many have wrongly named "flowers," but in the mercury of antimony. And this mercury of antimony, when it is brought to its perfection, is none other than the heaven of the metals, because its virtue is always vital, and nothing else than a perfect, pure quintessence. Therefore, even in the Deluge no virtue or efficacy was taken away from it; for the heaven, as though it were life itself, can be destroyed by no lesser thing. Its preparation I briefly subjoin here :—

Take antimony, purge it from dross and realgar, in an iron vessel, until the coagulated mercury of the antimony appears white and beautiful. And although it is an element of mercury, and has in itself a true, hidden life, yet all these things are potentially but not actively present.

If, however, you wish to bring it down to activity, it is necessary that you excite that life with what is like itself, such as living fire or metallic *acetum*, with which fire many philosophers have proceeded in different ways. Since, however, they agreed fundamentally, they all arrived at the destined end. One, with much toil, extracted a quintessence out of the coagulated mercury, and led down the mercury of the antimony therewith into activity. Others have discerned that a uniform essence exists in different mineral substances, as, for example, in fixed sulphur of vitriol, in magnetic stone, and thence extracted the same quintessence, and with that same have ripened their mercury or heaven, or have brought it into activity. And since they extracted their quintessence from a stony material, on that account they called that magistery a stone ; and, indeed, their opinion is right. Nevertheless, that fire, or corporal life, is found much more perfectly and sublimely in common mercury, which is testified plainly by its flowing, namely, that there is hidden in it a consummate fire and a celestial life. Whoever, therefore, desires to bring his metallic heaven to the highest grade, and to lead it to activity, ought first of all to extract out of the corporal life (the common mercury) the first liquid entity (as if celestial fire), the quintessence of the sun and a very sharp metallic *acetum*, by solution with its mother, that is, to mix it with the arcanum of salt, and with the stomach of Anthion, that is, with the spirit of vitriol, and he should dissolve therein the coagulated mercury of antimony, should digest it, and, lastly, reduce it to crystals, that it may be like a yellowish crystal, concerning which we have made mention in our manual.

CHAPTER VII

CONCERNING THE ARCANUM OF THE MERCURIUS VITÆ

Just as out of herbs, as the vine, a temperate essence is extracted, by which from every kind of herb or root their own essence may be drawn, even so that the mercury of wine does not shew its own peculiar nature, but the nature of that with which it is essentialised, in the same way is it with metals and minerals, for a like mercurius or spirit is extracted from the open and middle metal (mercurius), if the essence be extracted out of the perfect metals with that same spirit. Then, afterwards, that essentialised mercury is joined to the celestial balsam of the quintessence in a closed reverberatory, by means whereof it acquires life, and is on that account called the *Mercurius Vitæ*. The virtues whereof seem to us admirable, and must be kept silent and occult by us, lest they should be despised.

CHAPTER VIII

CONCERNING THE GREAT COMPOSITION, BEING THE CHIEF OF OUR SECRETS IN MEDICINE

In our Paramirical writings it is made sufficiently clear, that is to say, so far as it is necessary for a philosopher or a physician, if it be necessary that the whole human body, not only in its corporeal and earthly mass, but in its celestial balsamic part, should be preserved from and healed of all heavenly and earthly diseases—it is made clear, we say, that in the work of healing such a composition should be made as does not consist of a number of ingredients. For example, if any one should think that by pouring together water and wine a real mixture ensues, this is false, because one part can be separated from the other without injury to either, which is not the case in our great composition. For here is made a uniform and concordant mixture, so that two things, distinct in nature and properties, are united, and neither can be separated from the other without injury on account of their remarkable agreement, as occurs also in the male and female semen. If, therefore, such a composition is to be prepared conformable in its condition to man, through the due proportion of heavenly and earthly things, it is fitting to consider the name of the microcosm and that man is a little world. Wherefore if he is to be cured universally of all diseases, that must necessarily be done by his like. Concerning which Hermes Trismegistus said that it was necessary for him who intends to make this composition to create a new world; and as God created the heaven and the earth, so also the physician must form, separate, and prepare a medicinal world. And in order that he might point out to his disciples with sufficient fidelity from what thing or material this composition should be made, and how, also, the concordances of heavenly virtues are discovered by us in the Valley of the Shadows, he wisely and truly adds, a little after, that what is beneath is as that which is above, and that the inferior and the superior stand related to one another as man and wife ; and for the better understanding hereof he teaches that the heaven of itself agrees with the element of water, because it had its first spermatic matter in the water, and that the element of earth, coagulated and changed from its spirituality into corporeality and earthiness, is like the planets and the other stars, which also obtained their spermatic matter by their origin in heaven, and thence by separation passed over or were changed from a heavenly pellucid nature into a dense coagulated body.

In the primal creation, things above and things below, the upper and lower heaven, or water, the upper coagulated nature, or stars, and the lower terrestrial nature, were all mixed together and made one thing. But God separated the subtle from the dense, so that out of one water two were produced. The upper water was subtle, and to be considered as of the masculine sex, compared with the lower, denser, feminine water. But as God divided and separated still further the upper water, so that the subtle, airy part should be appointed for the stars, that thus the celestial bodies or stars may stand related to heaven as sons to a father, so, by parity of reasoning, has God appointed, together with this which is above, a separation also in the denser bodies, that is to say, in the female waters in the Valley of Shadows, and divided them into two parts. The seventh, clearer part He called water, and the other six dried parts, or coagulated portion, He called earth, which comprises in itself all the special fruits and planets which had their prime origin in water as their heaven ; as metals, minerals, and gems, which, in respect of the water, are reckoned as daughters in respect of a mother. So the superior heaven has a nature and properties like its own in its feminine nature, that is to say, in the inferior heaven or water ; and the superior terrestrial bodies, or stars, like the sons of a father—that is, heaven—have a similar agreement with and relation to their sisters, the earthly bodies. And just as by close relationship the higher heavenly bodies or stars are joined to their father, the heaven, so too, by a like and equal relationship, the lower earthly minerals and metals are connected with their heaven, the water, as with a mother. Whence the truth of the sentiment of Hermes is evident, which we commend to our sons of the doctrine in these words: that indeed the whole microcosm, so far as relates to the comprehensible mass, and to the living, moving, corporeal, generating spirit, ought to be collected and composed of these lower elements and dark waters which are their noblest essences. But as to the mental arcana, wherein consists the sound mind in a sound body, these should be attracted from the superior, heavenly waters, and their astral influences spiritually, in a mental manner, through the mind of the image of the Gamahela ; or, if these are not pleasing to us, they should be declined, as in our books on A Long Life we point out these things at length and with clearness. Since, both in other places and especially in the Paramira, we have included the theory of this grand composition, we pause at these words and add the practical method, namely, how the inferior world or heaven should be united and conjoined with its earth, or the sun with its heaven. But because we have premised already the preparation of the heaven and have taught it under the arcanum of the stone, we omit it here. And since of itself alone, like the male semen, it can bring no advantage in the body of man, but only restores the celestial parts, that is to say, the radical moisture, or balsam of life, therefore it ought to be joined to its terrestrial corporeal mass, and be brought into concord therewith, that so also the carnal element in man may be refreshed and restored, and not only one member, but the whole body, be restored to soundness. Therefore let a corporeal mass be taken, which in nature is equal to the sun above, and embraces in itself the properties of all the stars, since it is impossible for all the subterranean stars and the coagulated bodies to be included together in the number of the ingredients. This coagulated essence of heaven, that is, the sun, in its essence and temperate element, is so elevated and graduated that it also fixes with itself its own habitation, that is, the superfluous elements, that it cannot be destroyed by any element, and the inhabitant or corporal balsam hidden in it is able to remain eternal. If, therefore, as was before said, the whole microcosm is to be healed, then the corporeal coagulated balsam should be united with the spiritual celestial balsam and the discord between the elements of the sun should be reconciled, so that the superfluous elements may be separated from the fixed predestinated element and altogether die out and leave the fixed element, as their inhabitant, alone. If this dead body of the sun be after-wards cleansed from superfluities and brought into a volatile spiritual nature, then is perfected the true, sublimated, and resolved mercury of the sun, not that horizontal which many try to prepare with common mercury and sal ammoniac.

CHAPTER IX

CONCERNING THE CORPORAL BALSAM OR MERCURY OF THE SUN

In order that you may excite discord among the elements of the sun, or the habitation of gold, you must draw out Sol in a strong solution, by means of the phlegmatic fire, or quintessence of tartar, into its proper heat. By this method the element of air in Sol is very greatly increased, and by the air approaching the fixed element of the sun, as being its proper fire, it is so graduated that it can conquer and destroy the habitation of the other three. Putrefy this destruction with the quintessence of tartar and with struthio. Convert it by a proper sublimation into the matter of mercury ; and then the fixed mercurial element of Sol will remain alone without any habitation. But since this is still mixed with its superfluous tartar, therefore this must be re-moved from it. Dissolve it, then, in the circulated water of salt, corrupt it, and the tartar will be precipitated. Sublime the pure in a closed reverbatory of Athanor, dissolve it upon marble, and putrefy it. Thus is the mercury sublim-ated, graduated, and dissolved into the first matter of Sol, and is prepared in the highest degree.

CHAPTER X

CONCERNING THE COMPOSITION OF THE SPIRITUAL BALSAM, AND OF THE BALSAM OF THE COAGULATED BODY.

As is remarked in the Manual, this composition is made in the philosophers' egg. And so we put an end to this great work, in the name of God, to His praise and glory.

HERE END THE TEN BOOKS OF THE ARCHIDOXIES

THE MANUAL OR TREATISE
CONCERNING THE MEDICINAL PHILOSOPHIC STONE

PREFACE TO THE READER

READER. God, indeed, permitted the true spirit of Medicine to be brought into operation by Machaon, Podalirius, Hippocrates, and others, so that the true medicine, which shines forth from the clouds (where it cannot be fully and plainly known), should come forth into the light of day, and be manifested to mankind. By that same operation, too, He placed His prohibition on the spirit of darkness, so that it should not altogether overwhelm and extinguish the light of Nature, and that the mighty gifts of God, which lie concealed in Arcana, Quintessences, Magisteries, and Elixirs, should not be altogether unknown. God, therefore, has ordained certain means whereby, moreover, through the ministration of good spirits, research into such arcana and mysteries should be implanted in man, just as certain men have received angelic natures from that heaven which is familiarly acquainted with the angels. Such men have been able afterwards, as being endowed with a perfect intelligence of Nature, to search into Nature and her daily course more profoundly than other people, to compare the pure with the impure, to separate one from the other, and to adapt and modify what is pure in a manner that seems impossible to others. These, as being true and natural physicians, know how to supplement Nature, and by their arts to bring her to perfection. It must be, therefore, that all imperfect and diabolical operations give way before them, as a lie always gives way to what is true and perfect. We must, I assert, speak the plain truth if we would arrive at any happy result. And if it be lawful to grasp the truth by any means, no man ought to be ashamed to seek it in any quarter.

Let none, then, take it in bad part of me, that I myself have loved this truth and pursued it. I was forced to seek it, for it did not seek me. If a man wants to see a foreign city it is no good for him to stay at home with his head on his pillow. He must not roast pears at the fire, for in that way he will never become a doctor. No one will ever get to be a renowned cosmographer by sitting at table. No chiromancist ever became so in his chamber, or geomancist in his cell. So neither can we arrive at the true medicine save by investigation. God makes the true physician, but not without pains on man's own part. He says: " Thou shalt eat the labour of thy hands, and it shall be well with thee." Since, therefore, seeing goes before truth, and the things which are perceived by sight gladden or terrify the heart, I shall not esteem it toilsome, or deem it beneath my dignity, to travel about and join myself to such men as fools despise, in order that I may discover what lies hid in the limbo of earth, and that I may fulfil the duty of a true physician, by exhibiting Medicine according to the ordinance of God and for my neighbour's good ; and that it may not do more harm than good. An easy-going man will not take this trouble. Let him who will, then, sit in his cushioned chair. I like to travel about, to see and examine whatever God and the opportunity allow. For the sake of sincere readers, however, who desire to learn, and love the light of Nature, I have written the present treatise in order that they may know the foundation of my true medicine, pass by the absurdities of pseudo-physicians, and be able, in some degree, at least, take my part against them. For those distinguished fellows, of course, know all these things beforehand, and the asinine doctor has them all in his wallet, only he is never able to get at them. For a man must be a good alchemist who wants to understand this treatise, one who is not afraid of the coals, and whom daily smoke does not disgust.* Let those who will take pleasure in these matters, I force myself upon nobody. This alone, I say, the thing will not prove infructuous, however much my enemies, the sham doctors, blame and accuse me.

---

* I praise alchemy, which compounds secret medicines, whereby all hopeless maladies are cured. They who are ignorant of this deserve neither to be called chemists nor physicians. For these remedies lie either in the power of the alchemists or in that of the physicians. It they reside with the latter, the former are ignorant of them. If with the former, the latter have not learnt them. How, therefore, shall those men deserve any praise? I, for my part, have rather judged that such a man shall be highly extolled who is able to bring Nature to such a point that she will lend help, that is, who shall know how after the extraction of the health-giving parts what is useless is to be rejected ; who is also acquainted with the efficacy, for he must see that it is impossible that the preparation and the science—in other words, the chemia and the medicine—can be separated from one another, because should anyone attempt to separate them he will introduce more obscurities into medicine, and the result will be absolute folly. By this distinction all the fundamental principles of medicine will be overthrown. I do not think I need labour very hard in order that you may recognise the certainty of my reasons. I give you this one piece of advice: Have regard to the effect of quack remedies. They first destroy wounds which are already aggravated by a succession of processes, miserably torture the patients, and having after all accomplished no good, but removed all chance of recovery, they do the unfortunate to death. . . . I who am an iatro-chemist, that is, one who knows both chemistry and medicine, am in 'virtue hereof in a position to point out errors and to profitably reject all pestiferous remedies, relegating them to their own place. My ardent desires and ready will to be of use prompt me to this. – *Chirurgia Magna*, Pars. I., Tract I., c. 13.

THE MANUAL
CONCERNING THE PHILOSOPHERS' STONE

IN order that the Philosophers' Stone, which, for sufficient reasons, we call a perpetual or perfect balsam, may be made by means of Vulcan, it must first of all be known and considered in what way that Stone may be placed materially before our eyes, and become visible and cognisable by the other senses ; and, in like manner, how its fire may be made to come forth and to be recognised. In order, then, that this may be the more clearly set before us, we will take the illustration of common fire, that is to say, we will inquire in what manner its force shews itself and becomes visible ; and this is as follows :—First of all, by means of Vulcan, the fire is smitten out of the flint. Now this fire can effect nothing unless it meets with some substance that is congenial to it, and on which it is capable of acting, such as wood, resins, oil, or some other like substance, which, by its nature, readily burns. When, therefore, the fire meets with some such object it goes on forthwith to operate, unless it be extinguished or hindered by something of a contrary nature to itself, or unless the material wherein it should multiply itself be deficient. For if wood or some similar substance be applied, its violence becomes stronger, and operates in the same way until no more fuel is applied. Now, then, as the fire shews its effects in the wood, so is the same thing produced with the Philosophers' Stone, or the Perpetual Balsam acting on the human body. If that Stone be made out of proper material and on a philosophical principle by a careful physician, and due consideration be given to all the surroundings of the man when it is exhibited to him, then it renovates and restores the vital organs just as though logs were put on a fire, which revive the almost extinguished heat and are the cause of a brilliant and clear flame.

Hence it is clear that much depends on the material of this balsam, since it ought to have a special adaptation to the body of the man, and should so exercise its virtue that the human body should be safe from all the accidents which might occur to it from such matter.

Wherefore, not only much depends on the preparation of the Stone or Balsam, but it is of much greater importance, that before all things, the true matter adapted for it should be known, and then that it should be properly prepared, and above all that it should be soberly and prudently used ; so that such a medicine should have power to purge away all impurities of the blood and induce soundness in place of disease.

On that account, it becomes the true and honest physician to have good knowledge, and not to regard ambition and pomp, nor to order dubious or contrary things, nor to trust too much to the apothecary, but to make himself well acquainted with the disease and with the sufferer ; otherwise you will be constantly treated wrongly, and the only result will be that the sick man is deceived and defrauded, solely by the pride and ignorance of the unfit and unqualified practitioner.   For what else is it but a deliberate fraud, when a man asks money and fees for that about which he knows nothing, and tries to lord it, to his own infamy ; since many men think nothing of money if they can only get good advice.   If this is not given, and they lose both their bodily health and their money, still it is considered quite laudable to demand a fee. Let him who will trust them.   I, for my part, would provide marks for such a doctor in quite a different way.   It is quite evident that of such doctors, who in their own conceit are most highly learned, there is not a tenth part who possess an adequate knowledge of simples ; much less of what they order to be done, or how the medicine is compounded by the apothecary.   Hence it often happens that such a doctor orders some kind of simple to be taken, which he himself does not know, and the apothecary knows still less, and has not got in his store.   Yet this medicine is called perfect, and is swallowed as such by the sick man, often at a high price.   The result the sick man soon feels.   Although it furthers his recovery in no way, still it is profitable to the doctor and the apothecary for filling their purses.   If the doctor or the apothecary suffered from the same disease, they would not take the same remedy.   Hence it can be well understood how nefarious their conduct is, and how highly necessary is it that they should approach the subject in a different way, should amend their errors, and follow a better course of practice.   Still I fear that old dogs are very difficult to tame.

But to return to my subject, from which zeal for the suffering and solitary patients has led me astray.   In order to do it justice I would say that I do not purpose to romance or to boast about this Philosophers' Stone, but the nature of the case requires that it should be made of the proper material, and that it should be prepared and used with due caution.   You must know that many of the ancients in their parabolic writings have sufficiently indicated this material, and described the operation in figurative words, but did not altogether disclose it, so that unqualified persons should abuse it ; and yet they took care that it should not be concealed from their own disciples.*   When, however, few

---

* The congeries of chemical philosophy, which has been the subject of frequent reference and citation in the first volume, contains excerpts from the *Manual Concerning the Medicinal Stone of the Philosophers*, which offer in some cases considerable variations from the above text.   The most important passage is as follows :—Very many of the ancients have with sufficient clearness revealed this matter and its preparation for the ingenious, but still in parabolic and enigmatical words and figures, so that they might drive away the unworthy from so great a mystery at once of Nature and of art.   A very few, nevertheless, even of those who are fit for this art, have sought the perpetual balsam of Nature and the perfect Stone, on account of the vast labour and intricate difficulty which meet the investigator at every step.   Hence it is that sluggish and slothful dispositions stand aloof from this work.   The avaricious, whom the greed

persons followed their meaning, or properly undertook the operation, in course of time their instructions were forgotten, and in their place the nostrums of Galen were introduced. As was the foundation of these nostrums, such was the superstructure, and matters daily became worse. This you see, for instance, in their herbaries, how they fuss over these, and how the Germans mix up Italy in the matter, when it is quite certain that Germany does not lack those imported herbs, but possesses in itself a full supply of perfect medicine. In order, then, that truth may not give place to a lie, and that the obscurities of Galen, with his accomplices, may not quench and suppress the light of Nature in medicine, it is necessary that I, Theophrastus, in this book, should speak, not as a quack, but as a scientist who is not ashamed of his achievements in medicine, who also, by the grace of God assisting him, has had proof of this matter in many cases which you, O Galenist, would not have dared to visit ! Tell me, Galenic doctor, whence comes your qualification ? Are you not putting the the bridle on the tail of the horse ? Have you ever cured the gout ? Have you dared to attack leprosy ? Have you cured dropsy ? I believe you will wisely hold your tongue, and acknowledge that Theophrastus is your master. If, however, you want to learn, learn and note what I here write and say, namely, that the human body does not need your clumsy herbal, especially in chronic or long-standing diseases which you in your ignorance call incurable. Your herbs are too weak for these cases, and by their very nature are impotent to get at the centre of the disease.

You will effect nothing by your pills beyond merely purging the excrements. And even here, on account of their unsuitability, you often expel the good with the bad, and this cannot be done without severe damage to the patient. So, then, these pills must be left off. Furthermore, your draughts do nothing beyond causing nausea in those who swallow them by their foul taste, which irritates the sick person, and by and bye they cause gripings and danger

---

for gold and silver—vile things that they are—has inspired with courage, have persevered most diligently of all in the work, so much so that for it they have neglected their life and substance. But because so much worldly happiness was not intended for them by God, they have wasted their oil and their labour. There is in truth need of the keenest zeal and judgment for this matter, that from various comparisons and similitudes the meaning of authors who write about this art may be detected. In order, therefore, that the more intelligent may comprehend up to a certain point, we will bring forward a fitting similitude by which is prefigured the matter of the perpetual balsam, agreeing with a like balsam in the human body for the purpose of restoring and conserving its highest state of health and driving away disease. Take common natural fire for an example. This is invisible to us, wherefore it must be sought in the air where it is latent, and is to be found by the striking together a flint and steel. Now it does not owe its existence to these, but to the air, and is only retained by some dry object such as firewood. For the dryness immediately takes hold of its heat and that which is like itself, and both operate in the same way in the subject until all the humidity is consumed, and there remains only a dry and ashen body subject to death, being deprived of the fire and food of life. In no other way should philosophers investigate matter wherein the food and fire of life are chiefly present. These should be drawn out by preparations so that they may increase the vitality of human life when it has begun to fail. For as the flame and life of the fire revive the wood all but consumed if there be left only a few bits of coal or sparks of fire, by adding appropriate wood and fuel, so in the human body the perpetual balsam applied even to the smallest remaining atom of life, rouses this into a flame and into its pristine vigour. But a question arises as to the matter in which this rousing fire remains latent. It is not right, or even safe, to speak clearly and openly about this to every-body. Is not the grace of God sufficient for the children of light, whereby the faculty is given to them of bringing light out of the darkness of shadows, figures, and enigmas? The sons of darkness are proved in this way, since for them, even from the light itself, nothing can be elicited save the merest shadows. Nature requires a nature like itself, and takes pleasure in it. As iron, too, is attracted by the magnet, so darkness begets darkness, and light brings forth light. This Stone is hard—who will be able to extract the kernel from it ? What is harder than the stone and steel, except it be the diamond ? Yet this is worn away. But by what artifice the Philosophers' Stone is to be disclosed, so

through their unnatural action. But I quit the subject of your absurd and useless medicaments, since they are in direct opposition to Nature, and never ought to be used. Since, then, those of which I speak are the true medicines (and no such true medicine can be found in Galen, in Rhasis, or in Mesne, which attacks and purges those diseases I named from the very roots, as the fire purifies the spotted skin of the salamander), it necessarily follows that the curative process of Theophrastus differs altogether from the fancies of Galen, since it emanates from the fountain of Nature. Were it not so, Theophrastus would stand disgraced like the others.

If, then, we are willing to follow Nature, and to use natural medicine, let us see what substances amongst those used in medical art are most adapted to the human body for keeping it in soundness up to the limit of predestined death by means of their virtue and efficacy. If thought be given to the subject, I doubt not but all must confess that metallic substances have the chief adaptation to the human body, and that the perfect metals, in proportion to their degree of perfection, and especially the radical humour of those metals, can produce the greatest effects on the human body. For man partakes of that salt, sulphur, and mercury which, though hidden, enter for the most part into the composition of metals and metallic substances. Thus, like is applied to like, and this process is most serviceable to Nature if only it be dexterously applied. This is the great secret in medicine, worthy to be called its very arcanum. What marvel is it, then, if great, unheard of, and unhoped-for cures follow, such as the ignorant believed impossible? Not to delay longer, I am constrained here to set down what I determined to write in this treatise. I purpose to treat here more clearly than elsewhere of true medicine. First, however, it had to be pointed out how man derived his origin from sulphur, mercury, and salt, regarded as metals. This I have sufficiently indicated in the Paramirum, and it is not necessary to repeat it here.* I will, therefore,

---

that henceforth we have fire and life enough, we do not see. The eyes of the mind must be opened, and the first consideration must be what medicine is by Nature and art congruous with human life before all others, so that it may be conserved and preserved in health to its predestined end, and also may be safe against all corruptions. Nobody—at least, no true physician—will doubt that the metallic essences, especially those of the perfect bodies, are the most durable and least corruptible of all that Nature produces. If, then, life be the fire and heat of the natural form united to the humidity of its own matter by light, as is clear from Genesis, and the light lives more brightly nowhere than in bodies least liable to corruption, what will prevent the heat of the fire and the radical humour in the metals, each being incorrupt, from rousing in the organs joined to human life this vitality that is well-nigh dormant? For these are sleeping in metallic bodies alone, and in a state of repose, as a man overcome with sleep lies as if dead, and is only moved by respiration, but not in his body. As the spirit of metals, if it be liberated from its bodily sleep, will perform movements and actions as if its own in any body that is applied to it, none otherwise must we judge of human bodies. While these are sick the vital spirits in them sleep ; they are not able to breathe truly or freely on account of their corrupt domicile. But when the corruptions of darkness are removed from the body, not by an extraneous physician, but by Nature itself, fortified by medical aid, and with an accession of extraneous life, that is to say, of incorruptible metals, the vital spirits in men exercise their movements freely. It is no wonder, then, if miraculous cures are wrought by Spagyric physicians, otherwise impossible by the vulgar medicine of the Greeks, whence it came about that they considered those diseases incurable which they themselves were unable to drive away with their sleepy and, as it were, dead medicaments. Hence the distinction between the Spagyric and the Greek medicine is clearly seen. The latter sleeps with the sleepers ; the former, watchful and free from all slumber, rouses the dormant faculties of life. But returning to our investigation of the matter, this cannot be noted better than from the errors of those who inquire into this branch of the subject.

* ON THE ORIGIN OF DISEASES FROM THE THREE PRIMARY SUBSTANCES.—There are three substances which confer its own special body on everything, that is to say, every body consists of three ingredients. The names of these are Sulphur, Mercury, and Salt. When these three are compounded, then they are called a body ;

only shew how the Philosophers' Stone may be recognised, and according to what method it is prepared.

Know, then, for a fact, that nothing is so small but that from it anything can be made and can exist without form. For all things are formed, generated, multiplied, and destroyed in their proper agreement, and they shew their origin so that it can be seen what each separate thing was in the beginning, and what it becomes in its ultimate matter, while that which intervenes is a kind of imperfect condition which Nature intermingles in the process of generation. Since, however, these accidents can be separated by the action of Vulcan, so that they shall be rendered inoperative, Nature can, in this instance, be corrected. This is what is done in the Stone. For if you would make it of its proper material, which can be perfectly learnt from the circumstances pointed out, you must remove from it its superfluities, and you must form, multiply, and increase it as a separate thing in its adaptation, which without such adaptation cannot be done. In this instance Nature has left it imperfect, since she has formed, not the Stone, but its materials, which are impeded by accidents, so that it is not able to produce these effects which the Stone, after due prepartion, *is* able to produce. Such material, without preparation, is, so far as regards the Stone, a mere fragmentary and imperfect substance, which has in it no harmony whereby alone it could be called perfect, or serve the human body for healing purposes. You have an illustration of this in the microcosm. See a man who is formed by the Mechanical Power only as a man. He is not an entire and perfect work, since he lacks harmony, but is only fragmentary until the woman is created like him; then the work is entire. Each of these is earth, and the two at last make the entire human being, capable of increasing and growing,

---

and nothing is added to, or coheres with, them save and except the vital principle. Thus, if you take any body in your hands, then you have invisibly three substances under one form. We have now to discuss concerning these three. For these three substances exist under one form, and they give and produce all health. If you hold wood in your hand, then by the testimony of your eyes you have only one body. But it is of no advantage for you to know this. The clowns see and know as much. You should descend and penetrate beneath the surface, when you would learn that you are pressing in your hands Sulphur, Mercury, and Salt. Now if you can detect these three things by looking, touching, and handling them, and perceive them separated each from the other, then at last you have found those eyes with which a physician ought to see. Those eyes ought to see these three constituents as plainly as the clown certainly sees the crude wood. Now this example may make you able to recognise man, too, in these three, no less than wood itself. That is, you have man built up in a similar form. If you see only his bones, you see as the clown sees. But if you have separated his Sulphur, his Mercury, and his Salt, then you see clearly what a bone is, and if that bone be diseased you see where it is faulty, and from what cause, and how it suffers. So, then, the mere looking at externals is a matter for clowns; but the intuition of internals is a secret which belongs to physicians. Now if these visible things must exist, and beyond their mere aspect medicine fails, then their nature must be deduced so that it may lay itself bare and be exhibited. Moreover, see into what ultimate matter these things are resolved, and into how many kinds. You will find these three substances divided from one another into just as many kinds. Now the clown cares nothing about all this; but the physician cares. The mere experimentalist neglects it, but not the physician. The quack thinks nothing of it; but the physician considers a good deal. Before all else these three substances and their properties in the great universe should be understood. Then the investigator will find the same or similar properties in man also: so that he now understands what he has in his hands, and of what he is making himself master.—*Paramirum, Liber I.*

Morbific Effects of Sulphur, Salt, and Mercury.—With regard to Sulphur, its effect should be thus estimated. It never produces an evil effect by itself, unless it be astral, that is, unless a spark of fire shall have been cast into it. Then is power awakened by that spark. Is not the act of burning a virile one? Without it nothing is produced. So, then, if any disease declare itself from Sulphur, then, first of all, the Sulphur should be properly named. It is essentially a masculine operation. There are many sulphurs, such as resin, gums, botin, axungia, fat, butter, oil, *vinum ardens*, etc. There are some sulphurs of woods, some of animals, some of men, some of metals, as oil of gold, of Luna, of Mars; some of stones, as liquor of marble, of alabaster, etc.; some of seeds, and of all other things, each with their own special names. There is afterwards the fire falling upon each, which alone is their star, and this, too,

and this power is effected by indwelling harmony. So the Philosopher's Stone, which should renovate man no less than metals, if it be freed from its super-fluous accidents and established in harmony with itself, performs wonders in all diseases. Unless this be done, all your attempts with it are in vain. But if you wish to establish it in its harmony you must bring it back to its first matter, so that the male may be able to operate on the female, that the outer part may act on the inner, and the inner be turned outwards, and so both seeds, the male and female, may be enclosed in complete concordance; that by the action of Vulcan they may be brought to more than perfection, and be exalted in degree, so that each, as a qualified, tempered, and clarified essence, pours all virtue into the human body as well as into metals. Thus will they render each sound, will drive away defilement by the method of expulsion, and introduce what is good into the human blood by its power of attraction to a due place, so that the microcosm which is situated in the limbus of the earth, and is formed out of the earth, may by this medicine, as by something like itself, be radically, and not in mere imagination, but most surely, led to health and kept therein. This is the Mystery of Nature, and such is the secret which every physician ought to know. And this, too, every one can comprehend who is born of astral medicine. But, that I may more clearly describe the nature and preparation of so excellent a medicine, so that the sons of learning, who love the truth, may be initiated, know ye that Nature has given a certain thing wherein, as in a chest, are enclosed 1, 2, 3, the virtue and power whereof suffice abundantly for preserving the health of the microcosm, so that, after its preparation, it drives away all imperfections, and is the veritable defence against old age. This we name the Balsam.*

---

with its distinguishing name. And this operation is, in one respect, peccant matter. Moreover, with regard to Salt it should be known that it exists of itself as a material humour, and introduces no disease unless it be joined with its star. Its star is resolution, which gives it a masculine power. Salt, no less than the spirit of vitriol, tartar, alum, nitre, etc., exhibits itself tumultuously if it is resolved. Now, whence can such a nature be infused into humours, except by a star? About this physicians have formed a conspiracy of silence. Even if they had been guilty of no other blunder save that in all their causes and cures they had omitted the star, that would have been quite sufficient to prove that they had built their house on a foundation of moss and sand. You should know, also, that salts are manifold. Some are limes, some ashes, some arsenical, some antimonial, some marchasitic, and others of a similar sort. And from all of these are produced and begotten peculiar diseases according to the body of the salt; which diseases thereupon take the name and nature special to each. So, too, understand concerning Mercury. This of itself is not virile unless the star of Sol sublimates it. Otherwise it does not ascend. Its preparations are many, but there is only one body. But the body of this is not like Sulphur or Salt, which have many bodies, from whence come different salts and different sulphurs. This has only one body, but the star changes this in different ways, and into various natures. So, then, from the same source many diseases are produced. Thus its masculine nature comes from the star, and by this nature it has its morbific effect. If, then, all diseases are comprised under these three heads, each with its own name and title, know that you must reduce to Sulphur all that is sulphureous, in the sense that it burns. What is Mercurial should be reduced by sublimation if it be adapted thereto. What is from Salt should be reduced to such salt as is appropriate. These, then, are the three general causes of disease, as we have grouped them together above.—*Ibid.* Some further extensions of the philosophy of Paracelsus concerning the prime principles in their relation to disease will be found in an appendix to this volume.

* We must know before all things that a balsam resides in our body by the virtue of which the body is preserved from putrefaction. That singular balsam is present in all members of the body, for there is one in the blood, another in the marrow, a third in the arteries, bones, etc. While this balsam remains entire and uncorrupted it is impossible for any opening of the skin to take place. But when by means of the preparation of salts it happens to be corrupted, then the principles of corrosion begin to act. For cure in this case the whole attention must be devoted to the restoration of that which has been removed from the balsam by corruption. This is effected by the balsam derived from other elementary bodies. This, I say, is found in external elements, by which also other generated things are preserved from putrefaction. Hence it is reasonably called the mumia of the external body of the elements, that is to say, of their fruits. Such balsam is found congenital in the radical matrices of all growing things. Every body pro-

In what substance Nature has placed such a number you must know beforehand ; since I am unable, for many reasons, to describe it more clearly. How it is prepared, Galen, Rhasis, and Mesne were ignorant, and it will not be reached by their followers. For this medicine requires such preparation as mere pill-sellers do not compass, and understand less about it than a Swiss cow. Moreover, it involves, as it were, celestial and special modes of operation. For it purifies and renovates, so to say, by regenerating, as you can read more at length in my Archidoxies, and at the same time study the origin and essence of metals and of metallic substances, together with their powers. He, therefore, who has ears to hear, let him hear, and see whether Theophrastus writes about lies or about the truth, and whether he is uttering mere inanities from the devil, as do you, O Sophist, utter your trifles—you who are from the devil and are surrounded by lies and obscurities, and call nothing good unless it be comprehensible by your fool's head, and that makes for your broth without any previous labour. For you, with your one eye, wander about at random. and cannot get straight to your kitchen window. By all means, twist your intricate thread, and try to find the centre of the labyrinth, by means of an obscure star ; this will in no wise offend me. But if you would only use your foresight sometimes, and see whereupon the art of Theophrastus is founded, and how feeble your patchworks are, Theophrastus would not be so opposed to you. But what things I now write briefly, and shall write, so that my astral disciples may be able to perceive them, and enjoy them, and boast of them, these, through the diligence of one who is not ashamed to learn, can be easily understood ; since there is nothing so difficult but that it may be understood and learned by labour and study. The practical method of this work, then, is as follows :—

THE PREPARATION OF THE MATTER OF THE STONE

Take some mineral electrum in filings. Put it in its own sperm (others read, Take the immature mineral electrum, place it in its own sphere), so that all impurities and superfluities may be washed away from it, and purge it as

---

vided with life is vital by reason thereof. The first thing to be considered in its extraction is the quantity present in the given body, for our intention regards not the body, nor the form, but solely the inherent balsam by which it lives. The extraction of these balsams is almost accomplished by separation, that is to say, while the arcanum is being removed from the body which was sustained by the balsam. You must know that elementary balsam is nothing else than that which we are accustomed to call mercurial liquor in the three principles. Whence it follows that all curative efficacy resides in Mercury. Mercury of this kind usually appears most manifestly in thereniabin and nostoch, as also in minerals of water, the fruits of the earth, and in the stars. Hence it is clear why antimony is so efficacious in the cure of ulcers, namely, because it contains mercurial liquor more abundantly than the other species of marcasite. In like manner the curative virtue of gold must be understood, for no body produced from the element of water contains a more copious or more subtle mercurial liquor. Thus also, among things which grow out of the earth, there is not a more arcane remedy than the mercurial liquor of cheirine and realgar. The same is to be understood of chaos and the firmament according to the method of their operations. But for the understanding of every mercurial liquor it is necessary to know that metals can be transmuted into a certain matter of this kind—as iron into crocus of Mars, Venus into flower of copper, tin into spirit of Jupiter. Yet these are not true liquors, but need a more exact separation according to their canons. There are extant various forms and descriptions of balsams according to the categories of the four mineral elements in the writings of the ancients. Some are wont to describe them according to the form of a plaster, others of a powder, others of the oil of water, etc. But I think that these descriptions ought to be completely rejected as purposeless, devoid of skill, plainly repugnant to Nature, and dependent on a dietary regimen which is the foe of Nature. The true and only balsam perfectly purges out and removes that which is separated, restoring the body to its original condition by the accession of fresh balsam.—*Chirurgia Magna.* Pars. III., Lib. V.

completely as possible by means of stibium, after the manner of the alchemists, so that you may suffer no harm from its impurity. Afterwards, resolve it in the stomach of the ostrich, which is born in the earth, and is strengthened in its virtue by the sharpness of the eagle. When the electrum is consumed, and, after its solution has acquired the colour of the calendula, do not forget to reduce it into a spiritual pellucid essence, which, indeed, is like amber. Then add half as much of the spread eagle as the corporal electrum weighed before its preparation. Frequently extract from it the stomach of the ostrich, by which means your electrum becomes continually more and more spiritual. But when the stomach of the ostrich becomes weary with labour, it is necessary to refresh it and always to abstract it. Lastly, when it again loses its sharpness, add tartarised quintessence, but so that at the height of four fingers it may be deprived of its redness, and may pass over together with it. Do this for so long a time and so often until it grows white of itself. When, now, it is enough (for you will see with your eyes how it will gradually fit itself for sublimation), and you have this sign, sublimate. Thus the electrum is turned into the whiteness of the exalted eagle, and with very little labour it is transferred and transmuted. Now, this is what we seek in order to use it in our medical practice. With this you will be able to succeed in many diseases which refuse to yield to vulgar medical treatment. You will also be able to convert this into a water, into an oil, and into a red powder, and to use it for all purposes for which you require it in medicine.

But I say to you in truth that there is no better foundation for all medicine than lies hid in electrum. Although I do not deny, nay, I write it in my other books, that great secrets lie concealed in other mineral bodies as well; yet they require greater and longer labour, and the right use of them is not easy, especially to the unskilled; so that if any one attempts them he causes more harm than good. For this reason it is not advisable that every alchemist should seek to practise the art of medicine when he has no acquaintance with it. Some prohibitory method should be devised which might debar such imaginary physicians. I, indeed, will not take their blame, or recognise them as disciples, since they do not pursue the truth, but I hold them as known deceivers, and lazy vagabonds, who take the bread out of the mouth of true disciples, and do harm to such deliberately, and think nothing of conscience or of art. In the prepared electrum we have described there lies hid so much power of healing men that no surer or more excellent medicine can be found in the whole world. The Galenists, indeed, those drug-selling doctors, call it poison and oppose it, not from any knowledge of it, but from mere pride and folly. I grant, indeed, that there is a poison in its preparation, as great as, or greater, than that of the Tyrian serpent which forms an ingredient in theriaca. But that after its preparation the poison remains has not been proved. For Nature, though by some blockheads this is not understood, always inclines to her own perfection, and much more, therefore, may be brought to perfection by proper methods. What is more, I grant that after its preparation it is a greater and more potent poison than before; but it is such a poison as seeks after its like, to find out fixed and other incurable diseases, and to expel them. It does not suffer the disease to speed its course and do injury, but as if it were an enemy to the disease it attracts the kindred matter to itself, consumes it from the very roots, and washes it as soap washes the spots from a foul rag, along with which spots the soap retires also, leaving the rag pure, uninjured, clean, and fair to look upon. So, then, this poison, as you say, has a far different and better effect than your axungia which you employ to cure the French disease, anointing more frequently than the currier does a hide. This arcanum, which lies hid in this medicament, has within it a well-proportioned, well-prepared, and excellent essence, which can be compared to no poison, unless one understands it, as I said before; and it is as different from your quicksilver, which you use as an ointment, and from your precipitate, in virtue and efficacy, as heaven from earth. It is, therefore, called, and really is, a medicine blessed by God, which has not been revealed to all. It is better compounded than that dirty medicine which the slowly walking doctor has in his gown, or has filtered through his fillet or hood. Moreover, this blessed medicine has thrice greater force and virtue for operating in all diseases, however they may be called, than all the drug shops you ever saw. But I did not find this in idleness, by sitting still, and by sloth, nor did I find it in an urinal, but by travelling, or, as you say, by wandering, and by much diligence and labour it was necessary for me to learn it, so that I might know and not merely think. But you suck your medicine out of the old cushion and bolster on which some old witch has sat who has covered your celestial head with a blue hat for medicine and breathed on you. So, then, I shall never regret my travels, and I shall remain your master and follow in the steps of Machaon, which, indeed, spring from the light of Nature as a flower from the warmth of the sun. But in order that my proposed task may not be delayed and remain imperfect, observe further how the thing must be done, and what power and property of medicine Nature has given to the Philosophers' Stone, and how it leads on to its end.

Here follows the remainder of the Preparation

Having destroyed your electrum, as aforesaid, if you wish to proceed urther, in order that you may come to the desired end, take the electrum which has been destroyed and rendered evanescent, as much of it as you wish to bring to perfection, place it in the Philosophic Egg, and seal it closely so that nothing may evaporate. Stand it in Athanor until, without any addition, it begins of itself to be resolved from above, so that it looks like an island in the midst of that sea, gradually decreasing every day, and at last being changed into the resemblance of blacking. This black substance is the bird which flies by night without wings, which the first dew from heaven, with its constant influence, its ascent and descent, has changed into the blackness of a crow's head. Then it assumes the tail of a peacock, and subsequently acquires the wings of a swan. Lastly, it takes the highest red colour in the whole world, which is the sign of its fiery nature, whereby it drives out all the accidents of the body, and warms again the cold and dead limbs.

Such a preparation, according to the opinion of all philosophers, is made in a single vessel, in a single furnace, in a single fire, the vaporous fire never being allowed to cease.

So this medicine is as if heavenly and perfect, or at least can become more perfect than the moon, by its own flesh and blood, and by the interior fire being turned outwards and prolonged, as was just now said, whereby all the impurities of metals are washed away, and the occult properties of the same are made manifest. For this more than perfect medicine is all-powerful, penetrates all things, and infuses health at the same time as it expels all disease and evil. In this respect no medicine on earth is like it. In this, therefore, exercise yourself and be strenuous, for this will bring you praise and glory, and you will not be a mere quack, but a real scientist, and, moreover, you will be compelled to love your neighbour. For such a divine secret no one can perceive and understand without divine aid, since its virtue is unspeakable and infinite, and by it Almighty God is known.

Know, too, that no solution will take place in your electrum unless it thrice runs perfectly through the sphere of the seven planets. This number is necessary for it, and this it must fulfil. Attend, therefore, to the preparation, which is the cause of the solution, and for your glorified, destroyed, and spiritualised electrum use the tartarised arcanum for washing off superfluities which accrue in the course of preparation, so that your labour may not be in vain. Of the arcanum of the tartar nothing will remain, only you must proceed circularly with it according to the number mentioned. Thus, easily of itself is produced in the philosophic egg and the vapour of fire the philosophic water, which the philosophers call *aqua viscosa;* it will also coagulate itself, and represent all colours, so that at last it is adorned with the deepest red. I am forbidden to write more to you on this mystery, such is the command of the Divine Power. Assuredly is this art the gift of God. On this account it is not all who understand it. God gives it to whom He will, and suffers no one to extort it from Him by violence. He alone will have honour in this work; Whose name be for ever blessed. Amen.

Here follows the Use of the Stone

And now it is necessary that I should write concerning the use of this medicine, and concerning its weight. You must know that the dose of this medicine is so small and so light as is scarcely credible. It should only be taken in wine, or something of that kind, and always in the smallest quantity on account of its celestial power, virtue, and efficacy. For it is only revealed to man for this reason, that nothing imperfect should remain in Nature, and it has been so provided and predestined by God that its virtue and arcanum should be produced by art, in order that to man, made in the image of God, all created things should be made to do service, and that before all else God's almighty power should be known. Whosoever, therefore, has intellect from God, to him this medicine shall be given. That ignorant Galenist, Beanus, will not be able to comprehend it, nay, he will turn from it in disgust. For all his works are works of darkness, while this work takes effect and acts in the light of Nature. And so, now, in brief but true words, you have the root and origin of all true medicine, which no one shall take from me, even though Rhasis with all his base progeny should go mad, and Galen be bitter as gall. Let Avicenna gnash his teeth, and Mesne in a word measure the length and breadth, it will be too high for all of them, and Theophrastus will stand by the truth. On the contrary, the lame works of the ointment-people, and the unctions of the physicians and doctors, with all their pomp and authority, shall go to utter destruction.

There is only one thing more to say, since to many this description of mine will appear obscure. You will say, O my Theophrastus, you speak too briefly and abstrusely to me. I know your discourses, how correctly you impart your subjects and your secrets. Wherefore this description will be of no avail for me. To this I answer that pearls are not for swine, nor a long tail for a goat. Nature has not seen fit to bestow them. Wherefore I say that to whomsoever it has been given by God, he will find sufficiently and above measure, more than he wished for. I write this by way of initiation. Follow with foresight, and avoid not study and labour, nor let the parade of fools lead you astray, nor the diligence which is necessary turn you away. By constant meditation many things are found out, and this is not without its reward. So use to a good purpose what I here give you ; let it be to you a fountain, and then you will have no need to drink from the troughs of the pill-sellers, nor will your lot be with the body-snatchers, but you will be able to serve your neighbour well, and to give praise and honour to God.

These things I determined to set down briefly, in my book on the Philosophers' Stone, lest men might think that Theophrastus cured many diseases by diabolical methods of treatment. If you follow me rightly, you will do the same, and your medicine will be like the air which pervades and penetrates all that lies open to it, and is in all things, drives away all fixed diseases, and mingles itself radically with them, so that health takes the place of disease and follows it. From this fountain springs forth the TRUE AURUM POTABILE,* and nothing better can anywhere be found. Take this as faithful advice, and do not annihilate Theophrastus before you know who he is. I am unwilling to set down more in this book, even though it might be necessary to say something, and to philosophise somewhat concerning the *aurum potabile* and the

---

* The lungs, the liver, the spleen, and the reins may all be sustained and nourished by potable gold, which preparation all physicians ought to have by them, because no physician is of any consideration who is without it. I am acquainted with its preparation and am possessed thereof. It would not yet be advantageous to publish it, but time will perhaps reveal it.—*Chirurgia Magna*, Part I., Tract I., c. 17.

*liquor solis.** I only wished to note these things here, and if you prepare them rightly they are not to be despised in their powers. But since other of my books treat of these secrets and make them sufficiently clear, namely, what a true physician ought to know, I will let the matter rest, hoping that this book may not be altogether without fruit, nay, that it may be of use to the SONS OF THE DOCTRINE. May God grant His grace, to His own honour and glory! Amen.

---

* According to the treatise entitled *De Male Curatis Aegris Restituendis*, Lib. III., c. 30, Paracelsus in some cases made use of the *Liquor Solis* for the cure of leprosy.

HERE ENDS THE MANUAL CONCERNING THE MEDICINAL STONE OF THE PHILOSOPHERS

A BOOK CONCERNING LONG LIFE*

SEEING that certain medicines are discovered which preserve the human body for a second and subsequent periods of life, which also protect it altogether from diseases, corruptions, superfluities, and other diminutions of its powers; nay, even when these infirmities and corruptions have broken in upon it, take them away, every physician must carefully study these medicines, and learn them from the very foundations. Indeed, numberless wearisome diseases and accidents of all kinds are taken away and extirpated from the very roots by this conservation of life.

So, then, when we propose to write concerning the preparation for a long life it must be understood that we do this in two ways. The one is theoretical, the other practical; for the subject of long life can be understood in both of these ways. And no physician ought to wonder that life can be prolonged—this also for two reasons. One is, that no fixed limit has been laid down for us, so that on some pre-determined day we must die; nor is this left in our own power. The other reason is, that we have medicine from Him who has made us, by means of which we can keep the body in that state of soundness wherein it was created, and expel from it all diseases whatsoever.

Now from this it may be gathered that death brings with it no disease, nor is it the cause of any. On the other hand, no disease causes death. And although the two coexist together, they are still no more to be compared one with the other than fire and water. They are no more akin one to another, nor do they agree better together. Natural sickness abhors death, and every member of the body avoids it. Death, then, is something distinct from disease. So far as relates to our present purpose what we wish is to speak to our own disciples from experience. We would speak to men who know those properties of things which are discovered by the highest art and by daily practice. To the mere pretenders and presumptuous men of medicine all these matters are perfectly occult and unknown. So it is that we write only for our own disciples, and not for those others. It is more

---

* The Geneva folio contains a considerable literature on the subject of long life, and this, as it will be readily understood, is full of repetitions and unimportant variations. At the same time, there is much in it which is only partially represented by the treatise translated in the text, so much, indeed, that it exceeds the reasonable limits of foot-notes, and it has, therefore, been thought better to print the *Book Concerning Long Life*, which pretends to be complete in itself, without annotations of any kind, and to reserve the additional materials to be dealt with in a separate appendix.

certain than certainty itself that the restoration and renovation of the body do take place, and that by these processes the whole frame can be transmuted to something better. In like manner, we see with our eyes that all metals are in their bodies purged so that they are protected from rust, and that even wood and the dead bodies of corpses are embalmed so that they suffer no further decay.

If, therefore, such things as these are possible to Nature, why should any one shrink from our writings, supported, as they are, by examples, although put in a brief space, or stand aloof from them because we draw a comparison between the bodies of metals and those of men? We do not suppose them to be one and the same. We know that they are altogether different; but in both we see the same method of conservation, and this is just what would have been inferred from experience, for if a dead body, by means of embalming, can be conserved, by how much more can a living one be kept from decay?

First, then, it should be known that this conservation, so far as it refers to the body of man, is to be considered in three parts: firstly, with regard to early youth; secondly, with regard to middle age; and thirdly, with regard to old age. Hence it can be understood why we begin severally from youth, from middle age, or from old age, since the end cannot be expected at any determinate period of life. The young life is sometimes destroyed in the mother's womb, sometimes in the cradle, sometimes during the period of growth, and sometimes, again, by the inordinate use of food or drink, whereby the nature being depraved and the strength diminished, it fails to reach the true limit of human existence. In a case of this kind the physician cannot but compare this period of life to old age, because it is equally defective and stunted in its nature. When, therefore, as often happens, children are born diminutive and sickly from their mother's womb, directly they come to the light they ought to be imbued with these conservating substances, by smearing the mother's nipples therewith, as is more fully explained in our practical method. By this method life and strength can be promoted, just as they can in the case of old men who were not so treated in their infancy. But in the matter of bodily power there is considerable resemblance between the old man's last span of life and that early period in which children are sometimes so broken down that they cannot attain to old age unless that first period of existence be fortified for them.

That is considered middle age when the body has ceased to grow and gray hairs begin to appear. At that time should anything occur to weaken or break up this period of existence, such as excessive work or debauchery, a similar method of treatment should be adopted before old age supervenes. If there be too long delay old age may never be reached at all, because that aid came too late.

The final period of life may be said to begin when the hair grows gray, and it lasts up to death. If this stage be complicated with weakness and decay, again the same method must be adopted in good season. But where the powers are sound and strong the method of conservation need only be used in proportion to the requirements. This, then, is the three-fold division of life to which we alluded ; and in whatever period the regimen of conservation shall have begun, in each alike existence starts afresh from youth, and goes through the different stages according to its limits.

Now, although we are supposed to be talking with our own disciples, yet this argument, amongst others, might be brought against us by empirics. Since we have such a means of attaining long life and driving away disease and death, how comes it that so many princes, emperors, kings, and other great persons die premature deaths ? Why do they suffer from infirmities when they could purchase immunity for money, and even ward off impending death ? We answer those who make such enquiries thus : We have never read or heard of a prince or king who used these remedies, with the single exception of Hermes Trismegistus. Many others have existed, as we point out in our book on Restorations, but they are unknown to the dwellers upon earth. And, though we do not consider them as a solution of the difficulty, yet still it must be re-membered that the physicians of these emperors and princes understand less about medicine than the clowns who spend their lives in the fields. Hence it happens that they more frequently promote and conduct their princes and heroes to death than to life. Acting upon the advice of their medical men, it would be quite impossible for them to reach their proper period of life, and this on account of the ignorance of the faculty, who are physicians only in name. The argument, we repeat, is not thus answered ; but we might also put forward the irregular life of princes, who often shorten their own existence, and this more as a punishment than through devotion. Then, too, there are many princes to whom these and the like remedies are altogether unknown.

By these three cases we consider that we have answered the argument advanced against us.

In what has been just said, we desire to point out not that this three-fold division of the periods of life is alone to be considered, but also something else which is by far more important, and this in two ways : one, namely, where bodies lay themselves open to diseases by irregular life, whence proceed dropsy, jaundice, gout, the falling sickness, pleurisy, and other like diseases, chronic or acute. Not only the division of the periods of life, but each disease in particular has to be taken into account by itself.

In the other case diseases occur in conjunction with irregular life through the seasons or by means of accidents, as in the instance of pestilence, mania, and other visitations which are destructive of life. These two methods involve the process of long life. In this way we shall have to understand what diseases arise naturally and what from external causes connected with Nature ; then those which derive their origin from something beyond Nature, as from incantations and superstitious practices, about which last there will be much more to say. For where such a disease occurs it must often be considered as a punishment, and on this account is not curable. We admonish our disciples to abstain from such practices. But so far as the visitation proceeds from Nature, this we teach how to cure and guard against by natural means.

Although we know and do not forget the effect of rings, images, and the like, which guard life against death, still we set down nothing on that subject here, but pass it by, because it has to do with astronomy. We speak, however, of these matters elsewhere, and proceed by another method, which is well worthy of all attention.

We first set aside diseases which do not arise from the bodily constitution, or from debility, but from some other source, before we speak of means of conservation. So, too, secret diseases will have to be examined. If they were diseases like gout in the feet or the hands, falling sickness, etc., they would not have to be set aside on these grounds, but would be cured by methods of conservation. So, then, there are three different classes of disease : one of a long-standing character, such as fevers, hyposarcæ, jaundices, and the like ; another of diseases proceeding from within, as pestilences, pleurisy, abscesses, and so on; and a third of chronic diseases, as gout, falling sickness, and others of like nature. The first we must pass over, for the second prescribe remedies, and in the third case suggest methods of conservation, so that they may be guarded against.

Then, again, we must not omit those diseases which have a mental origin, and arise from the sufferer's own imagination, or from that of another person, as in the case of incantations, superstitions, and the like.

With such a division as this, it must be understood, by way of conservation, that those diseases which have a mental origin are in like manner removed by mental treatment : those which proceed from the patient's own ideas, by some objective mode of treatment ; those which arise from imagination, by imaginations ; and those which arise from incantations, by counter-incantations ; while those which come from superstitions have to be relieved and cured by counter-superstitions. Then, after ailments of this kind have been removed from the body, there follow at once seven means of preservation, one directed to the natural strength or weakness ; another against casual or accidental diseases, and the like, with the view of guarding against them for the future ; a third against mental attacks ; a fourth against incantations ; a fifth against imaginations ; a sixth against the patient's own notions ; and the seventh against superstitions.

Marvel not that no mention is here made of temperaments, because the preservation of the whole body does not depend upon these, or upon their proportions, but rests on the virtue of Nature alone, from which all other excellencies arise. For the sole virtue is that which resuscitates and re-kindles the humours, which are four in number, but of which no account need be taken in medicine. The physician who bases his treatment on the natural temperaments may be fitly compared to a person who extinguishes a fire and leaves coals still burning. It is better worth considering how to preserve the root than the branches of a tree, because from the root the strength of the tree issues forth. If by chance anything happens to the branches—as may occur to the root of life from the temperament and the humours—this is merely an accident which in no way brings health, but rather takes it away. That we said was to be guarded against, when we were speaking of the three classes of disease, with the idea of resuscitating and re-kindling the life of the root, and, together with the branches, deriving its own nutriment thence, according as the complex qualities suggest their own remedies. When we seek to compare a short life with a long one it is necessary to know what life is, where it is, and how it may be diminished or augmented, as also whether it is something analogous to sight, touch, taste and smell. Judging by Nature, it would be impossible for us to know what it is which produces sight, and how it supplies it, as we have set down in our treatise on the Bodily Faculties, or which of those powers possessed by the body can be investigated by natural means. It is not so with life, because life does not derive its origin from natural seed, but has a spiritual source, though its origin is still natural. It is just as when the flint is struck by the steel, the fire leaps forth. There is no fire in either, and yet fire is produced from them. They are not fiery by nature, neither in their elements nor in their combinations are they combustible bodies; they rather resist fire more than other stones and metals do, as is seen in their transmutations, wherein no fire is elicited or can be obtained. None otherwise life takes its origin from something wherein is no life, as from the seed, the root, the sperm, and the like, just as the spark comes from the flint by the force of the *ens*, not of the *esse*. So, also, the life of man proceeds from the *ens*.

And then one has to consider whether this life can be produced, improved, or strengthened, seeing it is an incorporeal, volatile something, like a fire to which the more wood you supply the more fiercely it burns. In like manner, life, the more of the humour of life it has, the more the spirit of life abounds in that life. *How* the fire bursts forth from some substance in which is no fire, it is impossible for us to know. Some think that it is due merely to hardness; but how this can be the cause of a substance supplying that which is not contained in it, is not a subject for discussion in this place, where we treat solely concerning life.

Since, then, we see that life is, as it were, a burning and living fire, in this way it is illustrated to us in a material fashion and in a sense which comes under our ocular observation, that life belongs to the same category as fire, and we know that fire lives in wood, as also in resins and oily substances. None otherwise life according to its goodness or evil is strong or weak. Whatever of good or evil fire there is in it, as in the wood, we are able naturally to amend, or to add that in which it delights, or by which it will be rekindled; we can regulate it at our pleasure, can supply something stronger and stronger when another thing fails, substituting something else in its place exactly as the fire is renovated with fresh wood. Though it had been well nigh extinguished down to its last spark, it is again kindled, and burns more brightly than before. It is increased by those replenishments, or by any other mode of regulation we see fit to adopt.

Nor let us think that we must die on some precise day, sooner or later, or that it is derogatory that a Christian should believe it possible to prolong life by medicaments created by God for that purpose. Nay, it is altogether a mark of idolatry or of idiotcy not to believe that this is put in our power, just as we said above of fire. Our only defect is that we do not know the special kinds of wood by which we can kindle our life. It is not against Nature that we should live until the renovation of the world: it only passes our comprehension. For the most part we lack wisdom, so that we are unable to judge what there is in existence which is useful for us, especially since we altogether ignore our powers in this respect. Adam, whom we think the wisest of mortal men, had perfect knowledge of these matters. Although he was deprived of the tree of life, this is not a matter of theology, but of medicine. There *is* a natural tree of life —the tree of the soul.

Since, then, we are writing and teaching about long life, we should, in due course, at this point consider that kingdoms, districts, states, and valleys more or less contribute to the prolongation of existence. Some are more useful for the purpose than others. They afford more joy, more fresh air, and humours which are healthful to life. Among these exist causes which we ought to understand, connected with the land, the elements, the winds, the stars, and so on, which can make life longer than the common span of human existence. From the earth is produced everything that nourishes and supports our life; and also that which destroys it and makes it perish. Hence we can gather that evil arises to our bodies not only from such things as poison and opiates, but from those things by which life is chiefly supported, as from gold and balm. Nor should it be thought that medicaments were created only for use against disease, as, for example, the Tyrian remedy against poison, nor should it be inferred that these have no operation in healthy bodies. For we should know that to keep the body in health just as many things have grown up as for taking away health. We can, by our daily food and drink, at one time injure our bodies, at another benefit them and keep them in health, according to our use or abuse of these things. Whatever we use of these elementary things becomes useful or hurtful to us according to our reception of them. For whatever the fire in us consumes as it burns, that the fire restores to us when it is extinguished. Whatever loss is brought about by the one or the other the air itself compensates. It never refuses life or deserts us; it is we who do this for ourselves when we withdraw from the air. In this way, whatever one element takes away from us, some other element supplies in the way of conservation; and by an alternation of this kind those elements are regulated, so that they may do us as little harm and as much good as possible.

And although the air may sometimes destroy us by the infection with which it is occasionally contaminated, still we are able to discover this, and even in another direction turn it to our advantage.   The air is both particular and universal, good in one place, bad in another, just as one land is better than another.   In like manner, it is compared with water, which, like the air, is sometimes to be avoided by us, and the same is the case with fire.   For from one kind of wood a better and more healthful flame proceeds than from another.   So, also, influences and the stars above ought to be understood, since they have the power of breaking up our health in different ways, and of destroying our health.   In like manner, it must be remarked, that among these some have the power of conferring health and of prolonging life.   For there is no section, species, or kind so bad but that some good also exists therein, though we are not able to twist and turn this, or to regulate it for our advantage, as, for instance, that we should be able to appropriate to ourselves Jupiter and to reject Mars, or to select for ourselves other useful and convenient stars, as we do with herbs, some of which we choose and others we reject.   We have not the same kind of power over superior objects of rejecting the one and selecting another; though even here we have some power when we bring it about through some means that they exert their influence on the inclinations, as when we use the rings or images of the planets, and other methods which we will explain more fully and clearly as we write.   Furthermore, too, it must be considered that sometimes our life depends upon the aforesaid causes, and similar ones, either to be diminished or prolonged.   We will say at greater length what good or evil can result to us from these, so that we may be able to understand the powers of God and His creation; not that we would presume to scrutinise these from their foundation, or fully describe them from the beginning, but only give some fragmentary thoughts concerning them.

More at length we have to consider in what essential way such things as diet, medicines, places of habitation, and the higher bodies exert their influence over us, so that we may be able to avail ourselves of the virtues contained in them, and to extend our life, and understand what their influence is, as also how they are able to do us good and help us.   Their virtue far exceeds ours, because it is less subject to any sensible virtue, and it may help us just as a log is not able to burn unless some oiliness survives in it.   The virtues of the things we have spoken about by far surpass ours in their consummate strength, and this in the following way : Their essence and property, their nature and quality, are incorruptible and permanent, since they do not die like mere sensible bodies, or as man dies, who is deprived of life and stripped of all his powers when his body becomes a corpse.   Yet herbs and like objects do not perish essentially.   For the substance of these things, even when the material body putrefies, remains, an incorrupted essence indeed, in the earth, passing on in turn to its own like until the remaining earth is consumed. They are taken up for its use, and their essence passes through their decaying bodies into the earth.   But in the case of man, his virtue, essence, and properties do not remain with his corpse, but withdraw themselves from decay.

Hence it is gathered that we are not able to receive into our bodies the essence of herbs and the like, so that those bodies may be increased and nourished. by the virtues of those herbs, simply because we receive the properties of the herbs, so that if one takes a pound of the quintessence of hellebore it relaxes the body as that herb does, or if a man eats gold his body shall become golden. By no means. Here is the difference : the one is appropriated, and the other remains in a material form within the body with all its virtues abiding in itself. This is an essence, not a property, but a certain *esse* from which is produced an essence endowed with all its properties and appropriate natural gifts. Such an essence, when it has entered the body, mingles with the humours for sustaining the spiritual life. This essence is a humour, and constitutes the life of the thing, and for this reason the two humours combine in one, and agree in their intermixture so that the inner receives the *esse* of the outer. The one is united inseparably with the other, just as one wine is mixed with another. Moreover, it must be remarked that consumption goes forth from that body because the humours of man's body are always mortal, and liable to decay ; whence it occurs that essences are deteriorated and weakened for the future. There are, then, in this one two contrary things, whereof one is corruptible as the humour of life, but the other, which is fixed, consists of the essences of things, their humours, and liquids. That which is fixed gives assistance until it is consumed, just as talc does, which fire has no power to injure, but, nevertheless, in process of time it is consumed in another way. With regard to long life, we complain bitterly, because we are deprived of the power to understand what we should take when weakness seizes on the fixed substance, and we become weak, like a lion worn out with a long fight, and no more abounding in strength, or when we have no wood to add to a fire that is going out.

We know that there are some medicines and some regions in which there is no death, in others it is slow, and life is very long because the harmonies which prolong life are in these places most abundant. Although there are in these places some who are mortal, still they live a very long time, concerning whom I do not purpose at present to write more fully. We should know also that some persons are immortal, so that we ought to learn what is mortal and what is not, as also what the grades of the universe are, what vast natural powers exist therein, and how supremely man may be beautified by a long life. This can be understood in two ways. That is mortal which cannot remain until the destruction of the world. That is immortal which will await such destruction, though these very things are themselves mortal. This lower world is not going to remain for ever in its substance. What is born therein is divided into mortal and immortal, so far as natural things are concerned only in common speech. For the world is not one and undivided, but partitioned off into regions, namely, Paradise and the outer world. In the latter we live with such hope as we can. But in Paradise, which is the other world, there is no death. This is no miracle, but purely natural, because the nature of that world allows it. Just as our gold preserves one from leprosy, so does Paradise preserve us from death, not by a miracle in either case. Nor must it be understood as though our dead body was to rise, which would be a miracle, as we have said in our book on Heaven and Earth. It is most necessary we should know that in Paradise is born a being immortal in its substance. Moreover, in the essence of the outer world we have nothing but a fixed essence and a corruptible body, which are both preserved in Paradise. But to write much about this belongs not to our experience, beyond what the earthy essence teaches, which affords a centre. Nor ought we to discuss at length about these matters, since they far exceed our imagination, and every faculty which seeks to learn whilst on earth the order of Paradise. We speak of these things after a spiritual manner, rather in a dream than waking, and only for this reason to shew that the life there is enduring, up to the consummation, or perhaps beyond it, but this is to us occult.

Whatever, then, is from Paradise is able to render our life immortal, and not the Nile itself could wash away that attribute or despoil us of that virtue. We are by our materiality, by our place of abode, and by our pursuits deprived of a certain amount of power, and so we will not write further about these matters. Still, we will endeavour to teach in the sequel about a long life, so far as it shall be in our power to look into the matter, leaving those things which we are not able to grasp, and rather considering the predestination of each in particular.

In order to write more fundamentally concerning long life, weighing everything that conserves and defends it, it should be known that mind and being are the two methods by which soundness is either taken away or secured. In these two the health of our whole body, and everything that we do, is established. Out of the mind those matters regulate us which are therefrom produced, which also specially belong to it, as incantation, imagination, thoughts, and influences, all of which operate in the mental sphere. From the being those things govern us which belong to that department, such as the complexions, the qualities, the limbs, and the bones. Furthermore, this must be known and understood, that incantation, estimations, imagination, and influences are submerged in our mind when our thoughts are so plentiful and so powerful as to master our reason and our mind. So, too, from the most powerful forces of imagination, estimation, and incantation, the reason is submerged like the fragrance from the rose, whence it happens that it introduces syncope and madness amongst its influences, and so rules in the mind that it masters it and dominates it, not because it is produced in the mind, as the three preceding things are, but it has such a power of inclination, that it can turn it to good or evil, just as the sun shining through some glass tints according to its own nature whatever is contained in it. No otherwise does our mind exert a transmuting power according to its natural life, like ice melting in the sun. Those things which proceed from the being bring to us bodily infirmities from our complexions and qualities, which complexions originally arise from the being, and this is conserved by the humours, according to the bodily organisation.

It is in no way necessary to consider diseases of this kind, or the origins, beginnings, and essences of the mind and the being, nor again to inquire what the cholera is, or the phlegm, or the blood, or melancholy, but only to proceed generically, as will be shewn below. First, then, we will approach our practical method, because those other matters have no relevance to mental affairs, or mental things to them. So we will assign to bodily things each its own substantial medium, and, while remaining silent about mental things, we will point out the effect which these others have in the curative process. Understand, then, that the bodily substance must be conserved in the humour alone, as being that wherein is the life, and from which the complexions, and so on, are regulated. It is not the complexions or the qualities that have to be conserved or purified, nor need the losses of the liver or the spleen to be taken into consideration. Conservation itself removes all these things, and blots out all defects of a similar kind, together with all that is superfluous in the body, and thus reduces it to a state of equilibrium. Conservation of this kind does not operate thus by means of some grand specific of its own, but by means of its altogether incorruptible essence.

We determine, therefore, to forget for the present higher matters, and to point out the practical method in its due order. Here the regimen must be first of all preserved, afterwards the disposition of the body must be considered, and thirdly, medicine. On these three the conservation is to be established. Regimen, from which we start, as regards regions, scarcely comes into the question amongst other matters, as it needs the least precautions, and only the convenient order and disposition of medicine. We put, therefore, our chief trust in that medicine which not only conserves man in youth, but also affects brutes. The chief essence of the same, which is present in natural objects, kindles the body with so excellent a virtue that no strength or virility can be wanting to it. At the same time no disease can affect it, since, through the conservation of youth, no decay is suffered to take place, no putrefaction can occur, no disease arises. It does not even admit the excrescence of superfluities, whence it follows that no disease proceeds from superabundance, whilst, at the same time, it prevents the body from being corrupted, and so no evil can arise from corruption. Finally, it so protects the body that none of those diseases can enter which flow out from imagination or impression. In like manner, those diseases which proceed from tartarus are unable to adhere to it, and if any has fastened upon it beforehand it is removed. When you have once received what we lay down concerning a long life, there is nothing left which you can wish to understand from common sense, since we disclose our meaning only to those who have a good and extensive knowledge.

Now, then, we will explain our method and practice as briefly as possible, though they will not be intelligible by common persons. We write only for those whose intelligence is above the average.

Its practical use is that one dram of this conservation should be taken once every month in the best wine, if the nature, up to that time, has not become debilitated. If decline has commenced, a dram should be taken every week. In the same way, if ninety or a hundred years have been reached, the same quantity mentioned above should be taken every three days. It should be carefully noted that if any one has been deprived of virile powers, the weight should be increased as necessary. For the natural power of man is virile in its own degree, and more potent for suffering in one than in another. All this must be taken into account in cases of conservation for the reason we have mentioned in the words preceding. Sometimes the constitution of a man is so weak that he cannot be conserved, as in the case of those who from birth have no good foundation or root. As a sponge cannot admit fire in the same way as wood does, so these persons do not admit of conservation in the same way as others do. So, then, though the medicine be perfect, still, on account of defect in the subject, in the way of constitution and perfection as required by Nature, the operation is weakened.

Concerning the female sex, too, it should be understood that, so long as this method of imagination is applied, the menses will not fail for another period of life, nor yet fecundity for conceiving and bearing children; provided only that the women, by their natural constitution, are qualified for this. For the spirit of life which is in them expels every contrary influence and conforms everything as it ought.

So, too, of the yet unborn fœtus. If at birth it be thus conserved it acquires a thoroughly sound constitution, proof against all diseases and preserved for a long life.

A long life depends on three things: the vital spirit which conserves life in us all: after this arise the different temperaments and qualities, like the trunk and branches from a single root. We place the reception of conservation herein, and endeavour to defend the rational spirit, by our conservation of the nature, from all those accidents which could be occasioned by temperament, such as excessive sorrow or joy, which arise from the humours. These, again, are fourfold, varying with the temperament and the qualities, and so on. For these no special medicine need be used. If the vital spirit only be conserved, these vital humours are conserved also, and if these exist without any defect, no perverted elements, such as temperaments, can arise. So, by this method, the body, and all that is therein, is rendered sound and healthy. Hence, too, it happens that no evil can occur to the rational spirit through the aggravation of the body, but it is conserved in a good essence and nature, just as it can easily be understood that it will be conserved with all its faculties, such as sight, hearing, taste, etc., in due proportion. The process of conservation, then, which we have already mentioned, we describe in two ways, one by means of simples, the other by arcana. Yet, we would not wish to separate altogether simples from arcana, but, on the contrary, join them together, since simples of themselves have such marvellous virtues and powers that some of them conserve for forty years, others for a hundred, and this, so far as conservation is concerned, we place on a par with the powers of the arcana. The essence of such simples which secure so long a life we consider in no way inferior to an arcanum. Take, for instance, the leaves of the daura, which prevent those who use it from dying for a hundred and twenty years. Its virtues we describe in the treatise on the Nature of Things, and here, for the most part, we preserve silence about them on the account of the incredulity of men. In the same way the flower of the secta croa brings a hundred years to those who use it, whether they be of lesser or longer age. There are many more of these no less worthy of regard which we do not set down here. The arcana, it is true, have greater powers of conservation than the simples we have named, because they fortify and nourish the radical humours more than the simples do, just as flesh and herbs nourish men, but in unequal degrees, since one has more power than the other, but Nature derives satisfaction from this rather than from that, and partakes of it with greater advantage. We give the recipe, then, for conservation during two or three periods of existence, for men or for women, as follows :— Take of cut flowers ʒj. ; of leaves of the daura ʒv. ; of essence of gold and of pearls each ʒss. ; of the quintessence of saffron, chelidony, and balm, each ʒv. Mix them all together according to the artistic method, and keep as a compound for use, in glass vessels as above. This medicine is also sufficient to cure and to drive out any accidental diseases which may subsequently occur in the subject.

Now we have to speak of mental affections and assign to each its special remedy. Mental diseases, either those which proceed from the mind or come into the mind from without, may occur in different ways, as we have before said. And though we cannot ascertain exactly the origin and nature of these latter, since they are invisible to our eyes in their seats, and consequently in their powers, still we speak only of their effects which we can see, and of the operations which they shew us. On these we base our practical treatment, saying that their effects are produced in the spirit of the mind (as we have elsewhere remarked), as our mind is merged in them.

We neither wish nor are able to hinder the course of the heavenly bodies, yet we have the power of resisting them, just as a strong wall can be assailed with bombardment and with engines. The sun impresses its influence on a stone. If that stone be thrown into the water, the sun can no longer bring its powers to bear on it, and thus the stone is preserved. Similarly, influence can enter a man in two ways : in one, when it is impressed upon him by ideas. This cannot be hindered, as the grass in the meadows cannot vegetate and grow without the sun. The cause of this we set down in another place. In like manner, there is one kind of influence which conserves and sustains us as nutriment does the body. Such is the influence of the constellations. The other is accidental, and this does us injury by hindering the former so that it cannot exercise its operations upon us. We know this of planets and the like.

We adopt measures of preservation against them so that their effects may have no power upon us. Such methods of conservation against the planets consist of rings and images specially constructed against their influences. What we have said on the subject of imaginations should be carefully noted. The rings of the planets have power of defending us from accidents, and so affect long life. But they are not satisfactory, for several reasons. And here is something to be known, namely, that the mental influence should be directed to some other object. For example, if Mars should be disposed to destroy me, and there be a mental inclination from him in my mind, which might induce mental disease, I construct my homunculus, that the operation of Mars may be directed to this image, and I may get off safely. It is easier to affect the homunculus, and so the planet is able to work its will more gently and without resistance. It takes the easier course, and leaves the more difficult one. The material is the same by means of the opening that has been made. To this material Mars was inclined with the view of injuring my nature ; but he begins to operate on that which is easier, is satisfied therewith, and remains therein. Other reasons may be assigned for this, but we do not put them forward, since they have no bearing on the conservation of long life.

In this way there is produced a conservation of the mind, that is to say, the heavenly bodies are distorted towards something less resistent, so that the mind may be freed from the heavy yoke of those heavenly bodies by means of which death is often inflicted.

Against incantations, too, we would prescribe something, so that a long life may not be taken away by means of these. Remedies of this kind we have mentioned in several places, not as conservations, but simply for a cure of incantations. But the same is to be understood concerning them as in the case of the malefic stars. Incantations, that is, are to be guarded against in the same way as mental influences transferred to a homunculus. A similar operation holds good in its reversal to that which binds our minds and mental organs and our beings, the seat whereof is in the mind, as we remark in our treatise on Incantations. It must be directed to some other subject, and not to that which we have from the stars, but to its own incantation in the following manner : I construct a homunculus of wax to serve my purpose, and this I put in its place. Then, whatever attempt is made against me by way of incantation will be fulfilled on this image. For that proceeds from my mind, and the incantation from his mind, so that the minds meet, and on neither side is any harm done, or any effect produced. Under this form we have made clear the mode of resisting and preventing incantations by means of images.

So, too, imagination destroys an imagination directed against us. It may happen that I am being killed by the imagination of another. Such an imagination, which should be fulfilled in me, should be diverted elsewhere, lest health and long life be taken from me thereby. This I illustrate by the following example. If anyone feels great enmity against me, so as continually to wish for my death, and is not an incantator, but still a most persistent enemy, and if I do not know this, but his evil disposition is hidden from me, I, on the other hand, settle my mind in the greatest possible repose, by this method of protection being hostile to none, hurting none. By means of this piety, such great envy which is directed against me cannot be carried out; for piety is the principal and most consummate means of preservation against bad imaginations which can possibly be devised.

Estimation, in like manner, which is in ourselves, may operate in the same way. When I think more of myself than my reason warrants, such estimation is a submerging of my reason, so that I may lose my reason by the inordinate mixture of self-esteem, as we have said elsewhere on the subject of estimation. Against such estimations as these the best method of conservation is not to impose more on the reason than it can bear, which is the way in which such estimations are fulfilled. Here is another method of conservation. For instance, where we make false estimates of God, a subject which surpasses our reason, we betake ourselves to a homunculus directed towards the stars, which homunculus is of the same nature as ourselves, and so our speculation, which might corrupt us, passes to the homunculus. For there is no resistance to such speculations being fulfilled without loss, by a subdivision of the senses existing in me and in the homunculus, with a difference of perfection and imperfection.

By means of estimations and imaginations many superstitions are fulfilled, which are not impressions, or incantations, or estimations, but simply superstitions capable of being understood by a similar example. I suppose myself imbued with the superstition that when I hear crows chatter on my roof I esteem it the sign of some one's death (there are many similar examples which need not be quoted). This superstition may make me ill myself, or kill my sick friend, the cause whereof I set down in my treatise on Superstitions. My method of conservation is this : that I regard such chattering as natural, and that it does not operate according to the foolish idea of men and of myself. By this means my superstition is destroyed, for it is rendered doubtful when I attribute it to my folly. Nothing destroys superstitions more thoroughly than considering them to be follies. Thus all harm is taken out of them. It is the consensus that leads to action, and this consensus is at once removed if I think of my own simplicity and the folly of such empty credulity. And this does not happen only in the case of crows, but with many other things, which it is not necessary, nor would be useful, to rehearse in a treatise on long life. Enough has been said on the subject of conservations.

As to the fact of our being hard to understand, that happens only to those who understand neither us nor Nature herself. Yet is what we say of some moment. For those who do not understand us on the subject of long life, we decline to teach them more plainly. For those who have any depth, we are conscious that we have written of our processes at sufficient length, and have disclosed them with ample clearness.

Concerning the regimen of food and drink, we will only set down so much as bears on long life, reserving the rest to be treated in another place and by another method. First, it should be known that foods ought to be prepared according to a process, just as medicine, so that they may expel all the superfluities of humours which lie hid and grow in the body. These humours, by means of conservation through food and drink, are completely removed. Next, it is necessary to know that the cures of all internal diseases are accelerated and promoted by suitable medicines, and also by Quintessences. As to the origin of diseases and medicines, we say nothing about these matters here, because they are extant in many parts of our Physics. This only we say, that for a long life the following is the best regimen : moderate diet. We do not give this in detail, because it is well known to every physician. One thing we will add, that food ought to be separated from all its impurity, as we teach in our treatise on the separation of the elements. The use of such separated food with moderate diet exhibits a wonderful sustentation and conservation of life ; it restores the flesh and blood, so that no disease can do harm to the body through external foods, which is not the case with other foods and drinks. The use of such separation of foods for the foundation of a long life we set down for those who wish to live long ; and we do this for many reasons, not necessary to specify here. This we do affirm here, that this is a regimen by which the complexions of the blood and the flesh are nourished. Nevertheless, conservation belongs to the vital spirit and its humour. Thus we believe we have said enough on the regimen necessary for a long life.

We describe the separations of foods and drinks in the treatise in our Archidoxies on the Separations of the Elements, the pure from the impure. In like manner, in the book on Regimen, in the same place, we set down what may be best and most conveniently adopted for a long life. We do not repeat here what we have written at sufficient length elsewhere.

What air, earth, and elements are best, we will indicate in the same manner. In the element of fire no watery disease can arise, and *vice versa*. So, in like manner, we must understand of the earth and the air. Whatever lives in the fire is free from other infirmities except those of fire.

What has to be considered, then, and assumed necessarily for practice, is a certain desired equilibrium of elements, wherein long life consists. This is the work of wisdom alone, in which work are comprised the operations of the elements with the perfection of their forces, so that the fire shut up in man shall avail to take away all diseases, drive them out, and keep them at the greatest possible distance. We look upon this work of wisdom as another earthly Paradise, in which no disease could germinate or remain, no venomous animal could dwell or enter, and health could never be destroyed. This we could wish conceded to us by the Lord God, that we might write freely, without the contempt of idiots, what experience has taught us about this work of wisdom. But, on account of those idiots, one has patiently to hold one's tongue with regard to the miracles and marvels of that work of wisdom, wherein is reserved the earth of the wise. Since, then, I must be silent about this, I determine to describe it only among my secrets, that it may remain buried within me, though without any end of life.

Thus far have I written on the subject of Long Life for our own and other disciples who are endowed with a happy and subtle intelligence.

HERE ENDS THE BOOK ON LONG LIFE

THE BOOK CONCERNING RENOVATION AND RESTORATION

By Theophrastus, Philosopher and Physician, of Germany, called
Paracelsus the Great

———

FIRST of all we must understand what Restoration and Renovation are ; what those things are which restore and renovate, and also what that is which can be renewed and restored in the creation of things. All minerals, indeed, are thus brought back to youth, are renovated and restored, so that rusted iron can be again brought back to new iron, and the rust or flower of copper into its copper. So, likewise, minium can be brought into lead, and Saturn into Mercury (or, as others read, the calx of Jupiter into tin). In this place, therefore, renovation and restoration signify that process which brings back a destroyed, or rusted, or consumed substance to its youth and its perfect essence. But still this renovation which we have here introduced cannot be in any way compared to that restoration or renovation which we purpose to expound. For though rust and verdigris are not a metal, still none the less in their essence the essence of the metal has not yet perished or been consumed. So, then, that renovation cannot, in this place, be taken for an explanation of restoration and renovation, because in the human race such rustings and ablutions do not occur ; and hence it is that men do not require a renewal of this kind.

In this manner, if a decrease be understood to happen to an old or decrepit man, like a kind of rust in his substance, then, in like manner, his body will be capable of being brought back from its state of decrepitude to youth, and there is a restoration from any kind of disease to health, but concerning this we have no desire to write at present. This, too, may be accounted a kind of restoration when out of salt, sulphur, and mercury a metal is made naturally. When this perfection is completed and brought to the actual metal, that metal may again return to its three primal elements, so that the salt, sulphur, and mercury again appear as they were at its first generation, insomuch that the metallic element altogether passes away, and there is no longer any metal present. So also it may happen that the matter of the three primal elements may return again to a metal as before, namely, if from the three primal elements of copper, copper be again made, and so on. This, too, in the case of metals is a restoration or renovation, when a kind of regeneration takes place from a metal previously complete into a metal again, perfect and complete. But that is not to be thought of as renovation or restoration if it be referred to man, because we cannot be brought back to our three primal elements, or reduced to our sperm, from which, again, we might be again renovated and restored, as in the case of metals quoted above. For then we should have the power of improving ourselves by a second generation better than the first was; even as iron, which is reduced into its three primal elements, and afterwards to silver or gold, and becomes incorruptible by this very process; or as Saturn, which is again reduced to its mercury, is at length changed to an incorruptible metal. None otherwise should we, too, be able to produce or create from ourselves an incorruptible creature, whereas we have no power to do this. For we lack that primal matter,* and are unable to go back, so as to be constituted an irreducible mass, but we must progress as we have begun, and in no way are we able to recover or to possess that out of which we proceeded.

Restoration, then, or renovation, is twofold. One, as applied to metals, we have introduced and made clear. Another, when an old picture is renewed with fresh colours, so that it appears recent and new as it had been before. But we must not, therefore, understand in this place that a new matter is made out of an old one, but that the old picture is so draped that it may appear new. Wherefore, again, this kind of restoration cannot be cited with reference to the restoration and renovation of man. But restoration and renovation must be understood in this way: that man's radical moisture, acting upon and energizing the spirit of life, shall not be diminished or driven back, but rather shall be increased in its powers and pushed forward, as a tree to which aid is given for the production of its flowers and fruits, so that when these drop off and are done with others are again procreated as before. But, although the example here quoted does not in every detail illustrate our theory, nevertheless, it affords the means for understanding how to promote the radical moisture of life just as we shewed in the case of the tree. We intend renovation and restoration to be understood thus: that they are not produced in the radical moisture,† but in that which is generated from the aforesaid moisture and derives its origin materially and corporeally therefrom. For as a bell made by fusion does not receive its sound from the tone, but from the body, so restoration or renovation does not receive its operation in the spirit of life, but in that which makes this same spirit; that is to say, the one is material, the other substantial. But when all this substance in which the radical moisture is

---

* It should be noted that while all things are constituted in the three prime principles, they cannot be separated without the destruction of the simple matter itself; for in separation the virtue of Mercury, of Sulphur, of Salt, vanishes and goes back into the first matter, as may be seen outside the Microcosm, Mercury being transmuted into soot, Sulphur into oiliness, Salt into alkali, whence it is manifest that the first matter cannot pass into the ultimate matter in the absence of a medium.—*Chirurgia Magna*, Tract III., Lib. III.

† For the conservation of the radical moisture in its proper quality a medicine is required which is also a material humidity, and while this is administered, no disease can be contracted. *De Morbis Metallicis*, Lib. II., Tract IV., c. 5. In the same treatise alum is said to contain an elementary humidity against the fire of the Microcosm.

present shall have been purified, its tone will be also improved, and the better the tone is the better the body will be. And when we say that the radical moisture proceeds from the body and the members, we understand it in this way, that the radical moisture itself, and that which proceeds from it, are just like the root and the tree, of which the one cannot live or subsist without the other.

Equally must it be understood here that these two things are so intimately united and conjoined as to be incapable of being separated. The radical moisture and the spirit of life,* together with the moisture of life, are in bodies and in limbs, just as in metals is the tone, which is not seen, only heard. For the spirit of life and the radical moisture are truly in bodies. It would be idle, therefore, if we endeavoured to purify it or to renovate the body by its means; but it is just that the body and the matter which are born and have their origin therewith should be restored and renewed. Hence, it may be gathered that restoration and renovation are transmutations of members existing super-fluously in the body; so that all which proceeds from the body, and not from the radical moisture, falls away, and something new is born in its place, just as we perceive in a tree, from which all the leaves, the flowers, the fruits, and the fungi fall away, and are born again, yet the wood itself is in no respect changed, so as to fall away and be re-born, but it remains. So, too, does the radical moisture remain. This is the life in the body; and when bodies reject from themselves hair, nails, teeth, and such things, these are presently re-born. This is restoration and renovation, whereby that very thing which should be restored and renovated is restored and renovated. For every re-storation and renovation occurs in superfluities, and in those things which have their origin and are born from the substance. By what method the body is able to be restored and renewed may be sufficiently understood from the demonstrations which we have made, and from the superfluities which do not form excrescences on material effects, as the hair, the teeth, the skin, and the nails; but are in the body as something in excess, which are not absent from matter or corporeal substances, but remain in their essence as four complexions†

---

* The spirit of life is a spirit situated in all the members of the body, however they may be denominated individually. In all and each of these the said one spirit abides, and it is the sole virtue indifferently of them all. It is that highest and most noble grain whereby all the members have their life. But being extended and diffused it manifests in various ways, according to the variety of its seats. . . . Nevertheless, its potences are one. The virtues which sustain the bones are in no way feebler than those which nourish and fortify the heart, nor do they abound more in the brain than in the marrow, although the opposite may seem to be true. There is the same necessity of the marrow as of the brain, and the virtues of both are the same. A like law prevails through all the members. Some of them may appear to be of greater importance but, nevertheless, one spirit of life is the moderator, virtue, power, and operation of them all. The spirit of life originates from outside causes or generations, not from those which are natural according to the flesh. While the generation of other things is twofold, that of the spirit is simple. – *De Viribus Membrorum*, Lib. I., c. 1.

† Concerning the four complexions—cholera, blood, melancholy, and phlegm, we would by no means be identified with that opinion which asserts that they are or do derive from the stars or the elements. We do not regard this as true even in the least degree. The principle or beginning of cholera is from bitterness; melancholy is acidity; phlegm rises from sweetness for every sweet thing is cold and moist. Blood is from salt; whatsoever is saline is sanguineous, that is, warm and moist. The four complexions, therefore, are acidity, sweetness, bitterness, and saltness. If salt in any man predominate from the ens of the complexion, then is he sanguine; if bitterness, then he is choleric; if acidity, he is melancholic; if sweetness, he is phlegmatic. Thus, therefore, the four complexions exist in the body as in a certain garden, wherein flourish amarissa, polypodium, vitriol, and salt nitre. And all these may coexist in the body, but so, nevertheless, that one alone will prevail. – *Paramirum*, Tract III., c. 10.

(otherwise humours), whereof one proceeds from coldness and humidity, which is retained in the whole body, and is born, not having any special place, nor any beginning or initial point from whence it proceeds, as is proved concerning the four complexions. Another springs from the exact contraries of the former, namely, from heat and dryness, which, too, are similarly in the body and have no special abiding place or origin, and also produce liquid. The third is cold and dry, deriving its birth in the same way. The fourth is warm and moist, itself also proceeding as the others did.

And here it must be noted as happening that these four humours* do not all exist in all bodies at all times, but sometimes one only, sometimes two, sometimes three, at other times four. Of them it must be remarked, too, that, in the process of renovation and restoration, they are consumed and expelled, for this reason, that Nature and the life of man are able to exist without them, and stand in no sort of need of them, since they exist only as superfluities, like the dregs in wine, or the froth flowing from it at the time of vintage.

Concerning the four complexions thus displaying themselves in man, this, too, is to be noted : that these are not renovated or restored, because they spring from no one of the members, either greater or lesser. Neither are they in the blood, nor in the flesh, nor in anything like these. Nor, again, is it true that the sanguine complexion proceeds from a liver abounding in blood, or the melancholy from the spleen, the choleric from the gall, and the phlegmatic from the brain, and the like ; since the aforesaid members do not supply their complexion to man, but those complexions come at birth itself and last right up to death. These points we do not undertake to discuss in this place, because they are too remote from our text concerning renovation and restoration.

Since, then, no one of the four complexions has its place or origin in the bodies we have spoken of, but exists in the spirit of life and in the radical moisture, complexions cannot be renewed or restored. But when the body shall have been clarified, their nature, too, shall be made clear.

In like manner do we point out as foreign to our text the division of complexions according to age, region, and regimen, because no complexions are impressed on a body by these three. It may, indeed, happen that old age brings sadness to bodies, but this is not a complexion. In like manner, the dwelling place may induce phlegm, but the complexion is not on that account phlegmatic. Bile may make one acquire a yellow colour, which need not be discussed here since it is made clear when we treat of the Construction of the Body.† For a division of this kind a special phase of intelligence is required,

---

* There are four humours contained in man—blood in the veins, moisture in the flesh, viscosity in the nerves, grease in the fat. These four have each their natural purpose. – *De l'este, cum additionibus*, Lib. II., Tract III. At the same time, the doctrine of the four humours as commonly expounded at his time was rejected by Paracelsus, because it was a thing hard to believe, founded upon faith only, whereas medicine is established, not in faith but in sight, and nothing in such a matter should be accepted upon faith, except the diseases of the soul and eternal salvation. — *Paramirum*, Lib. I., c. 1.

† Paracelsus has a treatise on jaundice, which will be found in the first volme of the Geneva folio. As in so many other cases, there is no work which precisely corresponds by its title to that named in the text.

since it must be remarked that they are not only humours, but sometimes minerals too, and sometimes corruptions, which all exist as superfluities contrary to Nature and virtue.   In like manner it must be said concerning the principal members which resist renovation and restoration, that is to say, in this way, that they do not perceive it, for they do not receive them into themselves, but they take up everything that passes through them, and is prepared with them, just as they take up food, not a medicament.   But wherever by any chance humours or superfluities are produced in them, they would be expelled.   So, also, of the other members, too, it must be equally understood, namely, of the bones, the marrows, the brain, the heart, the liver, the lungs, the kidneys, the spleen, the stomach, the intestines, the cartilages, the muscles.   And of the blood, too, it should be known that corruption or superfluity exists in it, though it be only an accident.   And so equally of the flesh. This accident is, as it were, purged away in the process of renovation and restoration.   Not, indeed, that another blood is produced, but that what is depraved is removed from it, and the good is preserved and predominates. The same judgment, too, is passed concerning the flesh.

To explain briefly what are those things that can be restored and renovated : leprosy, falling sickness, mania, pustules, gout in the foot, or in the hand, or in the joints, and other like ailments are removed in renovation and restoration, unless, indeed, it be some disease taking its origin from birth. This will not be removed.

But concerning leprosy, or any more severe disease which may exist, it is well to know that it undergoes transmutation in the body, not, indeed, that there is a separation of the pure from the impure, but that the leprosy is converted into health, as copper or iron are transmuted into gold.   And no one ought to be staggered by this conversion, for renovation and restoration consume, none otherwise than fire in silver or in gold, its falsities and impurities, and leave it pure.   In the same way the falling sickness and gout are taken away; for all things which are in the whole body are forthwith renewed, the blood and the flesh, with the other things which are embraced in it.   For, as alkali purifies mercury into the very best silver, so also renovation and restoration transmute the body into a good essence, as has been said above.

Renovation and restoration, then, expel whatever is superfluous and incongruous with Nature in the body, and change all this which Nature does not want, or which was of no account, into something good.   In this way it re-establishes everything and causes it to grow again, as we have said above, brings back the entire body to youth, and so on ; for this cause, that nothing can resist those things which are in the nature itself.

And now we must see by what method the body is restored and renovated, namely, by the kindling of the renovating and restoring medicament, which it has in the spirit of life and in the radical moisture, by which kindling the operations are prescribed like the burning forces of a nettle.   For who is so sagacious that he can investigate rightly forces of this kind when they do not appear to us in natural act, but are sensibly apprehended? In this way, also, renovation and restoration are accessions to Nature produced by forces which we cannot express. This, however, is openly known to us, that every visible thing is cleansed and purified with fire. Nature, indeed, demands that this one process shall be accomplished by fire, and that it shall not be possible by any other means. We understand, therefore, a two-fold fire, a material and an essential fire.* The material fire operates by flame; the essential by means of the essence and the virtues, like cantharides, burning the skin and raising it in pimples as a very violent fire does; yet still it is not fire, nor is felt as fire. A flame and a nettle produce the same effect, as we have often said.

It is in like manner certain that renovation and restoration in this way accomplish their operations when they come into the body or are joined in union within it, because a like operation takes place thereupon as is the operation of Saturn or Mars in Mercury, which are cast into the fire with their realgars, and although neither of them is warm or fiery, they are, nevertheless, burnt up like wood, and in the bottom the perfect metal is found which before appeared altogether leprous.

And, again, who can altogether trace or investigate how it happens that when a migdalio has been powerfully liquefied by means of vitriol, it becomes copper, in all respects and throughout its entire substance like copper, though before it had no likeness to copper at all. None otherwise must renovation and restoration be thought of by us than that they accomplish their operations like lime, which is extinguished by water, and purifies itself, so that all its powers and acridities are taken away and removed by its essential fire.

The renovation and restoration of our nature are none otherwise than in the case of the halcyon, which bird, indeed, is renewed in its own proper nature; and many other like animals are found which have the power of doing this, of which we have made mention in different ways in our Archidoxies, and still more in our Secrets,† from which could be quoted more examples were they not too far removed from our text concerning renovation and restoration, where the demonstrations we make come to be understood equally in this place concerning renovation, while we say again and again that we cannot sufficiently know how the fire operates, though we see it consume the wood, because by its excessive heat it overcomes and consumes everything else. But, leaving this, let us betake ourselves in another direction.

Since, then, we have spoken at sufficient length thus far concerning the beginning of renovation and restoration, let us now point out those things which do renovate and restore. We have, indeed, in our Archidoxies taught

* Fire in its nature is four-fold, that is, the sun and moon govern one part in water, the second and the third, which are resident in air and earth, are ruled in like manner by the sun and moon, and there is hence conjoined in all creatures that magnetic virtue concerning which there must be nothing more openly spoken, for herein is the knowledge of the labour of Sophia, the mother and fountain of the Magi; I have said.--*De Pestilitate*, Tract II., s. v. *De vi magnetica mumiæ in homine.*

† The reader of Paracelsus may not unreasonably be disposed to think that his secrets are synonymous with his whole philosophy. At any rate, there is no individual treatise under this title.

in writing how to prepare them, and entitled them with their proper names so that they may be known and marked. Now it is our intention to lay down the composition thereof, and first of all the processes. But while we teach concerning simples and arcana, it must be understood that their operations are brought about in different ways. For some are found which violently purge leprosy by means of renovation and restoration, but beyond that touch no other disease ; yet, nevertheless, they are perfect in renovation and restoration. Beyond these, in the distinctions of these kinds of diseases, are the Quintessence, the Magistery, and the Element of Antimony, which purifies the body from leprosy, none otherwise than silver and gold liquefied therein, and in these it leaves no trace of impurity.

In like manner the element of the sun and its quintessence, as also the oil of the same, and aurum potabile, take away leprosy, together with all diseases, renovate and restore. So also the quintessences of hellebore, of chelidony, of balm, of valerian, of saffron, of manna, and of betonia renovate the body, with the exception of those diseases which we have mentioned above, for these they in no wise diminish.

The quintessence of pearls, too, or of unions, of smaragdines, sapphires, rubies, granates, jacinths, renovate also the body and bring it to entire perfection. They take away tartareous diseases, as the calculus, gravel, gout in the hands, feet, and joints, together with congelations and coagulations, and similar diseases which arise from tartar. So also the quintessence and magisteries of minerals and of liquids renovate and restore the whole body from any defect, and free it from the falling sickness, syncope, suffocation, and all diseases which happen with a deprivation of the senses, such as mania, vitista, and the like.

The magisteries and the essence of tartar and alkali also renovate the body with perfect restoration, take away all abscesses, and remedy the putrefactions and the grossness of the humours.

In like manner, the essences, extractions, and magisteries of the greater drugs renovate and restore the whole body, take away fevers, both quotidian, quartan, chronic, and ephemeral. Likewise the first elements of the sub-margarites can renew and restore the whole body, and remove all diseases from women, with their accidents, and also render mankind fruitful, both the husband and the wife. The same arcana in like manner take away all long-standing and incurable diseases by renovation and restoration of the body to its highest powers.

None otherwise, too, the quintessence extracted from balsam renovates and restores the body. It takes away pleurisies and pestilences by its admirable operations and power of perfection. Of this class, too, are many more things which are also enumerated elsewhere, of much greater virtue than can be attributed to them.

In these matters, however, it must be understood that the compositions have to be carefully watched. For though there are many of them, still no one suffices generally to cure all diseases by itself, but such diseases are to be expelled by the medicaments of renovation. Finally, therefore, we purpose to demonstrate the manner and the practice of our intention, though we may not set down all the processes, for this may not be necessary. He who understands us will perceive the drift of our writings. He who does not understand us is not capable of being taught by our writings, yet none the less we will set down all the processes in sufficient detail. In truth it would be a heavy task for us to write those things which have been written by many or are already known beforehand. This doctrine cannot be conveyed better than by the first entity, in which there is a singular nature for operating upon the body and transmuting its essence. For that first entity, indeed, is an imperfect compound predestined to a certain end and to corporeal matter. And because it is not perfect it is able to alter everything with which it has been incorporated as Mercury can, which is like a primal imperfect entity, in respect of its own imperfection. Although this be determined and limited, nevertheless it is not changed from imperfection, but still it is limited thereby.

Mercury, also, has the power of renovating the whole body because there is in it a powerfully laxative force, and an alterative as well, which cannot be sufficiently explored. Yet, nevertheless, it is as a whole imperfect and useless in its operation, because, so to say, it is mercury, and its first entity should not be predestined into another body. For such as it is itself, such is its perfection. Nevertheless, we speak of a first entity, which is perfect for renovating and restoring the whole body, as is the first entity of gold, and for this reason, because it embraces altogether the spirit of gold and is most subtle, far more subtle, indeed, than the true body itself, that is, than gold.

Hence, also, the first entity of the sun, or gold, is penetrable, even as mercury in metals, and does not contain within itself the spirit of salt, whereby it may be coagulated. For the spirit of salt coagulating the first entity takes so much power away that the gold becomes not by a hundredth part so powerful in its effects as its first entity is. In the same way, generous wine coagulated by frost never again returns to its pristine power.

Now, in order that we may speak and write perfectly concerning renova- and restoration, it should be known that the first essence, that is, the first composition of gold which exists as a liquid not yet coagulated, renovates and restores whatever it lays hold upon, not only man, but also all cattle, fruits, herbs, and trees. And this must be understood, just as of the universal form of any metal, which is endowed with far greater virtues than its own metal, seeing that there is in the mineral form the spirit of arsenic and salt of sulphur and of mercury, all of which are lost in the purgation of the metal, the said metal remaining in one essence alone.

The very same thing is, in like manner, to be understood concerning the first entity of marchasites, as of antimony, which ought to be known no less than the first entity of gold. In the first entity of antimony, indeed, there exists such a virtue that of itself, by its own special nature, it transmutes everything of which it takes hold, none otherwise than antimony itself does by fire. For the virtue of that same separates from the body everything which is generated by the radical moisture, and altogether renovates the said body from its very foundation; for its first entity becomes so fixed in that predestination that such an essence proceeds out of it, and goes forth from it as heat does from a fire. No otherwise must it be understood concerning the first entity of resins. The first entity of sulphur is an entire transmutation of the body into certain renovations and restorations; and it is so vehement that it tinges all the first entities of metals into its own essence, takes away their operations, reduces them again to their first matter, and afterwards brings them to a perfect and new body. Indeed, the first entity produced from sulphur has such power over the body of man that it renovates all the radical humours in it, in all its places and parts.

In like manner, also, may we speak concerning the first entities of gems, which, indeed, by their primal essence most powerfully reinstate the whole body in its pristine powers, cleanse it from all its impurities, and renovate and restore it none otherwise than the fire changes lead into purest glass. For the first entity of the emerald regenerates and renews itself since it exists from the first as a perfect body. So, too, green marble, which, from its own predestination, has such a nature that it renovates itself from all uncleanness and impurities, and coagulates a second time until it becomes pure. Sometimes it renews itself thus a third and a fourth time, and returns to its youth; and the oftener it regenerates itself in this way, the purer and more enduring does it become. So far, then, as the virtues of first entities are known to me, these certainly far excel all the rest.

So, also, concerning the first entities of salts it should be remarked that, like their spiritual virtues, they are by far greater than in their perfection. Thus, the first entity of vitriol transmutes all white metals into red, and overcomes and masters all the perfections comprehended in them. It renovates and restores all the imperfect bodies of the metals, as tin into its first entity, and again into tin, in which there are more virtues than in the original tin.

In the same manner, it brings back to the radical moisture whatever proceeds from those radical moistures, and renders that same renovation and restoration more perfect, more plentiful, and more full: for nothing else operates so powerfully on the radical moisture.

In no way different are the first entities of herbs and trees, which, before they have received their body, that is, their stalk or trunk, are a thousand times more powerful than when incorporated. Even so, the first entity of balm renovates and restores the body far more powerfully than seems possible to be done in natural things. It should be known, that the halcyon is not thus renovated or restored by its own nature, but because its nature is such that it should be nourished and live on first entities in this way: When it eats the bodies of herbs, or of seeds and the like, its stomach, by digestion, reduces them to their first entity, and afterwards, by means of that first entity, it perfects the operations of its own renovation and restoration. For its digestion was by Nature predestined only to first entities, whence it happens that it first of all transmutes all its food and drink into a first entity. On which account, also, it eats those bodies only which regenerate and restore, with which from the very time when it is hatched it is furnished and nourished by the parent-birds. In the meantime, this is its own nature, that after death it is renewed and restored; on this account, forsooth, because all these first entities cannot have their progress in the bird whilst it lives, since the life of this bird takes away all the power of those entities by turning them into blood and flesh; but when it is dead it flourishes according to the annual seasons; and just as first entities put themselves forth in the earth, thus, in like manner, do they then put themselves forward in the bird itself, and so renovate and restore the dead flesh, which, indeed, in Nature itself is a wonderful argument for its powers and virtues. If these things did not lie open to our eyes, they would be incredible, though they should be written down by many persons. From this cause, also, it happens that halcyons renovate themselves at irregular times, some sooner, some later, as they have earlier or later eaten the first entities; for some are born and come forth sooner or later than others. Also many worms are renewed and restored for this reason, that they are fed and nourished on first entities while they are still imperfect in the earth. There are many marvellous things which are occult, and far more than are known or openly investigated, concerning which one could write more copiously, were they not too far removed from the text of a book on renovation and restoration.

And although, as we have previously written, we cannot very well take first entities, or have them in the same essence; still it is possible for us to do so. For if we know where the mineral of gold lies hid, we shall also find its first entity in the same place if we come before its perfection. For there are certain signs by which it may be known how the form of the metal is placed. So whilst it is in its first entity it makes trees fruitful, and renders fertile its foundation, that is, the earth. It renovates old trees which for twenty years have borne no fruits; for when the first entity of gold has seized upon them, or upon their roots, they begin again to live and flourish as before. But although there are many more particulars concerning the first entity of gold which are worthy of our admiration besides those we describe, these suffice for a demonstration of the first entity of gold, namely, that it is there.

But when flames and sparks are seen, then it must be inferred and noted that the metal is being produced out of its first entity, which metal has betaken itself to the process of coagulation. These are reckoned as signs concerning the origin of minerals which apply to gold, silver, or the other metals; for the signs of the other first entities as to their origin are like those whereof we have spoken.

When a sign of this kind has been seen or found it must be understood that this same first entity is not gathered up in one heap, as it is when it lies in its perfection, but that it is spread over the whole of the land in that district. Wherefore, this land is in the power of the first entities, and out of it these are drawn, as it is with the celandine when it is not yet composited. Its first entity is in the earth, where it has its position. For this reason similar earth should be taken, and from this at length an extract should be made, as we have pointed out concerning the virtues of celandine. It must be noted that between the first entity and the perfection there is this difference, namely, that the first entity has power to renovate for the reasons before mentioned : but when perfected it has only the virtues of the natures, so that it tends in that direction, but not perfectly. Thus it may be gathered that if it be wished to have from these virtues of the same kind as those of the first entities, it is necessary that they should be removed from their coagulation, and should be corrupted, as is pointed out in arcana and quintessences. But that every thing has greater powers in its first entity, let not philosophers wonder, because even out of any earth in which a herb is produced essentially, before it is incorporated, all the virtues of that herb can be extracted, so that the virtues may be preserved and the earth be put back into its own place, so that for the future it shall be mere earth, and have no fruitfulness in it, because its first entity is taken away, which had lain hid in the earth.

In this way it often happens that the power of a first entity of this kind may be enclosed in a glass, and be brought to such condition that the form of that herb grows of itself without any earth, and even when it has quite grown it has no body, but something shaped like a body, the cause of which is that it has no liquid of the earth. Hence it happens that its stem is nothing more than a mere apparition to the sight, because it can be again rubbed down to a juice by the finger, just like smoke, which shews a substantial form but is not perceptible by any sense of touch. In growing things of this kind the quintesssence is entirely incorrupt and in its highest state of perfection, as in the earth.

Wherefore, from the first entity of gold is produced, in this way, the finished gold, which to the touch is like red water, and is stirred up and exalted like gold.

So far concerning these things. Now, let us next in order betake ourselves to the practice of those things which renovate and restore, if they be prepared by the power and rule of art. These things, although briefly written by us, are sufficiently patent, nevertheless, to intelligent men, that is to say, those who have good instruction in medicine and philosophy.

First of all, then, must we know those things which renovate and restore, as we have pointed out, and their first entity must be extracted, and by that the work of renovation and restoration must be done. As a deduction from this argument we set down four mysteries, that is to say, of minerals, gems, herbs, and liquids, as follows :—

THE FIRST ENTITY OF MINERALS

Take of mineral gold, or of antimony, very minutely ground, one pound, of circulated salt four pounds. Mix them together, and let them digest four months in horse-dung. Thence will be produced a water, whereof let the pure portion be separated from that which is impure. Coagulate this into a stone, which you will calcine with cenifiated wine, separate again, and dissolve upon marble. Let this water putrefy for a month, and thence will be produced a liquid in which are all the signs as in the first entity of gold or of antimony. Wherefore, with good reason, we call this the first entity of these things. In like manner, it will have to be understood concerning mercury and other things.

THE FIRST ENTITY OF GEMS

Take of emeralds, well ground, ʒj., calcine them in dissolved salt until they be turned to a white colour. Afterwards let them be dissolved, and enclosed in a phial hermetically sealed, and placed over an open fire. Let the matter be suspended on high in a naked glass vessel, so that it shall not touch the bottom, and let this be continued until, from its spiritual nature and condition it falls down to the bottom into a body like the liquor of honey. This displays the virtues of the emerald. Wherefore, it may rightly be called the first entity of the emerald.

THE FIRST ENTITY OF HERBS

Take celandine or balm ; beat them into a pulse, shut them up in a glass vessel hermetically sealed, and place in horse dung to be digested for a month. Afterwards separate the pure from the impure, pour the pure into a glass vessel with dissolved salt, and let this, when closed, be exposed to the sun for a month. When this period has elapsed, you will find at the bottom a thick liquid and the salt floating on the surface. When this is separated you will have the virtues of the balm or of the celandine, as they are in their first entity; and these are called, and really are, the first entities of the balm or of the celandine.

THE FIRST ENTITY OF LIQUIDS

Take the mineral of sulphur and of dissolved salt; let them be completely resolved into water by themselves, which distil four times. First will ascend a certain whiteness which displays all the virtues of the first essence of sulphur. Therefore with good reason we can have it in the place of the first entity of sulphur, and so name it.

Now that the four first entities have been thus described generally, it must be further remarked in what way they are to be utilised, so that their virtues may be perceived. This is the method. Let either of those first entities be put into good wine, in such quantity that it may be tinged therewith. Having done this, it is prepared for this regimen. Some of this wine must be drunk every day about dawn until first of all the nails fall off from the fingers, afterwards from the feet, then the hair and teeth, and, lastly, the skin be dried up and a new skin be produced.

When all this is done that medicament or potion must be discontinued. And again, new nails, hair, and fresh teeth are produced, as well as the new skin, and all diseases of the body and mind pass away, as was declared above. Herewith we would conclude our little book on renovation and restoration.

HERE ENDS THE BOOK CONCERNING RENOVATION AND RESTORATION

A LITTLE BOOK CONCERNING THE QUINTESSENCE *

By the Great Theophrastus Paracelsus, Most Excellent Philosopher and Doctor of both Faculties

———

ANY have written concerning the quintessence of those things which either lie hid in the bowels of the earth or grow and sprout therefrom, concerning the quintessence, namely, of metals, salts, saps, stones, trees, herbs, roots, quadrupeds, fishes, and other animals, etc. To few, however, has it been vouchsafed to point out the method or process whereby the quintessence can be extracted from the said things, so that a certain hope might, indeed, be left for the human race, not only that a certain quintessence of this kind resided in things, but also that the glorious and honourable body of man might become partaker of the same essential blessing. Wherefore, I am now about to teach in part, avoiding the method of those loquacious triflers, who merely threaten with railing, but do not strike. It has often been stated that an implanted quintessence inheres and dwells in all things; but I shall, in addition, indicate how it is to be extracted, not only from metals, stones, salts, and saps, etc., but also from roots, herbs, animals, and all other things which have anywhere been formed by the Creator.

All essential and created things contain in themselves water, oil, and salt; hence it will be a matter of very great moment to separate the salt and the oil from the water. During the process of distillation the waters first appear, next the oils, lastly the salt. However, frequently the oil, the water, and the salt remain conjoined together. Thus, if juniper oil be extracted from berries in which salt is present, then the oil is separated by reason of its levity from the water, which is somewhat heavier. No one, notwithstanding, has hitherto succeeded in separating from it the salt, which is the most precious of all. Moreover, I am convinced that it is a true quintessence of juniper.

Further, no fresh water proceeds from sulphur, but only oil. Yet, under that oil lie hidden salt and concealed water. Nor can it be properly designated oil unless the salt and the water have been previously removed and separated

---

* This treatise is not included in the Geneva folio, and is derived from the Basle octavo of 1582. As the fourth book of the Archidoxies has already discussed at considerable length the subject of quintessences, and much further information is scattered through previous books, there is no need for annotation at this point.

from it. Moreover, oil and water are present in common salt, but it can not be properly designated salt until the water and the oil have been separated from it.

In plants and things growing in the earth each has its own water, salt, and oil, yet in a distinct manner. Thus, the warmest things have the most salt, the coldest the most copious supply of water. The lukewarm contain equal proportions of oil, salt, and water ; each in its own grade and kind. Thus they differ from one another in colour, odour, and savour. Also the subtlest and purest spirits which are in these things are extracted, and justly and rightly deserve to be described as salutary to the life of man.

Growing metals, as are gold, silver, iron, lead, tin, copper ; also arsenic, marcasite, lapis lazuli, rust of copper, green of the mountain, calamine, vitriol, are the first and chief. These are followed by gems, as the emerald, carbuncle, amethyst, which is also called pyrops, cyamus, hyacinth, which is also called chrysolite, corals (white and red), also pearls, etc. And, indeed, the last possess the highest virtues of them all, provided that they be duly prepared ; although in addition to the above-mentioned stones various stones are found, as tiles, wherein no small efficacy for preserving the health of the human race is discovered—alabaster, bolus armenus, and others. But before the rest we desire here to speak of the more excellent. In the third place, order demands the mention of vegetables growing out of the earth, as trees, shrubs, herbs, roots, and similar vegetables, their marrow, and other things under which both the saps and the liquors which proceed from growing things are comprehended. Thus, wine, various kinds of oil, as of olives, of nuts, of flax, of the berry of the laurel, of the nutmeg ; as also gums distilling out of trees, shrubs, and from the stalks of other vegetables, as Chian turpentine, myrrh, mastic ; also from the cherry and heliotrope, from the plum tree and the blackthorn, and others innumerable, each of which has been formed by the Creator of things for the profit and use of mankind. In the fourth place follow the animals, in which no small virtue and healthy operation lie hidden in many ways. Firstly, they are in the blood, which is most certainly the soul of every brute ; secondly, in the marrow or fat. Also, there is a force lying hidden in the flesh of animals provided with blood ; and equally in fishes, which have it not, from all which things the quintessence can be extracted to the unspeakably great advantage of man. Moreover, that I may at once shew by means of what things and methods this is effected, I will take the essence of the blood of a stork as an instance. This is a most excellent remedy against any poison which has been taken, a result which arises solely from sympathy and compassion. Moreover, the blood of man when distilled abounds in these powers, because if its essence be kindled, it burns without ceasing, as long as the man whose blood it was, lives, and is extinguished when he dies. If those great doctors do not hold it as a nigromantic exorcising, the cause will be that they do not understand the natures of these things. But there exists not a more excellent cordial than that which is extracted from pearls or margarites, which possess such marvellous powers that by its means I could restore dying men who are already in the agony of death, revive them, and enable them to speak again.   It has also given me no small aid in raising up and restoring a man both paralysed and convulsed, who (in the judgment of all the bystanders) was solely in God's hands.   The man afterwards spoke, wrote, and begat children.   Nor do the other gems possess less efficacy, yet each after their kind.   Concerning herbs, roots, and the family of the same, there is no doubt that even with the doctors who follow Galen they possess unspeakably great power and efficacy.   What shall I say with regard to the excrement of men and of other animals?   If so many virtues reside in this cast-away and refuse thing, how much greater virtues must dwell in the noblest metals, in gold, silver, antimony, etc.   But how great a virtue must reside also in marcasite, since, without the addition of any other medium, it is a most salutary medicine and purge for all fistulas, for cancer, and similar ulcers?   I will say nothing here of vitriol, antimony, and other things which abound in virtues in no wise inferior to the former with reference to diseases of this kind, yet each according to its mode and genus.   The same also is (perhaps the essence) of gold thus prepared, whereby I have frequently purged and cured an exceedingly foul leprosy and elephantiasis.   I pass over the fact that I have cured gout of forty years' standing, contractions and relaxations (of the sinews) and other (ailments) by means of this and other adjuncts.   Nor are marrows and fatnesses to be altogether rejected, especially that of a man, the badger, pigs, rams, bulls, stags, apes, and similar animals, living either in water, air, or land, none of which things has grown or has been formed except as a special boon (to man-kind), as is proved by the fact that the minutest Spanish flies, and the dung of the fly and the moth, have their respective properties, all of which matters we have thoroughly enquired into, have examined completely, and have ascertained that nothing remains which has not received its powers and its virtues for the use and profit of man.   But to approach our theme.

CONCERNING THE QUINTESSENCE OUT OF OIL, ALSO OF THE SALT OF THE SAME

Take of the purest gold, reduced to minutest grains and lunated, or of gold calcined together with plumbago.   Add to this 100 parts of the most excellent white wine discoverable, and 10 parts of the white pine.   Leave them to be macerated in a glass for 40 days.   Pour out the wine.   Pour on the same quantity of fresh wine.   Similarly macerate.   Do this a third time. Pour these three relays of wine into a glass.   Seal well, and distil in a strong fire, so that it may come forth strongly.   When it has been distilled, place the glass with water (liquor) upon hot ashes, being well sealed up with a blind alembic, the ashes being too hot for the fingers to bear.   Let them stand under till nine parts are consumed or dried up, and the tenth part alone remains.   Add the same quantity of the albumen of eggs to the water that remains.   Shake together.   Distil together, at first slowly; when white water comes forth, separate this.   Next distil more strongly until the bocia glows, and the matter comes forth in the form of a somewhat attenuated honey, of a strong (offensive) odour. Retain this, for this is the Quintessence of gold, the royal medicine. Place somewhat whitish and thick water in a glass. Cook until it be consumed. You will then find white and excellent salt. This is salt of the sun. However, the half of your gold will have gone, that is to say, the superior, which enters into the Quintessence. It is permissible to call this same essence the oil of the Sun.

Concerning the Oil or Quintessence of Silver and its Salt

Take silver reduced to the thinnest filings. Cut them up into the smallest parts in the form of a denarius. Do the same with particles of cinnabar. Place together in a glass, layer by layer. Let the glass be half filled and perfectly closed up. Then place it in a cupel for 30 days. Keep in a continuous fire. At last let it cool. It will then have extracted the soul or essence, and the silver will have been reduced to the form of a sponge. Purge the cinnabar with lead, and you will find the soul of silver in the residuum, and this is the most excellent silver that can anywhere be found. Reduce the said silver into the minutest grains, and pour on the strongest vinegar of wine. Let them stand in a glass, and the silver will become blue. Take away the blue and reserve it. Pour on other vinegar. Let it become a second time blue; remove this again, repeating the process of adding blue matters till the silver be totally removed. If there were two ounces of silver, take one ounce of camphor. Let it be dissolved in the said blue silver and vinegar. Distil it as follows, at first slowly, then stronger and stronger, until the water commences to be tinged with a swarthy hue. Then remove the water from its receptacle, place and retain in another glass. After pressing the residuum through a bag, distil with a very strong fire from the fæces until the bocia glows, and you will find a quintessence in the glass, like a dark or coarse beer of somewhat caustic quality. This is that oil and quintessence of Luna. Cook the water retained until it be consumed (as in the case of the salt of the sun), and it will become a green salt, which is salt of silver.

Oil of Mars

Pound up crocus of Mars into a most minute and delicate powder. Wash this with fresh water. Pour out the water of the lotion. Let it stand until it sinks. Then separate the water from that which sinks, and let the crocus be dried up. Take any quantity of this, and make a paste with the yolk of eggs. Let it be again thoroughly dried. At last beat it into a powder, and spread on a smooth glass slab. Place in a wine cellar, and it will be dissolved into a clear oil. This is the oil of Mars, suitable for all external ulcers.

Oil of Saturn

Take viii. oz. of spume of silver. Beat and pound extremely small. Place in a jar, which must lean in the fire to the side, as cinders of lead are usually produced. Stir the spume with an iron spoon. When it is sufficiently heated pour it out into an iron frying-pan. Let two measures of sharp and boiling vinegar be injected. The oil will then separate itself from the spume, when let the oil be again poured into the frying-pan. Let it be consumed until scarce one quarter of one measure remains. This oil has a marvellous sweetness. If you mix it with stale urine it will grow white like ceruse, and if it be boiled in an iron frying-pan it will subside and become like silver. Further, if a small quantity be placed in a vessel and it be left until it dries, there is produced from it a tenacious matter like thin (delicate) gluten. Let the matter which produces the oil be distilled through a well-sealed retort; you will then have the pure and clear oil which is called the Quintessence of Saturn.

OIL AND SALT OF JUPITER

Let Jupiter be calcined in the following fashion : that is to say, let him be placed in a crucible and cooked by means of a secret fire, that is, by descent, for forty days. You will then have the powder which they call tin cinder. Take of this ℥v., and of juice of lemons ℥lxv. Let these subside for the space of twenty days. Then wash away the ash. Afterwards wash the cinders. Pour on it again the said juice of lemons. Do this a third time. Afterwards let these be distilled, and let them be poured again over the cinder. Let them once more subside twenty days as before. Finally let them be distilled through the alembic, until at length pure water comes forth. If you perceive its colour to be red, apply a strong fire. The water which first comes forth is to be thrown away ; there will next issue a water mixed with oil, which is to be separated. You must then rectify the oil in the heat of the sun, and this is the oil of Jupiter. The second water is to be diminished by boiling to the tenth part. This you next allow to subside until it commences to have rays of such an extremely beautiful green colour as to almost surpass the colour of the emerald, and this is the salt of Jupiter, which also I judge as the spirit of Jupiter.

OF THE QUINTESSENCE OF ANTIMONY

This is the most excellent and most sweet matter of all things which has ever existed ; it also excels all things proceding from Art or Nature. I except potable gold, because it is surpassed by the quintessence of our Stone, which not unfittingly but truly, we call by its true name, the Stone of the Philosophers. Take, therefore, antimony, which breaks into long and sharp grains which glow. Let it be pounded and sifted until it becomes a powder most thin and most subtle. Let this be imbibed in strong and good wine for thirty days. Let the vessel be well closed. But notice that for a pound of antimony there must be taken two measures of wine. Let these be put in the sun ; afterwards let them be distilled in a fire, slow at first, then stronger, until at last the water coming forth commences to redden. Then bring out another receptacle. Keep the water, having sealed it up well in the vessel, sufficiently long to permit it to subside. Then, within the space of nine or ten days, something black will appear at the bottom, from which take the water which swims above it, and for every ℥v. let there be taken an ounce and a half of carline thistle cut up small. Let these again be distilled together. The other water, which is also red, needs in distillation a strong fire so that the lid may glow, whereupon you will see upon the water a red oil, which must be separated from it. In this manner is produced juniper oil. Each must be kept in a separate place. Also in this way are extracted those three waters out of antimony, which contain in themselves the Quintessence. The process for extracting it is as follows : While these three species of waters are subsiding during the period of 30 days, you will find something earthy at the bottom of the glass vessel, just as was done previously with the first water. Then let the waters be strained, and the clear separated from the turbid ; the former are to be retained, the latter rejected. Afterwards let them be distilled in the first grade of the fire only for 30 days, among cinders, until the matter be coagulated and become a hard stone, like to granite in colour. Let the stone be beaten up and dissolved in distilled vinegar ; afterwards let these be distilled through the alembic. The water which comes forth is to be placed in a glass vessel over cinders in the second grade of the fire. You will then find a red stone like to spinetum, and this is the quintessence of antimony. It was never previously known that it was useful to humanity, but it cures all leprosies and removes all fistulas, also the French disease ; indeed, all the incurable diseases of that type, as also all pellicles and small specks in the eyes. It is neither bitter nor acrid, but has a sweet savour ; its consistency is like oil, its colour that of red wine. It is the special cure of dropsy, for it quickly consumes (removes) the dropsy if the patient take the quantity of one pea of this medicine in the water of violets. It also heals paralysis, apoplexy, and epilepsy, if with three drops of potable gold the quantity of one third part of a scruple be taken, for this stone is dissolved therein. Lest it should perish, it must be kept in dry places, and preserved in seed of millet ; for if it be placed in humid situations it corrupts in four months' time.

Concerning Oil and Salt out of Marcasite

Pound as subtly as possible such marcasite as is found on the mountains. Pour over it strong wine ; then, in order that it may overflow, let it be stirred with a stick daily, and after three days let the wine be poured out, fresh being added. Let this be done until to 10lb. of marcasite an addition has been made of 20 measures. Let the wine strained or poured out be distilled with marcasite until water issues forth, which water is to be retained. Next close the aperture of the tube in the glass vessel, in an effectual manner, with lute. Apply a somewhat fierce fire, so that the matter in the upper part of the cucurbite, which resembles pure silver, may attach itself to the sides. Keep this and cast away the dross. Afterwards diminish it by boiling, so that a single measure only may remain. Put the white matter, which was in the upper part of the cucurbite, previously well pounded over a stone, into a cellar, and beneath let there be placed a small vessel into which oil flows. This is a most excellent remedy against fistula, cancer, and other diseases of this kind. That is called the quintessence which is extracted from immature metal. However, again distil the water cooked in this manner. Diminish it by boiling till half has departed, that is to say, till half remains. Let this subside for the space of 30 days. Throw into it straws and pieces of wood cut small, upon which there will collect a salt like to crystal, except as regards colour, which is slightly green. And thus you have oil and salt out of marcasite, which two are by no means the smallest aids to the preservation of the life of the human race and its health.

OIL OUT OF COMMON SALT

The confection of oil out of salt is not a useless and unproductive labour, since that wherewith it is prepared possesses the greatest diversity in nature from it, not so much in point of provocation of thirst, and removal of putrefaction, as of sweetness of savour. For oil painlessly eats into and consumes in a single hour all things whatsoever which are smeared therewith, and are putrid by nature, whether it comes in contact with them in wounds or other injuries. On the other hand, salt nourishes whatever is putrid, and prevents its consumption, and is sharp. Moreover, salt excites thirst, whereas oil repels it, as may testify several subjects of dropsy, who, having taken some, were free from thirst between nine and ten days. Lastly, as regards the savour, the oil is not sharp as the salt, but forcibly reminds one of the sweetness of honey, or of the juice of wild apples.

OIL AND SALT OF CORAL, ALSO OF CRYSTAL

Crystals of the first quality are found in the Helvetic Alps. Let these be reduced by pounding to the smallest powder, over which let there be poured the juice of lemons; let them be put into a cucurbite with a narrow neck on hot cinders as deep as the matter which is in the glass. Let them subside for 40 days. Then the crystals will be dissolved, and from them is produced a gross water. Add to these vinegar, equal in quantity to the juice; let them subside for 20 days again. Afterwards let there be added musk, which is good, in order that the matter may further ascend; then let all be distilled in a well-closed glass vessel. All the water which issues forth must be retained, for no oil comes out, and care must be taken lest the fire be too fierce, otherwise the matter would be burnt up. Next cook the water until half be consumed, and distil until it acquires a gold colour. Afterwards pour it out and distil it in a strong fire until there comes forth pure water of a yellow hue. But if it be turbid let it be preserved apart, and that which is coarse (gross) is to be distilled in a glowing cucurbite, wherein, while it cooks, you will find a yellow oil swimming on the surface of the water. This is the quintessence of crystal. Collect all the waters, the white, also the yellow, and that which comes out last, into a glass vessel. Put thereupon small and thin straws of the length of a finger. Seal up the glass vessel well, place in a cellar, and leave it for 40 days. Then upon the said straws there will grow a matter having rays like salt nitre. Dry these, and you will then have salt of crystal of a marvellous nature, virtue, and efficacy.

Oil and Salt of Pearls

Arabian pearls, and those which are fetched from India, are among others the best. They require to be purged with pure and warm water ; next to be dissolved in vinegar nine times distilled. You will be able to renew the vinegar every three days ; pour out into a glass vessel, well sealed up, and keep therein. And if they be altogether consumed and dissolved, the vinegar must be abstracted by means of distillation : that which remains at the bottom must be again dissolved into vinegar, and again distilled as before. Then a portion of the pearls or margarites will distil. Keep the distilled water until nothing remains of the pearls. Add to one measure of the same water half an ounce of camphor, which is of so great a virtue that it will not suffer any dross, but renders that which is earthy and heavy light and volatile, so that it can ascend. These are to be distilled again, and the vinegar consumed by boiling to the twentieth part. This must be done in an open glass vessel having a large and wide mouth. Let the remainder be distilled through the alembic until pure water flows. When this has commenced to turn yellow put another glass vessel beneath. Subject it to a fire which increases in vehemence, and then there will proceed a gross matter resembling thin honey. This is the quintessence of pearls, a true crown of human life. But salt is confected from the water which precedes the yellow colour. Let that be cooked until it becomes a salt. This salt is beautiful, white, and soft, yielding the sweetness of camphor.

Of the Essence and Salt of Things growing on the Earth

There exists no better method of extracting the essence of herbs and roots than to cut them up as small as possible and boil them in strong wine in a jar, well closed, lest any of the spirits should evaporate. Let the wine be separated by frequent straining. Also let fresh be poured in again until they lose their strength. Then no further process is needed except to collect all the wines and distil them together through the alembic. This can be done as frequently as you desire, but the aperture of the glass vessel must be effectually sealed up. It is then necessary to wait until the virtue which resides in the wine, and circulates throughout the vessel, collects into one place, for the smaller the quantity of the essence of these things, the better and more subtle the quality. But when at length the waters have boiled after the first and second distillation, salt is discovered at the bottom. Note, however, that in order to the extraction of oil from most herbs it is requisite that both herbs and roots be distilled. But in the case of those which are fat, their leaves alone, when exposed to the sun in a glass vessel, yield an oil, whence ultimately a quintessence can be extracted, and in the same manner also a salt. Wherefore it is impossible to assign to each herb its peculiar operation. Nevertheless, we shall distinguish between those which have oil in them spontaneously, and distil it of their own accord, for instance, nutmegs, the rind of a species of quinces, or golden apples, and the like. Eggs, nutmegs, berries of the laurel, and similar fruits must be melted up and roasted a little in a frying-pan over the fire, so that they may acquire a savour of burning. Afterwards press out the oil. This method of extracting is simple child's play, but there exists scope for skill in extracting from that oil something yet more subtle : just as no small art is required to extract oil from cinnamon, ginger, cloves, and the like, in which the method is quite dissimilar to the above. Indeed, these are distilled through the pores like juniper oil. Consequently, the same amount is not extracted from these as from other matters, although they be more excellent, whereas the berries of the laurel, nutmegs (they are also called *myristicæ*), and similar things out of which an oil proceeds without any process, as aforesaid, yield a larger supply than those things which are light and tender, or are even pressed.

The Method of Separating Oil and Salt from Oil of Olives

Oil which proceeds out of olives also contains salt, though not of special quality. It is extracted more abundantly out of old oil than out of fresh. Out of this is produced a red water, which, marvellous to say, is a most speedy remedy for the stone, whereas it is manifest that no oil contains an aqueous humidity, nor is of such a nature as to break the stone. Wherefore let those who exclaim " How can this or that be done ?" speak with more deliberation and modesty. They bring forward the argument that it is contrary to its nature and property, but they totally ignore the fact that cooking involves so great an alteration as to frequently effect what otherwise lies not in the essence or nature of a thing. Similarly, our silver, although it be not gold, is reduced to such an extent by coction as to arrive at the most excellent gold. Thus also iron is converted into genuine and most excellent copper. The matter of which we are here speaking takes place in the same way.

That Red Water out of Oil of Olives is made in a similar Way

Take lb. iiij. of oil. Place in a kettle of copper. Heat it so that it may commence to smoke ; let it cool again, and put in a bocia. Close up its neck most effectually with a sponge. Superimpose a helm well stopped up with lute, lest any vapour should escape. Distil the matter in a slow fire ; at length the virtue begins to issue forth. The fire is to be maintained at an equal temperature—not at one moment hot, and at another cold. The water will then become red. If it commence to thicken in the upper part of the receiver, it is necessary to cease operation and proceed no further. And this is the water which is extracted out of oil. It cannot long be kept entire by reason of the putrid nature it contains in itself. Notwithstanding, it dissolves in a moment, as it were, the stone in the bladder. For nine days daily let three spoonfuls be administered, one in the morning, one at noon, and one in the evening. After this the patient must fast a whole hour. The savour of this water is nauseous, and frequently affects a man so much as to give him a headache.

Salt is made in the following Fashion

Distil oil out of which the red water has been made ; abstract it through the alembic by means of a moderate fire, and if it flow in excess, remove the fire. If it cannot endure great heat you will find turbid matter at the bottom of the bocia. This is to be removed into a small glass vessel. There must be added to it as much pure fountain water as there is matter. The oil will then float on the top, and must be separated from the water, in which there is no fatness. Distil it again and you will have pure water, which must be cooked in a glass vessel until it be consumed. You will then discover at the bottom a somewhat black salt, most suitable for purging ; it is also a strong laxative.

Method of Extracting the Oil and Salt from Pepper

Beat the grains to a fine powder. Add thereto the sap of the alder, sufficient to cover the powder. Let it settle, exposed to a very hot sun, as during the dog-days, for nine days. Afterwards press through a bag. When the water has been distilled pour it again upon the fæces, and let it settle for nine days. Next, when the water commences to recover colour, let it be removed. Put out another vessel in the place of the first. Extract the oil with a strong fire. If any water has been mingled with it, it must be separated by means of a glass funnel. The last water must be added to the first, and again distilled, and consumed, till but a quarter remains, by coction. Let this settle for 30 days in a glass vessel well secured over a capella. A salt will then be found at the bottom, possessing the savour of pepper, most excellent as a pickle for food, by reason of the extremely excellent odour it yields. It is wonderfully hot, and adapted for cold limbs and nerves.

Oil out of Gums

Take myrrh, mastic, and gum, or like things, which must be pure and clear. Let them be pounded and sprinkled over fresh eggs, which have been cooked hard, split through the middle, and placed in sand in a pan, which must be put in a cellar until the powders are dissolved into oil. This cures all wounds and alleviates gout. A man's face is rendered fair if it be anointed therewith after a bath. In like manner it preserves the flesh from putrefaction so that the same can never ensue.

Preparation of Colocynth

Take an apple of colocynth and remove all the seeds. Cut them up small. Let marrow be taken to the weight of three coins of Ravenspurg. Let it be placed in a glass vessel. Add thereto of

Cinnamon Cloves Ginger Nutmeg Mastic	The weight of three coins of Ravenspurg.

Pound them all together and place in the glass vessel where the colocynth was put. Pour over five tablespoonfuls of good wine. Let it settle for eight hours. If you desire to use it, it must be prepared about twelve o'clock, mid-day, and must be taken at night. It is next necessary to lie down, then purgation will commence at midnight, or at one o'clock. Note, before taking this medicine, that it must be pressed through the bag, and it must be eaten. But be careful not to press it out too violently. Take only that which penetrates by itself, so to speak, spontaneously, about seven or eight drops, lest it become too violent. When, therefore, the purgation has occurred several times, take, about seven o'clock in the morning, some soup wherein peas have been cooked, but which contains neither salt nor butter, only a slight admixture of flesh. Take it as hot as can be borne, for two hours, without anything else being eaten.

HERE ENDS THE BOOK CONCERNING THE QUINTESSENCE

ALCHEMY: THE THIRD COLUMN OF MEDICINE *

THE third fundamental part, or pillar, of true medicine, is Alchemy. Unless the physician be perfectly acquainted with, and experienced in, this art, everything that he devotes to the rest of his art will be vain and useless. Nature is so keen and subtle in her operations that she cannot be dealt with except by a sublime and accurate mode of treatment. She brings nothing to the light that is at once perfect in itself, but leaves it to be perfected by man. This method of perfection is called Alchemy. For the Alchemist is a baker, in that he bakes bread; a wine merchant, seeing that he prepares wine; a weaver, because he produces cloths. So, whatever is poured forth from the bosom of Nature, he who adapts it to that purpose for which it is destined is an Alchemist. Hence you may understand the difference between this art and all others, from the comparison which has been set before you. For, consider, if any one should put on a rough sheep's skin for garment, how rude and coarse this would be compared with the work of the furrier and clothier. Equally rude and coarse would it be if one had anything taken straight from Nature, and did not prepare it. Nay, this would be much more rude than the former. For this is concerned with the body itself, its life and health. Hence it has to be handled and dealt with much more carefully. Now, these universal methods of treatment have rivalled Nature, and have so mastered her properties that they can express the nature itself in everything, and elicit that

---

* This treatise constitutes the third section of the *Liber Paragranum*, wherein are described the four columns upon which Paracelsus built his system of Medicine. These are Philosophy, Astronomy, Alchemy, and Rectitude. The first distinguishes between the false and adulterated philosophy of Aristotle, and that sure and genuine species expounded by Theophrastus himself. The latter alone enlarges that knowledge whereby the physician is instructed in the matter of all diseases. There is no other way by which the truth concerning the body of man and its nature can possibly be estimated. Outside it there is only pure imposture. Disease itself is of Nature, Nature alone understands and knows disease, and Nature also is the sole medicine of disease. The matter out of which man is made testifies to the physicians concerning that which is produced therefrom. Man, as the exemplar and type of all things, contains within himself all knowledge and wisdom which are required in Medicine. There are two species of philosophy—one is of heaven, the other of earth. The philosopher is he who is acquainted with the things of the lower sphere; the astronomer, on the other hand, is he who is familiar with the things of the sphere above. In their medical aspect philosophy is concerned with the earth and water in man, and astronomy with the air and fire which exist in the same subject. For man is heaven, air, water, earth. As there is a Zodiac in heaven, so is there a Zodiac in man; as there is a firmament in heaven, so is there a firmament in the body. Man has his father in heaven and in the air. He is the son, made and born from air and the firmament. Heaven operates in us. This operation can be understood by no one who is ignorant of the essence and nature of heaven. The star is the basis of celestial knowledge; he who is acquainted therewith is already the disciple of Medicine, and is on the way to understand the heaven in man. So far concerning the first two pillars of Medicine. But the fourth is the probity of the physician, which consists in the certitude of his art, and the rectitude, and sincerity of his faith in God, which forbids him to falsify anything, and makes him a fulfiller of the works of God.

which is the chief feature in each. But in Medicine, where it is most specially necessary, this power exists in the smallest degree; so that, in this respect, Medicine is most rude and unpolished. How can man be more rude than when he eats raw flesh, clothes himself in undressed hides, and has his dwelling in the nearest rocks, or is exposed to the rains? In the same way, how can a physician be more dense, or the preparation of medicine than what ointment-sellers use for decocting substances?* Nothing can be more objectionable than this method for pounding, subduing, and mixing medicines, or for polluting them in any other different way. As its own special art dresses the hide, so so does its own special art prepare the medicament. And since, in this place, the true basis of preparing remedies, in which lies the whole essence of medicine, is laid down and established, be well assured of this, that such a foundation must be extracted from the most secret recesses of Nature, and not from the imaginative brain, as a cook dresses a mess of pottage, according to his own judgment. For in this preparation the extreme and ultimate condition of things is posited. Thus, if philosophy and astrology—that is, the nature of diseases and their remedies, with all their combinations and conjunctions—be understood, it follows next in order, and is chiefly necessary, that you should decide how you will use and avail yourself of your knowledge. Nature, indeed, of herself teaches you on these subjects, and you should give her your chief attention in order that you may reduce your medical science down to practice. As summer puts forth the pears and the grapes, so should you do with your medicine. If you act thus you will assuredly compass the results you covet. And if it so be that as summer puts forth its fruits, so you do with your medicine, be certain that summer does this by means of the stars, and in no case without their aid. Now, if the stars accomplish this, take care so to regulate your preparation in this case also that it shall be duly directed by the stars. These it is which complete and perfect the work of the physician. Now, if the stars have this effect, it is right that medicine should be understood, and naturalised in all respects, with reference to them. Let it not be said: This is cold, this warm, this moist, this dry; but rather let it be said: This is Saturn, this Mars, this Venus, this the Pole-star. In such a way the physician proceeds by a straight road, especially if, beyond this, he also knows how to subject, conjoin, and harmonise the astral Mars with that which is produced from it. For here is situated the covering or nucleus which none of the physicians from first to last before me has ever arrived at. This must be understood, that medicine should be prepared with reference to the stars, so that they exert their astral influences. The higher stars weaken and cause death, but they also heal. If any of these effects is to be produced, it cannot be done without the stars. But if this is to be done by the help of the stars, it will be done after such a manner, and in such a way, that the preparation will be reduced to practice, so that medicine will be compounded and prepared by means of heaven, just

---

* The text at this point is unintelligible, as the comparison is not completed. It has been rendered literally in translation.

as prophecies and other acts are settled by heaven. That is, you see, that the stars presage and prognosticate unfavourable weather, that they foretell diseases, and the deaths of kings and princes; the stars, moreover, portend battles and wars, pestilences, and famine. All these things are indicated by heaven. It makes and produces them. What it produces it naturally predicts. All these effects come from this source, and from it, too, proceed all the branches of this science. So, then, if they are produced by heaven, and come from heaven, they will also be governed by heaven, so that all those matters which have been mentioned and pointed out will be produced at its will and pleasure. The occurrences which have been predicted from heaven come to pass at its pleasure; so that heaven produces and regulates them. Moreover, lay this well to heart. If Medicine is from heaven, without any contradiction it will remain subject to heaven, will accommodate itself thereto, and be regulated according to its will and pleasure. Now, if this be true, it is absolutely necessary that the physician should form an opinion concerning degrees and complexions, humours and qualities, and, whether he will or not, he must learn that Medicine is in the stars; that is, he must judge the nature of Medicine according to the stars, so that he shall understand the superior as well as the inferior stars. Since Medicine is worthless save in so far as it is from heaven, it is necessary that it shall be derived from heaven.* And this bringing down from heaven means neither more nor less than the abolition and elimination of every earthly element which exists in it. Heaven does not rule it except these earthly elements be separated from it. If you have effected this separation, then Medicine is in the power of the stars, and is ruled and protected by them. For instance, everything relating to the brain is led down to the brain by means of Luna. What relates to the spleen flows thither by means of Saturn; all that refers to the heart is carried thither by means of Sol. So, too, the kidneys are governed by Venus, the liver by Jupiter, the bile by Mars. And not only with reference to these, but in all other respects, this must be, in an ineffable manner, perceived. For of what use is the medicine which you exhibit for the matrix of a woman unless it be directed by Venus? What remedy would there be for the brain unless Luna gave it its origin? So judge with regard to the rest. Otherwise all remedies would remain in the bowels, and by-and-bye, being ejected through the intestines, would produce no effect whatever. Thence it happens that, if heaven does not aid your efforts, but refuses to direct your medicaments, you will profit very little. There is need of heaven as the regulator. Herein consists true

---

* The stars, therefore, have generated physicians constituted in the light of Nature, so that they might not deviate from investigating by their own skill the various arts. The first source of their discoveries was in the stars and influences, which, turned into alchemy, in no slight degree enriched the medical art. For alchemy is but a medical pyrotechny, whereby marvellous preparations, transmutations, transubstantiations, etc., of things medical are artificially produced. For such is the affinity of the firmament of the constellation with the nature of terrestrial bodies, that he who is informed with celestial doctrine is not debarred from the desire of the knowledge of terrestrial things, and when these are combined there is joined to them an influence from heaven, and so out of those three, thus united, there arises the true physician. A physician generated after this manner will never lack those remedies which suddenly become necessary for a pressing purpose. –*Chirurgia Magna*, Pars II., Tract I., c. 16.

art, that you should not speak after this fashion : " Melissa is a herb that acts on the matrix, marjoram on the head." Thus speak the inexperienced. The matter rests with Venus and Luna. If you wish to attain the ends you anticipate you must have heaven kindly and benignant to you, otherwise no effect will be produced. This is the source of that error which is so abundant in medicine. " Do you at least drink it. If it does you good it does, and there is an end of it." Any clown can practise this art. There is no need of an Avicenna or a Galen for it. You physicians, who have sprung from them, are wont to say that we must have directories for the head, the brain, the liver, etc. And how can you dare to lay down these directories when you understand nothing about heaven ? That is the sole director. Moreover, you forget one thing, which convicts you all of folly. You have, indeed, found out what things direct the brain, the matrix, the anus, the head, the bladder, etc.; but of what things rule disease you are utterly ignorant. Now, if you do not know what rules a disease, you are ignorant as to where the disease is situated. You do just the same with the principal parts which you say are affected as the sacrificers do with their gods. They put the whole of them in heaven, although very many of them lie buried in the infernal regions. So, according to you, all diseases arise from the liver or the lungs, though sometimes they affect the rectum.

This is because heaven rules by means of the stars, but not so the physician. So medicine must be reduced to air, that it may readily be ruled by the stars. Can a stone be lifted up by the stars ? No, unless it be volatilised. Hence it is that many, by means of Alchemy, hunt after a fifth essence ; which means nothing else than that the four bodies shall be separated from the arcana. Then, what remains is an arcanum. This arcanum, moreover, is a chaos, as easy to be deduced from the stars as a feather to be swayed by the wind. Such, then, should be the preparation of medicine, that the four bodies shall be taken from their arcana. To this should be added the knowledge as to what star is in any arcanum. Then it must be known what is the star of this disease, and what is the star in medicine that operates against this disease. Hence, at length, proceeds a direction. If you drink a medicament, then the belly, which is your alchemist, is compelled to prepare this for you. But if the belly can reduce the medicament to such a condition that it is received by the stars, then that medicament is directed. If not, it remains in the belly and passes off with the excrement. Now, what, I ask, is more worthy of a physician than a knowledge of the concordance existing between each star ? In this consists a knowledge of all diseases. In this respect Alchemy is an external bowel, which prepares its own sphere for the star. It is not, as some madly assert, that Alchemy makes gold and silver. Its special work is this— To make arcana, and direct these to disease. To this it must come, and here it is symmetrical. For all these things proceed from the guidance of Nature, and with its sanction. So ought Nature and man to be conjoined, brought together, and estimated one by the other. The whole principle of cure and sanitation rests in this. Alchemy perfects all these processes; and without it not one of them can exist.*

Now if all arcana belong to Medicine, and all medicines are arcana, and, moreover, all arcana are volatile, by what right, I ask, can that sausage-stuffer and that sordid concocter of the pharmacopœia give himself out as a dispenser and a veritable concocter? In undisguised truth he is a dispenser and concocter, but of mere trumpery. How great is the folly of those doctors who trick people by means of such clownish concocters with their electuaries, syrups, pills, and ointments, which are based on no foundation, or art, or medicine, or knowledge! Not one of you, if put on your oath, would dare to examine what works conscientiously and truly. The same, too, is the principle of your Uroscopy. From the urine you divine as to the blue sky, and you persist so strongly in your trifling as to confess that there is nothing but divination and conjecture in the whole matter, nor any coincidence except what occasionally happens by chance. In your surgeries you lie so consummately, and with your washings and your decoctions you assume such a magisterial air, that nobody could think otherwise than that the whole kingdom of heaven is affected by you, whereas you conceal nothing save the mere bottomless pit of the infernal regions. If you would put aside these your incapacities, and would examine arcana, what they are, what director they have, and how the stars rule disease and health, then at the same time you would learn that your whole foundation amounted to nothing but phantasy and private opinion. The ultimate and sole proposition is that the principle of medicine consists of these arcana, and that arcana form the basis of a physician. Now, if the sum total of the matter lies in arcana, it follows

---

* Thirdly and lastly, there is alchemy, wherein the physician should eminently excel. For if he does not thence take his preparation, his practice is nothing worth. Herein consists all the art of preparation. It is also the art which teaches how to separate the stars from the bodies, so that those stars obey the stars and firmament in direction, for the direction is not in the bodies, but only in the firmament. Hence it also follows that everything which the brain produces is a sign to the Moon through its course; that which the spleen produces Saturn attracts to himself; that which the heart produces is attracted by the Sun; and in this manner the external firmaments are the directors of the interior. So do they speak wrongly who say that Melissa is good for the matrix and sage for the cerebrum. For unless Venus and Luna direct them thither, they sink into the stomach and pass out through the intestines. Therefore that which does not separate in medicine is not directed by heaven, that is to say, the course of heaven is absent, and so nothing operates. So has every part its director from stars, and they are called stars. But heaven directs nothing except that which is separated from the body—that is, heaven directs only the arcanum, not the body itself; just as it directs reason to man and then reason directs the body - so heaven directs substances, which, if they be in the stomach are cooked therein, and then it directs one thing—that is, the arcanum. The stomach, indeed, is an alchemist, that is, one who fulfils the function of alchemy; but this takes place much more usefully without, before the substance sinks into the stomach, for then its operation will be much more powerful. Unless this be done, it will be like raw flesh, which the stomach digests much more thoroughly if it be assimilated after cooking. But if so much care be required over the preparation of food, how far more is necessary in the case of medicine. Many have despised alchemy as a senseless search after the confection of gold and silver, but it is not our intention to give a more prolix definition here. I have decided only to deal with the preparation - that is to say, how much virtue and efficacy there is in medicine which is devoid of a body. He, therefore, who contemns alchemy, herein despises that which he does not understand. Although I know well enough that not even apothecaries, barbers, and servants about the baths, will cease from their cookings; nevertheless, if you double the faith which you at present have in your medicines, the congeries of your recipes sufficiently proves that you are nothing but fools. It is evident enough whom ye cure, how, where, and when. Since, therefore, I am decided thus, lastly, to teach of alchemy, which is itself the fountain and pillar of medicine, it must be stated that without this art no one can be a physician, for he who lacks it has the same relation to a physician as his own cook has to that of the prince. All Nature therefore recognises alchemy, and desires that it should be understood by the physician, and that the same, being skilled therein, should not ever be cooking soups and colewort. — *Fragmenta Medica.*

that the foundation of all is Alchemy, by which arcana are prepared.* Know, therefore, that it is arcana alone which are strength and virtues. They are, moreover, volatile substances, without bodies; they are a chaos, clear, pellucid, and in the power of a star. If you know the star, and know the disease, then you clearly understand who is your guide, and wherein your power consists. So, then, these arcana prove that there is nothing in your humours, qualities, and complexions, that such terms as melancholy, phlegma, cholera, and the rest, are falsely imported into the question, and that in place of these should be introduced Mars and Saturn, so that you should say, "This is the arcanum of Mars, this of Saturn." In these, true Physic consists. Who of you, my hearers, will venture to reject and turn away from this foundation? Only your teachers do this : and in this respect they are like the old and case-hardened students.

If, then, it be right for the physician to know such things as these, it will also be convenient that he should ascertain the meaning of calcination and sublimation. And he should not only know this as a matter of handicraft, but as one of transmutation, which is far more important. For by these methods, as they are met with in preparation, there are very often produced such maturations as not even Nature herself is able to bring about. Towards this maturation the physician should direct his art. It is the autumn, the summer, and the star of those things which he ought to bring to perfection. Fire is the earth; man is the order; and the thing operated upon is the seed. And, although all these things are simply understood in the world, they are in result various and manifold. So also are they manifold in the locality of the result. And yet by our process all arcana are born and produced in the fire. That fire is their earth; and this earth is also a sun; and so the earth and the firmament, in this second generation, are one and the same. In this the arcana are decocted; in this they are fermented. And as the seed in the earth putrefies before it is reborn, and fructifies, so here also in the fire a dissolution takes place, wherein the arcana are fermented, lose their bodies, and, by means of ascension, go off into their exaltations, the times of which are calcination, sublimation, reverberation, solution, etc. ; and, secondly, into reiteration, that is, into transplantation. Now, all these operations take place by means of motion, which is given by the time. For there is one time of the external world, and another of man. But the operation, or force, of the celestial motion is truly marvellous. And, although the artificer may be disposed to rate highly both himself and his work, still here is the sum of the matter, that heaven, in an equally wonderful way, decocts, digests, imbibes, dissolves, and reverberates, while the alchemist does the same. The motion of heaven, too, teaches the motion and regimen of the fire in the Athanor. So also the virtue which is in the sapphire, heaven draws forth and discloses by means of solution,

---

* Alchemy indeed brings forth many excellent and sublime arcana to the light, which have been accidentally discovered rather than sought for. Wherefore let alchemy be great and venerable in the sight of everyone, for many arcana are in tartar, in juniper, in melissa, in tincture, in vitriol, in salt, in alum, in Luna and in Sol.—*De Caducis*, Par. IV.

coagulation, and fixation.   Now, if by these three methods the heaven is thus constituted in its operations, whilst it reduces them to this point, it necessarily follows that the solution of the sapphire shall also, in like manner, be made up of these three points.   The solution is of this nature, that bodies are thereby excluded and the arcanum remains.   For, hitherto, while the sapphire remained entire, there was no arcanum.   But afterwards, analogously to the life in man, so this arcanum has been infused by heaven into this matter. Therefore, the body, which impedes the arcanum, has to be removed,   For. as nothing is produced or begotten from the seed, unless it be dissolved, which dissolution is nothing but a putrefaction of the body, not of the arcanum itself, so, in this case, is it with the body of the sapphire, save in so far that it has received an arcanum.   But now its dissolution is brought about through the same processes which caused its previous coalescence.   The seed corn in the field has no little experience of the industry of Nature during its own progress to the corn-ear.   For there is an elixir, and a most consummate fermentation, which is retained in Nature beyond all other places.   Afterwards follows digestion ; and hence begins increase of the substance itself.   Whoever wishes to become such a natural originator must gain his end in this way.   Other-wise, he will be a mere cook, or scullion, or dish-washer.   For Nature demands that in all respects the same preparation shall take place in man as in herself ; that is, that we shall follow her as our guide and not the follies of our own brain.   But you, doctors and ointment-sellers, what do you ferment, or putrefy, or digest, or exalt ?   Nothing, save when you make up some medley of sauces which you serve out and shamefully palm off upon people. Who can praise a physician when he has not learnt the method and principle of Nature ?   Who will trust him ?   The physician should be nothing but the skilled advocate of Nature, who, in the very first place, knows its being, pro-perties, and condition.   If he is thus ignorant about the composition of Nature, what, I conjure you, can he know of its dissolution ?   Understand that dissolution is a kind of retrogression.   Whatever things Nature has gradually formed by composition, those things you ought to be able to dis-solve by a reverse process.   As long as you or I shall be inexperienced in, and ignorant of, this solution, so long, at all events, we shall act the part of robbers, murderers, rascals, or simple novices.

What, I would ask, can you produce from alum, in which are latent as many arcana for diseases of the body as for wounds ?   According to your method of proceeding, who ever, by following the pharmacopœists, applied it to that use in which it is chiefly powerful ?   The same may be said of mumia as of alum.   Where will you seek it ?   Beyond the ocean, among the heathen ? O, you simpletons, who fetch from afar what is before your very houses, and within your city walls !   Because you are ignorant of Alchemy, you are on that account ignorant also of Nature.*   Have you persuaded yourselves that,

---

* I also assign the greatest weight to experience, as most helpful in the attainment of a proposed end, especially in alchemy, by which things unheard of, and indeed scarcely credible, are produced, whence also fertile science and

because you disparage Avicenna, and Savonarola, and Valescus, and Vigo, that you are as capable as these men? These are mere trifles. Apart from this arcanum, nobody can inquire into the true composition of anything in Nature. Bring together into one spot all your doctors and writers, and tell me what corals contain, and what they can do. However much you may chatter, and whatever blatant nonsense you may talk about their powers, directly you begin to reduce it to practice, it is proved that you have not one particle of experience or of knowledge about corals. This is the reason why the process of the arcanum has never been handed down in writing. But if the process is accomplished, then its virtue is ready to hand. So great is your simplicity, however, that most of you think everything consists in pounding, and that it suffices if you write: " Let these things be strained and mixed. Make a powder with sugar." What Pliny and Dioscorides wrote about herbs they did not prove by experience, but gathered from the famous authors who knew many such matters, and then they filled many books with their feminine chatter. Dare to make the experiment for yourselves whether what they hand down is true. Will you never be able to reach the goal of experiment and proof? What do Hermes and Archelaus say about vitriol? They mention its vast virtues, indeed; and these are present in it, but you are ignorant what powers are in it, whether the green or the blue. Can you be masters of natural things and not know this? What you know, you have read, indeed, but you profit nothing and do no good by it. What do the alchemists and other philosophers teach about the potencies of mercury? Their teaching is copious, indeed, and full of truth. That you know truly, but how it is to be verified you know not. Cease, then, to shout. In this respect your academies and yourselves are novices and mere tyros. You skim over all these matters in your reading, and you say, " This property is in one thing, that in another; one is black, another is green. God is my witness, I know no more. So I find it written." So, unless it had been thus written, you would have known nothing about it. Do you think that I am wrong in laying and fixing my foundation in the Alchemical Art?* This reveals to me what is true, and that you are unskilled in proving the truth. Is not such art worthy to come into the light? And is not that deservedly termed the foundation of Medicine, which proves, augments, and establishes the knowledge of the physician? But what is to be thought of

---

notable experience are gathered in the light of Nature. I could state on oath that from such experience the greatest and most notable fundamental principle in medicine has arisen. Who without it will ever be a physician, or know and understand anything?—*De Caducis*, Par. III.

* We assume no person will doubt that the chemical art has been devised to supply the deficiencies of Nature; for although Nature supplies very many most excellent remedies, she has, notwithstanding, produced some which are imperfect and crude; for the perfection of these a separation must be effected, by which the pure is set free from the impure, so that it may at last fully manifest its powers. We desire the surgeon to be versed in this art, without which he does not indeed deserve his name. The preparation of medicaments is of great importance, so that they may be brought to their highest grade of action. God does not will that medicines should ready too easily at hand; He has created the remedies, but has ruled that they should be prepared by ourselves. The chemical art must not therefore be repudiated by the surgeon. As long as physicians are content with the preparations of the pharmacists, they will never accomplish anything worthy of praise. Furthermore, the alchemists themselves, despite the excellence of their remedies, will find their operations barren until the arts of medicine and chemistry are completely united. - *Chirurgia Magna*, Tract II., c. 9.

your judgment when you say "Serapion, Mesne, Rhasis, Pliny, Dioscorides, Macer, report about verbena, that it is useful for this or that purpose"? You cannot prove that what you say is true. What sort of a judgment can even you yourselves think this? Be yourselves the umpire: Is he not more powerful who is able to prove that true which is within? And this you cannot do without the aid of Alchemy. Even though you should read and know twice as much as you do, all your knowledge would be of no avail. Let any one read my work, and then how can he have the face to make it a charge against me that I lay these things before you and explain them to you? You do not reduce to action those powers and virtues which you parade and boast that you possess. Answer me: if the magnet fails to attract, what is the cause of the failure? If hellebore fails to make you vomit, why is this so? You know what causes purging and vomiting; but what are the arcana of healing just now spoken of? In this matter you are the very brother of Ignorance. Tell me in whom confidence should be placed as to the operations and powers of natural things? In those who have only written about such things without ever having tried or used them, or in those who have put them to the proof, though they may not have written about them? Is it not a matter of fact that Pliny has proved none of his assertions? Where, then, is the use of his statements? What has he heard from the Alchemists? And if you know nothing of these you can be at best but a travelling quack-doctor.

If, then, it be of such vast importance that Alchemy shall be thoroughly understood in Medicine, the reason of this importance arises from the great latent virtue which resides in natural things, which also can lie open to none, save in so far as they are revealed by Alchemy.* Otherwise, it is just as if one should see a tree in winter and not recognise it, or be ignorant what was in it until summer puts forth, one after another, now branches, now flowers, now fruits, and whatever else appertains to it. So in these matters there is a latent virtue which is occult to men in general. And unless a man learns and makes proof of these things, which can only be done by an Alchemist, just as by the summer, it is not possible that he can investigate the subject in any other way.

Now, seeing that the Alchemist thus brings forth what is latent in Nature, you should know that there is one kind of virtue in the twigs, another in the leaves, another in the unripe fruit, and yet another in the fruit when ripe; and that the difference between these is so palpable that the later fruit of a tree is altogether unlike the earlier, and this not only in form but in virtues. Whence the knowledge should be of such a kind that it shall extend from the first to the last. This is Nature. And since Nature thus manifests herself, so also does the Alchemist when dealing with those matters which Nature defines for

---

* I include Chemia in the circle of medical perfection for many reasons. It supplies true simples, magnalia, arcana, mysteries, virtues, powers, all things which pertain to the science of remedies, much more perfectly than ordinary pharmacy. But you object that alchemy is universally unpopular. I ask for kind words. Other arts also — astronomy and philosophy — are contemned, but are not the more imperfect for this.—*Fragmenta Medica.*

him. For instance, the genestum keeps the process of its own nature in the hand of the Alchemist. So does thyme, with its flower, and the rest. One thing does not contain a single virtue, but several. You see this in flowers. They have not one single colour, and yet they are in one thing, and are themselves one thing ; and every colour is severally graduated to perfection. So is it with the different virtues which are latent in these things. Now the alchemy of colours is so to separate Art and Nature, that this separation shall extend not only to the colours, but to the virtues. As often as a transmutation of colours takes place, so often occurs a transmutation of virtues also. In sulphur, there is yellowness, whiteness, redness, darkness, and blackness. In each colour there is a special power and virtue ; and other substances which possess these same colours have not the same, but different, virtues lying hid in these colours. And there is a latent knowledge of colours just as there are latent colours, and a latent cognition of virtues, as there are latent virtues. And the manifestation of virtues is the same as in form and colours, where are first the twigs, then the pith, afterwards fronds, flowers, and leaves, then the beginning of fruits, then their mid-period, and, lastly, their full development. If the virtues ripen by a gradual process of this kind, and thus increase, then the indwelling virtues are changed in degrees and in number every day, nay, every minute. For as time and not matter gives its purgative effect to the elder, so that same time confers its powers on other virtues, some in one way, and some in another. As time again assigns styptic powers to the acacias, which do not naturally arise from time, and as is the case with other wild growths, so time also in this case imparts the middle virtues before the final term. For these signs or intervals must be carefully noted in Alchemy, on account of the knowledge as to the true end of operations, and of the autumnal period by which the time of mature or immature virtue is defined ; and the same is carefully attended to in Medicine. So, also, these ripenings are divided into buds, fronds, flowers, pith, liquids, leaves, fruits ; and in each of these are their own proper beginnings, mid-periods, and ends, divided into three ways or principles, namely, into laxatives, styptics, and arcana. Those things which loosen and constrict are not arcana. And none of these is conducted at once to the final end, but they exist in the primary and middle virtues. How evident is this illustration in the case of vitriol, which is now everywhere very well known to all, and the virtues whereof are especially patent. Now, I propose in this place not to obscure its virtue, but to unfold and manifest it more widely. First of all, then, vitriol puts forth its laxative virtue, being the chief of all laxatives, and possessing the greatest power for the removal of obstructions. There is not in the body any member, external or internal, which is not penetrated and affected by it. This effect arises from the first time. The second time gives it a constrictive power. As powerfully does it now constrict as in the beginning of the first period it loosened. And still its arcanum is not yet at hand, nor have buds, fronds, and flowers burst forth. If it has sped to fronds, what is more effectual in the falling sickness? If it proceeds to flowers, what is more penetrating? It is like an odour which most readily diffuses itself on all sides. If it issues forth in its fruits, what is more excellent for promoting heat? And there are many other qualities in it, reckoned by its appropriate periods. At all events, so much we have endeavoured to explain, how the arcana in any one thing separate into many parts, and each part is borne on to its own special period, and that, moreover, the limit of periods in things is an arcanum.

So in the first transmutation of tartar; what excels this arcanum in cases of itch and scab, or other similarly disagreeable affections? In its second period, what is more effectual for the removal of obstructions—not in the way of a mere laxative? What, at length, is more powerful in the healing of wounds? Now, it is Alchemy which opens and manifests these qualities. Then, why not raise the foundation of Medicine hereupon?* Learn from this, at all events, and dismiss these dirty ointment-vending quacks, who do not know this process, but, together with their teachers, are double-dyed asses, and so mad as to presume to think everything of this kind false and impossible. They are so ignorant and experienced that they have not learnt even the be-ginning of a decoction, and yet health and safety for the sick are to be sought from such men as these. What else do you find in them but desire for money and thirst for goods? It is all the same to them whether their medicines do good or harm, whether they remove or increase diseases. Is it not right, then, that ignorance of this kind should be publicly revealed? I do not adopt this course from any hope that they will imitate me. They will feel no shame on this account; but rather hatred and envy will so take possession of them that they will persevere in their ignorance. Yet, notwithstanding, whoever wishes to pursue truth alone, will turn aside to my monarchy, and not to any other.

Mark, I beseech you, my readers and hearers, what a wretched and dis-torted process is adopted in the falling sickness, not so much by the ancients as by those writers who are contemporary with me; and this to such an extent that they are scarcely able to rescue or to cure a single patient. Do I act unjustly when I despise such writers and such teachers, who demand, as a matter of right, that people should use their remedies, though they are not of the smallest power? On the other hand, if any one investigates another method, by which help can be given to patients, they call him a vagabond, a chatterer, and a fool. What is rather true is that their prescriptions, like their diagnoses in the case of falling sickness and other diseases, are mere lies. This is proved by results. The patients themselves bear witness to it, while the nature of things cries out and proves the foundation on which Medicine must be built up. No one disease can they heal by a well considered and consistent system of

---

* If, therefore, it be the part of the physician to cure, and the foundation must be taken from the four things named above, how shall he conclude? By alchemy alone. What is alchemy? That which prepares Medicine, making a pure and precious remedy, exhibiting it perfect and entire, whereby the knowledge of the physician is completed. If such then be Medicine, and the knowledge of the physician must be acquired in this manner, how, I say, dare those square and knotty doctors and masters, without forfeiting their honour, blame me because I do not deal with trifles, but with truth itself, that I may establish science more fundamentally and exactly, for they adhere to their antiquity. — *Fragmenta Medica.*

Medicine, since God does not call and choose such uncertain and erratic men to be physicians, but rather well assured and experienced men. If He supplies an assured and experienced husbandman or quarryman, much more will He give a physician who is certain about his art and confirmed in its practice, since on him rests more responsibility than on all other men. But they call the foundation itself doubtful, and place it in the hand of God. So, then, the hand of God is stretched like a veil over their imposture and ignorance ; and they justify themselves, but accuse God, when they say that their art, indeed, is perfect, but that God interrupts it and stands in its way. What is impiety and imposture, if not this? But see by what unshaken argument I will establish Alchemy as the foundation of Medicine. I base it on this : that the most severe of all diseases, such as apoplexy, paralysis, lethargy, the falling sickness, mania, frenzy, melancholia or gloom, and other similar ones, can be cured by no concoctions of the ointment-sellers. As meat cannot be boiled with snow, so much less can this kind of medicine be brought to any successful issue by the art of the drug compounders. For as the magistery of each several substance is that to which it specially looks, so it should be noted of these diseases that they have particular arcana. Hence, they require special preparations. What I say of these special preparations I would have to be understood in the sense that particular arcana require particular adminstrations, and different administrations in like manner demand different preparations.* Now, in the surgeries there is no other preparation beyond some kind of kneading and decoction, such as one would see in a cookshop. By this kind of cooking the arcana themselves are stifled and result in no energy whatever. Nature must be kept under proper restraint and management. Thus, you see, there is one kind of preparation required for bread, another for meat, and so on. In the same way is it with herbs. By parity of reasoning it should be inferred that Nature never mixes up in disorderly confusion foods, drinks, meats, and breads in one mass, but deals with each separately and by itself. Now, this arises from no trifling causes, though to recount those causes here would be a work of unnecessary labour. Now, if Nature admonishes us hereby that in all things due order is to be kept, we are also in another way warned by the same mistress how to prepare medicines, and how to adapt the several medicines to their special diseases. The thirsty liver demands wine or water. But think how often wine is pre-

---

* If the seed, that is to say, the matter, be present, it requires preparation. But it is prepared by nothing save alchemy. This is not that which teaches cooking and preparation, which Nature has instituted for the benefit of man. Thus, Nature is acquainted with many virtues in S. John's wort, but for every virtue there is another preparation. Nature orders this to alchemy, saying, as it were : Prepare for one disease thus, and for another after another manner. Then arises the physician, and is famed in the medical art, for he knows the foundation, he knows also what cooking or preparation is here needed. But for what purpose do ye scintillate, ye men of Montpelier, of Leipsic, or of Vienna? Ye must turn hither, hither, ye who would know what is philosophy, and what is alchemy, what preparations Nature institutes, and by what methods she instructs her alchemists. Where, then, will ye remain, ye apothecaries and sordid cooks? For it is a shame and disgrace that your whole business is nothing but sheer fancy and wickedness, opposed to the whole art of Medicine. Strange, if all this become publicly known, that is to say, how many tortures are hidden in your golden gallipots and in your solemn concoctions—how great will be the measure of your opprobrium ! Yet the matter must be completely brought into the light. But though you multiply scorpions in my food, the venom will only operate in yourselves, and not in me, and will overflow your wily and mendacious designs, and will break your own necks, not mine.—*Fragmenta Medica.*

pared, and, as it were, reborn, before it satisfies the thirst of the liver. In like manner, the bowels require food. Here, too, notice how variously the food is changed and prepared. Believe it to be the same with diseases. Now, if you are going to undertake a pure and artistic method of cure, you will make no difference, but act just as if apoplexy were a thirst for which there was need of a particular medicine and a special preparation thereof. Or suppose that the bowel is falling, and that it requires for its restoration another preparation, and one which acts on the stomach. Imagine, again, that mania is like the spermatic vessels, and demands that its necessities shall be supplied by other methods. So you will come to the same conclusion as to medicines and their preparation in cases of mania. I admonish you, then, with due cause, that if you have chanced to meet with effectual remedies and arcana against diseases, you should not let them be tampered with and wasted by these mere decocters. Are these things not to be brought into the light? Truly, indeed, they are, in order that such errors may be avoided, and patients may advance to those sure arcana which God has designed for their use and requirement. You will gather from hence how necessary it is to act on my prescription rather than on yours. In this respect you have to follow me, not I to follow you. Though you fulminate ever so much against me, nevertheless, my monarchy shall stand, while yours goes to destruction. It is not in vain that I write at such length concerning Alchemy, but I do it with this purpose and for this end, that you may well and surely know what is latent therein, and how it should be understood. Nor should you be offended hereby because you get no gold or silver by it. Rather its result should be that by means thereof arcana shall be unveiled, and the seductions of the ointment-sellers shall be brought to light, since by these the ignorant folk are deceived, while they sell them for a florin what they would not buy back for a penny. So precious, in this sense, are their secrets!

Who will deny that even in the very best things a poison may lie hid? All must acknowledge this. And if this be true, I would now ask you whether it it not right that the poison should be separated from what is good and useful, that the good should be taken and the evil left. Such should certainly be the case. If so, tell me how it is separated in your surgeries. With you all these elements remain mixed. See your own simplicity, then, if you are forced to confess that a poison lies hid, and are asked how it is to be got rid of. Then you bring forward I know not what correctives, which shall drive out and take away the poison. Comfrey, for example, they use to correct scammony, which is then called diagridium. But what kind of a corrective is this? Does not the poison remain afterwards as before? And yet you boast that you have so corrected it that the poison no longer harms. Whither has it gone? It remains in the diagridium. Try it, if you are wise. Exceed the proper dose, and you will soon see where the poison is. You will not be long in finding out. So you correct turbith also, and then call it diaturbith. These are your methods of correction, fit only for clowns, and useful to drench

horses with ! Risk an excessive dose; you will see whether you do not feel the
poison. To correct is to take that which has been corrected. A criminal who
has broken the law is punished; but his correction is not extended farther
than the free will of the culprit lasts. Such are your corrections. The power
is in them, not in you. In such a case, all the physician thinks of is how to
eliminate the poison. This must be done by separation, For example, a serpent
is venomous, and yet it is used for food. If you take away the poison you can eat
the flesh without injury. And so it is with all other substances; only a similar
separation is absolutely necessary. If this is not brought about, you cannot
be sure of your work, unless, indeed, it come to pass that Nature supplies your
place, or some special interposition of providence favours you. You have no
protection in your art. If a sure foundation be necessary for the extraction of
the poison, that is afforded by Alchemy. It must be so arranged that if
there be Mars in Sol, Mars must be taken away; or Saturn in Venus, the
Saturn must be separated from the Venus. As many ascendants and impressions as there are in natural substances, so many bodies are there in them.
But when the bodies are contrary, it is absolutely necessary that one of them
should be taken away and removed, so that in this way all contrariety should
be eliminated, and so the evil which you are searching for should be separated
from the good. As gold is useless except it shall first have passed through
the fire, so medicine is much less useful unless it, too, shall have passed through
the fire. It is necessary that everything which is to benefit man shall have
passed by fire to a second birth. Should not this, then, be deemed the right
fundamental principle of every physician? A physician should exhibit not
poisons, but arcana. But all the preparations of your surgeries, how many
soever they be, do not contribute the smallest tittle of learning. They are
employed simply in correcting, which is just as if a dog should break wind in
a room and you should kill the stink with fumigations and juniper wood; but
does not the smell remain in the room as much after as before that process?
Although the smell is not perceptible by the olfactory organs, could anybody
say that the stink was separated and no longer remained? It is there, though
corrected by the fumigation. The stink and the fumigation enter the nose
together. Of such a sort also are the corrections of the drug-vendors who
disguise alöepaticum with sugar, that it may not offend the palate. The
sugar and the honey form a magistery in this case. So, too, they correct
theriacum with gentian. Are not all these operations instances of asinine
ignorance? And yet these people boast themselves the physicians of princes,
and sell their skill for money ! Yet who is so dense as not to scent the fact
forthwith that all this is worthless? What else can they trumpet forth about
their remedies beyond saying " This electuary is sweet, being compounded of
spices, sugar, honey, and other condiments, and is held in very great esteem ? "
And thus you mock your patient with your medicine, though all you can say in
its favour is that it is nice. Just think how idle it is to lay your foundation in
compounds of this kind, and to entrust everything to fatuous doctors. This

method differs as widely as possible from the true basis of Medicine; and is nothing but a worn-out and ridiculous phantasy.

So, then, up to this point, we have sufficiently discussed this Column of Medicine, that is to say, Alchemy, in which consists the fundamental principle of all Medicine. Whoever is not built up on this foundation is washed away by every wave, the wind blows away his work, the new moon breaks through it. Every new moon destroys that building, or the shower softens it and casts it down. In view of a system of Medicine built up on such a foundation, do you, reader, judge whether I am an irregular doctor of Medicine or a heretic, disregarding truth, and with a mad brain. Do I deservedly, or undeservedly, gird at my opponents? What right have they to rise up against me? Let who will care for their cudgel. When once it has grown warm in their hands they will not readily lay it aside. Any fools can do this; but a wise man should not imitate them. A far-sighted man throws away one cudgel and seeks another. What matters to me their persecution? I shall not try to stop them. What I shall do is to shew them up, because they rely entirely on fraud and impostures, and have no foundation save what is elaborated from their own phantasy and their own brain. Whoever is a trustworthy and honest man to his patients, whoever in his practice tries to imitate Nature, will not avoid me or turn away from my teaching. But those who live in this century do not follow Christ; in fact, they despise Him. Why should I expect such a privilege as not to be despised by any? At first, indeed, I ploughed by no means inactively in the same furrow with them. But when I saw clearly that from such art arose only murders, deaths, paralyses, mutilations, and other forms of destruction, I was compelled to retrace my steps and to follow truth by an altogether different road. Then they complained that I neither followed nor understood Avicenna or Galen, or knew their writings. They boasted that they understood all these things. Out of all this boasting it arose that on every side they injured, tortured, and murdered far more people than I, in my ignorance, did. This is really as much as to say that there is one and the same mode of operation for the one who understands and for the one who does not understand, and that neither the one nor the other is worth anything at all. But the more I contemplated the havoc wrought by them and by myself, the more I began to burn with hatred for the system, and I advanced to such a point as to perceive that it was nothing but a patchwork and a hotchpot mixed up with imposture. And I do not wish the matter to be concluded here; but in all my writings I shall make it clear how, and in what way, all these matters are combined with ignorance. Every day I grasp more and more that not in Medicine alone, but also in Philosophy and Astronomy, these people rest on no good and praiseworthy foundation, as I have already said. A vast tumult will be stirred up against me because I reject those who, for so many ages, have alone occupied the throne of glory and magnificence. But I confidently predict that the time will come when they will be cast down from that throne of glory and magnificence. Their force is nothing but phantasy; and I shall not end with the single expression of this sentiment, but shall assiduously bear the same testimony in all my writings. If the academies do not approve of me, what matters it? They will by-and-bye fall to the ground and be humiliated. Meanwhile, I will expose and oppose your errors with so much severity that to the very end of the world my writings shall be truthful and acceptable. Yours, on the other hand, shall meet with this fate, that, full as they are of bile, and venom, and the poison of asps, they shall be cast out by all like toads, trodden under foot, and shunned. I do not attempt to destroy you and level you with the ground in a single year. It is better that at some future time you shall decay and die by your own infamy and ignominy. I shall judge more of you when I am dead than now I am alive. You may demolish my body, but you will only destroy its refuse. Theophrastus will struggle with you even when he has no body!

But those who shall hereafter be physicians, I would admonish that they deal more cautiously with me than with their own teachers, and that they rather weigh our disagreement with due care and judgment than condemn the other and absolve the other without maturely thinking the matter out. Weigh carefully, I beseech you, with yourselves what it is you would aim at, namely, the healing of the sick. If this be your aim and the subject of your argument, tolerate me as your teacher, since my sole object is to lead you towards this healing of the sick. On what basis, and with how much seriousness, I do so has already been said, and shall every day be more copiously set forth. Let not my writings be an offence to you because I stand alone, because I make a new departure, or because I am a German. By these writings and not by any others, the Art of Medicine must be discovered and learnt. Above all, I would enjoin that you carefully read and consider the works which, by the divine favour, I am to finish. I would name particularly one volume on Medical Philosoph in which all the causes of disease will be investigated. Another will be on Astronomy, with a view to sanitation. A third, and last, will be on Alchemy, that is, the method of preparing medicaments. If you read and understand these three books; even you, who before disagreed with me, will become my followers. Nor shall I fix my limit here, but as long as the divine favour illuminates me I shall go on to regard the Monarchy in certain separate treatises published for this special purpose. Indeed, if I had not been oppressed by the unseemly hatred and envy of certain prominent men in Medicine these treatises would have already for the most part seen the light. I can already easily foresee that the astronomers also, like the philosophers, will noislly set themselves against me. It will be that they fail to understand me. They will raise a precocious clamour against me, but at last they will be silent, and betake themselves to their dens. Let not these things affect you, my hearers. Rather do this—read their writings while they follow mine in full cry. Thus you will find what you seek. At all events, I have set myself to write in what position, and on what basis I build up my system of Medicine, so that you may be able to gather what I wish to erect on this foundation. I lay it before you so clearly that you shall not be able hereafter, with any show of justice, to repudiate me at the suggestion of your fathers and teachers and professors. Take care that you be not led away by vulgar physicians, surgeons, or bath-keepers. These like to look great and powerful, and pour out their long words, which have no science in them, but plenty of ambition and boasting. These are like psalm-singers in a choir, who, indeed, chant the psalter, but understand not its meaning. Such are the physicians who constantly chatter and shout. And just as the nun sometimes understands a single word, but then turns ten pages without comprehending them, so the physicians sometimes make a hit, and then go astray again. Think over these matters with yourselves, and be your own witnesses as to what basis most of these people have for their studies. Even in Medicine it is no new thing for these accusations to damage any one. Medicine in their estimation admits any amount of rascaldom, and is directed only towards persecutions and injuries. All these are signs of doubtful and uncertain art. Those who make such professions give themselves up entirely to wallow in envy and hatred, and wherever one man can stand in another's way, he thinks he has reached the highest point in his practice. So the devil governs them. From him they have derived their discipline; this you cannot doubt. This is attested by their constant rendings and tearings of one another. The hand of God is not the cause of such things as these.

THE "LABYRINTHUS MEDICORUM"

CONCERNING THE BOOK OF ALCHEMY, WITHOUT WHICH NO ONE CAN BECOME A PHYSICIAN *

ANYONE who would become a physician must learn the book of Alchemy thoroughly by heart. Its name, no doubt, will prevent its being acceptable to many; but why should wise people hate without cause that which some others wantonly misuse? Who hates blue because some clumsy painter uses it badly? Who reviles a stone because it has been broken by the quarryman? In like manner, who will hate Alchemy, which is innocent? He deserves hatred who is guilty, who does not take in the Art, or use it properly. Does anybody hate him who has injured none? Who will blame a dog if he bites anybody who seizes him by the tail? Which would Cæsar order to be crucified, the thief or the thing he had stolen? I trow the thief. No science can be deservedly held in contempt by one who knows nothing about it.

Now, in good sooth, this same Science or Art is of great use and necessity. Into it is dove-tailed the Art of Vulcan, and we know how useful a work Vulcan can accomplish. Alchemy is an Art, and Vulcan is the operator therein. Whoever is a Vulcan, he has power in this Art. Whoever is not a Vulcan has no power herein. In order that you may understand this Art more thoroughly it is necessary to repeat, first of all, that God made all things out of nothing. Out of nothing, I repeat, he made something. Now, this something is the seed which gives the result of its own predestination, its own special office. And, although all things are created from nothing for their own end, there is, nevertheless, nothing which is entirely adapted to its end. That is, it is adapted to its end, but not wholly so adapted; and it is Vulcan

---

* The *Labyrinthus Medicorum Errantium* distinguishes eleven books out of which the physician ought to obtain his art and experience. Of these the fifth is alchemy. The others are wisdom, which is knowledge as opposed to surmise and guess work; the firmament, of which book the stars are the alphabet; the elements, which are all essentially present in man; the greater anatomy, by which the physical body of the microcosm is made known; experience, because the whole of medical science is nothing but a great and certain experience, and whatever acts or operates therein is founded exclusively thereon; the entire natural world, for this is the great storehouse of apothecaries and doctors: theoretic medicine, which must be founded in Nature, even as theoretic theology is founded in God; magic, because medicine should not be constituted in speculation but in manifest revelation, because disease and the medicine thereof are alike hidden, and magic is the science which makes manifest that which is concealed; the book of forms, for all medicines have their forms, of which one is visible and the other invisible, one corporal and elementary, the other spiritual and sidereal; finally, the book of the generation of diseases and their Iliastric and Cagastric seeds.

that must complete the adaptation. All things are created with this view, namely, that they should be placed in our hands, but not altogether perfect. Wood grows to its proper end, but not to coal; clay is created, but a vessel is not formed from it. The same reasoning applies to all growing things. Carefully study, therefore, this Vulcan. We will explain the matter by an illustration. God created iron, but not in the form it should afterwards assume; not as a horse-shoe, a sickle, or a sword. These modifications are entrusted to Vulcan, and so this Art is good. Unless it were good Vulcan would not bring about these adaptations. Hence it follows that iron must be first separated from its ore, and then wrought, for this the artificer requires. Now, this is Alchemy. This is the metal-founder, named Vulcan. What the fire operates is Alchemy, whether in the kitchen or in the furnace. He who tempers the fire is a Vulcan, whether he be cook or heat-producer. And the same is the rule of Medicine. It is created, indeed, by God, but not fully prepared for its final end. It is, so to say, hidden in the ore. Now, the work of Vulcan is to separate the ore from the medicine itself. What you saw about iron is also true of Medicine. That which the eyes perceive in a herb is not Medicine, nor what they see in stones and trees. They see only the ore; but inside the ore the medicine is hidden. First of all, then, the ore has to be removed from the medicine. When this is done, the medicine will be ready to hand. This is Alchemy; this, the special office of Vulcan, who superintends the pharmacopœia, and brings about the elaboration of the medicine. And as it often happens that gold and silver are found in a pure state, so medicine also is sometimes found in a state of purity, and its subsequent separation is then the easier, just as the pure gold needs only fulmination and fusion. When all that is necessary has been done, if in this way, by means of Alchemy, the medicine has been prepared and produced, then it is given to the sick as a remedy, or to the sound as food. Take an illustration from bread. The external Art of Alchemy cannot produce the ultimate material in the furnace, but only the intermediate substance. That is to say, Nature produces the first material up to the time of the harvest. Then Alchemy reaps, grinds, bakes, and cooks this up to the very time when it is taken into the mouth. Thus, the first and the intermediate matters are perfected. Then, at length, the Alchemy of the Microcosm begins. This takes up the first matter in the mouth, that is, it masticates it, which is the primary operation. Then it deals with it in the stomach, which is the second matter. It decocts and digests it until, at length, it becomes flesh and blood. This is the ultimate matter, though afterwards another Alchemy may intervene in the shape of weakness, which is a primary matter. To this succeeds decline, a secondary matter; and at last death, the ultimate matter. Moreover, then ensues putrefaction as a first matter. Next to this is decay; at last, dust and earth. Thus Nature deals with us by means of her creatures. And this makes good my position that nothing is created in a state of perfection for its ultimate matter. All things are created for their first matter. Then Vulcan is applied; and, thanks to the alchemical art, reduces this to its ultimate matter. This is seized upon by the Archeus, or inner Vulcan, who, by circulating and preparing, according to the nature and difference of each separate substance, by sublimation, distillation, and reverberation, puts the finishing stroke to the process. All these arts are prefigured and practised within the body of man, no less than without, in Alchemy. It is here that Vulcan and the Archeus differ. This, indeed, is Alchemy, which directs to its final end everything which has attained some intermediate end; by reducing lead ore to lead, and afterwards shaping lead into whatever it is designed to make. Thus there are Alchemists of metals, Alchemists who work with minerals, who reduce antimony to antimony, sulphur to sulphur, vitriol to vitriol, and salt to salt. Know, then, that this only is Alchemy, which, by preparation through fire, separates what is impure, and draws out what is pure. Though all fires do not actually burn, still they are fires and they remain fires. So, also, there are Alchemists of wood, such as carpenters, who prepare timber for building purposes, or statuaries, who take away from the block of wood whatever does not form part of the contemplated statue. So, too, there are Alchemists of Medicine, who take away from medicine what is not medicine.

Hence, then, it is quite clear what sort of an art Alchemy is, such an art, namely, as separates the useless from the useful, and reduces it to its ultimate matter or nature. The reason why I define these things more at length in this book is because most printed books contain no art at all, but are crammed full of elisions and senseless punctuations, so that swine would rather eat dung than taste such a concoction. Since such ill-digested mixtures are of no use or force, God has put in their place Alchemy, the true and sublime Art of Nature herself. That crass and rude preparation of medicines which the drug-vendors of Montpelier produce is not worthy to be called an art, but is mere cramming, and a most abominable concoction. Yet this is how they make up their syrups and laxatives, or compound other like matters. Those printed books of the pseudo-physicians teach the same artifice, or, at all events, they put it forward. Yet not even syrups or laxatives, such as the practitioners of Montpelier prescribe, should be prepared in this way, but as the science of Alchemy teaches Medicine. So has God appointed and arranged. This should suffice for every physician that, since God has created nothing in its state of ultimate finality, but has committed the finishing stroke to the Vulcans, he, too, should fully perfect his medicines, and not weld the ore with the iron into one mass. Take another illustration. Bread is created and given to us by God, but not in that shape which the baker confers upon it. Those three Vulcans, the farmer, the miller, and the baker, produce from that first matter a second, namely, bread. The same should be done with medicaments, and the same mode of reasoning applies to the Vulcan within us. So, then, the physician should not be ashamed of Alchemy; but in all things proceed according to the method which has been pointed out. Unless he does this he will not be a doctor, but just a freshman dubbed doctor—a doctor only to the same extent as that is a man which is seen reflected in a looking-glass.

HERE ENDS THE BOOK OF ALCHEMY FROM THE LABYRINTHUS MEDICORUM

CONCERNING THE ALCHEMICAL DEGREES AND COMPOSITIONS OF RECIPES AND OF NATURAL THINGS *

———

THEOPHRASTUS BOMBAST, EREMITE OF HOHENHEIM, DOCTOR AND PROFESSOR
OF BOTH FACULTIES, TO THOSE DESIROUS OF THE MEDICAL ART,
HEALTH IN THE LORD

SINCE Medicine alone among all branches of learning is necessarily accorded the commendable title of a divine gift by the suffrage of writers both sacred and profane, and yet very few doctors deal with it felicitously at this day, it has seemed expedient to restore it to its former illustrious dignity, and to purge it as much as possible from the dross of the barbarians, and from the most serious errors. We do not concern ourselves with the precepts of the ancients, but with those things which we have discovered, partly by the indications found in the nature of things, and partly by our own skill, which also we have tested by use and experience. For who does not know that very many doctors at this time, to the great peril of their patients, have disgracefully failed, having blindly adhered to the dicta of Hippocrates, Galen, Avicenna, and others, just as though these proceeded like oracles from the tripod of Apollo, and wherefrom they dared not diverge a

* The Geneva folio adds the two other dedications which here follow :—THEOPHRASTUS, EREMITE OF HOHENHEIM, DOCTOR OF BOTH FACULTIES, PHYSICIAN IN ORDINARY AT BASLE, TO HIS MOST FAMOUS D. CRISTOPHORUS CLAUSERUS, THE MOST LEARNED DOCTOR OF THE PHYSICIANS AND PHILOSOPHERS OF ZURICH, GREETING. It is the most excellent and the best sign of a true physician to be acquainted with medical truth, and to know whether he possesses the secret or not, exactly as you, O Cristophorus, most eminent of the physicians of Zurich, do nothing in your medical capacity which is contrary to your judgment and your most tender conscience, to which thousands rightly appeal. But understand this authority which I exercise in this our Monarchia. There is inborn in me a medical virtue derived from the soil of my fatherland. For even as Avicenna was the best physician of the Arabs, Galen of the men of Pergamon, and Marsilius of the Italians, so also, most fortunate Germany has chosen me as her indispensable physician. You know well that experience is the very mother of all physicians, yea, also of our whole Monarchia. But since each country is autonomous and a foreigner cannot be properly identified with it, but on the other hand an alien can well be compared with the man who corresponds to him, so this observe, that you may compare me to Hippocrates and Averroes ; you may compare Rhasis with us three together, each according to his country. Thus. the Arabs, the Greeks, and Germans stand on the same level, even as a triple horebound, and they equalize the amber of Germany with the Greek amber, with storax, turpentine, balsam, and mumia. Nor are you unaware that each country contains within itself the matrices of its own element, and produces that which is needful for itself. So amber is amber to its own country, and though perhaps there can be no comparison of the Chaldean rose to that of Arpinum, what has this to do with diseases, since each rose is for its own country ? Exactly in the same way every nation brings forth its proper and peculiar physician ; and that from its own Archeus. For every want gives work to an artificer, and the same necessity is the teacher and parent of every physician. Therefore the Italians can dispense with the Greeks and the Germans with both, since each of these have their own need, and its own minister, one for the nature of each nation. There is no call that any one should copy the mind or morals of the Arabs or the Greeks. If there be error at home, there is arrogance abroad. For this takes place at random, as by a dream, and without any reason— and hence a physician must be generated out of these

finger's breadth. From these authorities, when the gods please, there may indeed be begotten persons of prodigious learning, but by no means physicians. It is not a degree, nor eloquence, nor a faculty for languages, nor the reading of many books, although these are no small adornment, that are required in a physician, but the fullest acquaintance with subjects and with mysteries, which one thing easily supplies the place of all the rest. For it is indeed the part of a rhetorician to discourse learnedly, persuade, and bring over the judge to his opinion, but it behoves the physician to know the genera, causes, and symptoms of affections, to apply his remedies unto the same with sagacity and industry, and to use all according to the best of his ability. But to explain the method of teaching in a few words, I must first speak of myself. I, being invited by an ample salary of the rulers of Basle, for two hours in each day, do publicly interpret the books both of practical and theoretical medicine, physics, and surgery, whereof I myself am author, with the greatest diligence, and to the great profit of my hearers. I have not patched up these books, after the fashion of others, from Hippocrates, Galen, or any one else, but by experience, the great teacher, and by labour, have I composed them. Accordingly, if I wish to prove anything, experiment and reason for me take the place of authorities. Wherefore, most excellent readers, if any one is delighted with the mysteries of this Apollonian art, if any one lives and desires it, if any one longs in a brief space of time to acquire this whole branch

---

things. But he who in spite of this randomness and slumber is raised up as a physician by the need of his country, he at length becomes the perfect physician of his nation, and is plainly its true Hippocrates, Avicenna, and even Lully. However, in this place, I cannot praise the men because they were raised up by this necessity, since their own country will not permit that I should pass over their errors in silence. For how, I ask, did Rhasis benefit Vienna? What good did Savonarola do to Friburg or Arnold to the Swiss? What did Gentilius or the commentaries of Jacob de Partibus and Trusanus to the physicians of Meissen? What did Avicenna confer upon them all, since the health of the sick is the one thing to be considered? This, therefore, is the faculty by which I write, which also my fatherland gave me, and this by that necessity whereby I said that I was born. Hence I dedicate the whole of this book to you. But I am persuaded that some ignorant person will at once reply, and I again shall make answer; so is it manifest and clear on both sides that the whole duty of every physician is concerned with the health of the sick. But those whom I love most dearly will perhaps give interpretations of some obscure places herein, though not my oldest friends of all, namely, the foxes. My crowd of physicians is divided into two parts, the false of tongue, and the false both of heart and tongue. Now you understand what I wish. I will shortly send you some prescriptions, together with my improvement of the oil of colcothar. Act as a friend always, and be careful. Farewell.—*Given at Basle, the fourth day of the Ides of November*, 1526. THEOPHRASTUS OF HOHENHEIM, DOCTOR OF BOTH FACULTIES, AND PHYSICIAN AT BASLE, TO THE MOST EMINENT ASSEMBLY OF THE STUDIOUS AT ZURICH, HEALTH. Alas, how wretched is the estate of mortals, because there is scarcely any joy which is not presently followed by sorrow, a most fine company of helpers! Hitherto I have not fully perceived my blindness, for I did not consider in the present that the wise man must most diligently observe not only those things which are at his feet, but those which are behind him, like a two-headed Janus, and those also which are in all directions around him. The reason is that your most delightful assembly, which I lately enjoyed, and do still recollect with gratitude, had so enchanted my heart and eyes that I forgot all about the future. My mind presaged no disaster; I thought the whole matter was well managed and deemed that joy would be obtained and perfected without the company of grief. Now when I see those things which I ought to have foreseen, how, I say, shall I restrain myself from grief and mourning, since the dearest friend I had at Basle, whom I left in health and strength, has been killed by the accident of a sudden fall from an upper storey, where he was accustomed to sleep? He had been freed by me from the heaviest chains into which he had been thrown by the petty doctors of Italy; by me was he restored to health, of which fact Erasmus of Rotterdam is a witness, with all his family, as the epistle written by his own hand sets forth. Now when I was thus feasting with you, and taking life easily, he died whom I had left in good condition; he, I say, whom I loved as my own eyes; being snatched away by the accident I have mentioned, namely, John Frobenius, the parent and tutor of all learned and good men, being himself also wise and good, the most diligent promoter of all kinds of learning. Wherefore also have I need to fear the same suddenness in death which has overtaken him. What shall I say to myself? Death is common to all. Wherefore be warned. Watch, most excellent fellow-learners, and if to any extent we fail in our office, attribute it to that severe grief wherewith I am now tortured, and can find no relief. Farewell, most sweet companions. Love your Theophrastus.—*Basle, from our Library, the third Ide of November*, 1527.

of learning, let him forthwith betake himself unto us at Basle, and he will attain to far other and greater things than I can describe in a few words. But to make it clearer to the studious, we do not, for instance, shrink from submitting that we in no wise imitate the ancients in the method of complexions and humours. The ancients gave wrong names to almost all the diseases ; hence no doctors, or at least very few, at the present day, are fortunate enough to know exactly diseases, their causes, and critical days. Let these proofs be sufficient, notwithstanding their obscurity. I do not permit you to rashly judge of them before you have heard Theophrastus. Farewell. Look favourably on this, an attempt at the restoration of Medicine. —*Basle, the Nones of June, 1527.*

CONCERNING THE ALCHEMICAL DEGREES AND COMPOSITIONS OF RECIPES AND OF NATURAL THINGS *

By Theophrastus, of Hohenheim, Doctor of both Medicines

BOOK THE FIRST

CHAPTER I

BEFORE I begin to treat of Degrees, two complexions of Nature should be noted : one is hot, the other cold. Moreover, each of these has in itself a certain inborn diathesis : for everything which is hot is dry, and that which is cold is moist, nor can heat or cold be alone. So these two natures, heat and dryness, are one thing, and in like manner, cold and humidity. Hence, therefore, degrees are easily determined, how each and every thing exists in its own degree, and what degree each thing respectively occupies. At this point, no doubt, those who are suffering from cataract and have familiarity with works of darkness will cry out that there are four complexions, hot, cold, dry, and moist, from which they gather that cold is present in moisture and in dryness, and in like manner heat is conjoined with both. According to this opinion they have arranged everything, that is as much as to say that the cold may be dry, and heat may be moist, which is a contradiction of terms. If they had approached more nearly and made a more searching investigation into Nature they would have found our arguments, which here follow, to be nearer the truth. They did not sufficiently understand that these four are two only, and so they falsely ascribe to the four elements those which are nothing less than they are elements, as philosophy clearly demonstrates.

CHAPTER II

But in order to more clearly understand what I have said about the two complexions, take the following. Whatever the elements have produced in

---

* While there is some matter in this treatise which is outside the purpose of the present translation, it has been thought well to include it in the section devoted to Hermetic Medicine because it enters at length into a subject, or, more correctly, a class of subjects, to which there is frequent reference in Paracelsus—that, namely, of degrees and complexions. It is not very clear in itself, and it adopts an arbitrary terminology, which will be dealt with in the Vocabulary at the end, but it will help to illustrate the obscurity of previous references, and may perhaps give a little light indirectly.

the nature of things is either cold or hot. If cold, it has in itself a certain innate individual humidity. Where there is humidity there is cold, and so what is hot is dry, for dryness subsists in heat alone. It cannot come about that the cold is dry and the heat is moist. For these are elemental conjunctions which come from Ares, as is clear from the example of a man and a woman. A man has in himself what is warm and dry; a woman that which is cold and moist; but they contribute to the complexions only according to their several degrees. From the very first, therefore, it must be remarked what is moist, what coagulated, and what, lastly, is resolved dryness. For hence arises a common error which is apt to spread even amongst the chief physicians. For example, take a crystal which appears cold, dry, and arid, since it dries and renders arid, but this appearance is delusive. For the most arid force of the crystal is a coagulated moisture, and in its action it masters everything, transmutes and forcibly changes it into a coagulated moisture, which is finally dissolved like ice. Similarly in petroleum, the dryness is not resolved as it appears to be, for the dryness is resolved in the substance of its own body. Wherefore I lay down this definition in place of an epilogue, that degrees must be observed in a twofold manner, the warm and the cold. Moreover, the dry is double and the moist is double, that is to say, dry *per se* and a resolved dryness, moist *per se* and a congealed moisture. The remainder of what is requisite at this point is contained in the Philosophy itself.

CHAPTER III

Although in this place more was to be said on the subject of degrees than I have so far set down, still, since these matters are well established among those who are any way skilled in medicine, I pass them over in silence here, and speak of those subjects which have up to the present been put forth falsely, and with a certain amount of pervading error. This is what should be accepted. In the first place, it is not only necessary to observe the sum of the elementated degrees, because this only serves in the case of elementated ailments; but attention must also be paid to those things which concern mundificatives, incarnatives, laxatives, constrictives, repercussives, diaphoretics, narcotics, cicatrizers, and other things of this kind. For this purpose it is of prime necessity to acquire a full knowledge of diseases, and, moreover, of the special degrees of each kind of disease. In the case of wounds, one has to know the degrees of incarnation; in hyposarcha, the degrees of drying; in gutta, the degrees of strengthening; in epilepsy, the degrees of specification; in cachexy, those degrees which arise from commixture. If you have thoroughly examined all these matters, then at last approach the composition of recipes.

But we should not omit to mention that for the perfect knowledge both of diseases and of degrees there is required not only the medical but the astrological profession; and, moreover, the Spagyric form. All these require perfect, and, moreover, a prolonged experience; since thereby alone, and not by constant reading, or by a judgment, however exact, the scope of this book is made clear.

Lastly, if you miss anything in this place on the subject of degrees, seek it in daily practice, to which I relegate you all, so as to learn the virtue of anthera, more particularly of tereniabin, which is remarkably ennobled; and, lastly, as respects the flower of cheiri.

CHAPTER IV

Before, however, we come to the degrees themselves, we must observe certain rules of the degrees, by which method the degrees are at one time intensified and at another relaxed. In the first place you will observe this method. Whatever proceeds from the elements of the earth, that occupies the first degree. Of this kind are the lettuce, the violet, the anthos, etc. In like manner, whatever is of the air, such as pestilence, pneumonia, fever, is in the second degree. That which is produced from the element of the water holds the third degree, such as lead, sapphire, topaz, etc. But those things which spring from the element of fire, as ice, crystal, snow, claim the fourth degree, either hot or dry. It must be noticed, therefore, that whatever sensitive thing comes from an element is the same as the element, as the frog whose sperm is in the third degree; in like manner, camphor. That which is of the earth, as man, is in the first degree, as Rebis. What comes from flying things is in the second degree, as vua. What comes forth from the fire, like the Salamander, is in the fourth degree. In what respects one excels another will be made clear in the following chapters.

CHAPTER V

Furthermore, in order that the degrees may be more clearly marked in their points, take the present example. As the degrees of herbs have been divided into four, so all of them, how many soever there may be, are referred to the first degree, but still not all on an equality. For one is sometimes more than another as to the beginnings, middles, and end of both; but still so that whatever descends from the element of earth remains in the first degree, and must not be placed outside that same. Among you the nenuphar occupies the fourth degree, and with you Saturn is allotted the third, though in coldness it exceeds nenuphar by almost eight degrees. So, then, they cannot be arranged in the same degrees. So, too, whatever exists in the second degree, there also the first point excels the fourth degree of the element, which is of the earth, and the fourth point is higher by four degrees than the last point of the first degree. In the third degree the same judgment must be formed, and likewise in the fourth. Thence are gathered sixteen points, which mount as if by stairs to the true degrees, even to the six hundred and sixty-third. Deservedly, then, we say that those have been in error who collected camphor, the sperm of frogs, nenuphar, and alums into one degree ; since from these a true and certain degree could not in any way be taken in recipes, as will be shewn in the following modes for compositions.

CHAPTER VI

But in order that you may have in particular both the degrees and the points of those things which induce heat, remark : Whatever makes ashes or lime or glass is in the fourth degree of fire, as fire itself, mercurial water, aqua fortis, etc. So whatever produces a biting effect and brings things to an Ischara, so as to cause putrefaction, occupies the third place. Of this kind are colcothar, arsenic, sal ammoniac, borax, pigment of gold, and others of that kind, as alkali. But as to the virtues of these things in which some things excel others, that is a matter of points, not of degrees. Moreover, whatever produces scars or blisters belongs to the second degree, of which kind are rabeboia, cantharides, flammula, melona, and others of that genus ; for although flammula be in the first degree, nevertheless, in another way it affects the second, because the spirit of salt in it reduces the flammula so that it is just comprised in the first point of the second degree. Lastly, whatever warms, but does not attain to the signs above mentioned, such as ginger, cardamum, abrotanum, and other things of that kind, exists in the first degree, together with its higher and lower points. But it is to be observed in this rule that the degrees are not arranged according to the nature and proportion of the elements, but are, independently of them, condensed into the present rule, for this reason, because the present rule is taken from the first three principles and serves for them, namely, those which predominate in salt, mercury, and, lastly, in sulphur, wherefore, in this place care must first be taken not to use the present rule in elementated diseases. For they are only, as it were, gathered from these, and serve for diseases which can be healed by the first three principles.

CHAPTER VII

But in order that you may ascertain the degree of cold, apart from that which belongs to the elements, take the following : Whatever congeals humours belongs to the fourth degree, of which kind are those things which are born of the element of fire. But whatever refrigerates (to use a common expression), yet does not injure the vital spirit when administered as a remedy in a proper dose, as narcotics, anodynes, sleeping draughts, the sperm of frogs, hemlock, belongs to the third degree. Whatever extinguishes unnatural heats and allays paroxysms is in the second degree, and, lastly, whatever prevents a disease from breaking out into a paroxysm is of the first degree. This rule does not differ much from that one which applies to heat, for these offer a direct enantiosis to the aforesaid. But whatever degrees they occupy which are of the elements, that same remains, according to what has before been prescribed, together with the present degree ; so that now there is produced a double degree of Nature, and it operates exactly according to the proportion and nature of the elements.

CHAPTER VIII

Moreover, the rule concerning colours must be noticed, since these, at the same time, indicate the nature of the things in which they exist. For instance, the centaury, which is red, is therefore of a warm nature ; the lily, which is white, is for that reason of a cold nature. But of colours which are external, nothing certain can be defined, except in this way. The rose is red, yet of a cold nature, on account of the anther lying in it which attracts the heat of the rose. Again, wherever there is any yellow in a red flower, there is the heat, but the redness is judged to be of a cold nature, and so must it be concluded with regard to other flowers in like manner. Moreover, there are flowers which, though by nature they appear warm, are nevertheless cold ; among these is the minium. Others, again, seem cold when they are warm, as copper is. In order to ascertain these things, observe the following rule : Whatever is green, as soon as it is gathered from that with which it may be mixed, is warm. So, too, is the body under which these colours lie hidden. Silver is, by its nature, cold, and keeps the colour of a cold body, for finally it passes into the colour of lazurium. Mars is naturally of a cold colour, and admits of being transmuted into a warm nature ; but, nevertheless, it preserves the force, and so the universal virtue of its own proper nature. Black colours are of no special nature, for they are nothing but sulphur, which is burned, and nothing underlies this, but it belongs to the elements. Whatever is white, livid, black, and hyacinthine, is cold ; the other colours are warm. Whatever is variegated belongs to one nature, presumably that of its principal colour. So, also, in green, though cold be present, yet it is comprised under its own head.

CHAPTER IX

Whatever is fat, and, moreover, moist, is cold, even although it exists in something green ; for the greenness is changed into a cold nature. Whatever, on the other hand, is dry, assumes a warm nature. Moreover, whatever comes from Sulphur, Mercury, and Salt falls under each nature, the warm and the cold, and that on account of the three principles. Summarily, whatever burns is sulphur, and of a warm nature, unless it exists in warm colours. But whatever goes into sublimation or calcination admits a warm nature. So whatever resolves itself or is brought to an alkali is warm. Moreover, whatever is austere is cold. Sweet and bitter assume a warm nature unless affected by the former rules. Whatever dries the skin is warm ; that which constricts it is cold. If you would judge by the odours of these things you can define nothing accurately, except so far as they retain the same nature as the body. Lastly, there are other rules which are to be admitted, so long as they do not oppose those given above.

CHAPTER X

Moreover, it should be noticed that there are certain things wherein, besides those which are natural, degrees are concealed in two ways, and that according to two bodies, as is the case with metals, gems, and stones. According to this view, mercury is chief among the metals, and embraces in itself a certain peculiar nature, warm and cold, nor can this be taken away from it. Now, if from thence be generated a metal, in iron or lead, beyond that nature it acquires another, and so two natures will be in one substance. Wherefore, from henceforth lead will be in place of mercury, if the leaden nature which it has acquired, together with its own, shall be suitable to your affairs. In the same way it must be judged concerning tin, silver, iron, and copper, because they return into their own body. Thus the liquid in gems remains in its own nature, and that a mercurial nature. If, now, it be congealed into a gem, it equally puts on a two-fold nature, because the constituent parts are reduced into their primal liquid. So, too, must it be judged concerning common stones. In certain herbs, too, a similar nature is present. Wherefore, read, and read again, and finally recall for experiment whatever is committed to your memory concerning the nature of things. And so recall it that you may now not merely think, but know exactly each of these things ; for in this lies the essence of a true and sure philosopher.

CONCERNING
DEGREES AND COMPOSITIONS IN ALCHEMY

———

BOOK THE SECOND

CHAPTER I

ALTHOUGH I have written concerning the relollea of Nature, according to its reason and nature, that it belongs both to the cold and the hot, together with its innate essence, still there are other things which the natural Ares has produced which in many respects excel what I have treated of in my former book. And, to begin from this point, if you wish from the beginning to speak exactly concerning accidental complexions, you will find that in this place the former relollea of Nature are little approved, and for this reason there are two natures universally in things which are both together in one substance, although only one of them appears. These are the innate accident and the elemental accident. Moreover, everything in its own nature is warm. The first matter of things is warm *per se*, nor does it change the innate accident, because all the three principles in the complexion remain even to their ultimate matter ; that is, in whatever nature they are found before the relollea in that same nature do they still remain until the relolleum departs. In whatever way, therefore, experience compasses the end, in that same way the beginning is manifested in itself. But before we pass on to those simples which are in the degrees we must observe that neither heat nor cold is an innate accident of them, but rather an elemental and external accident.

CHAPTER II

Nature sends forth absolutely nothing from herself, as the man experienced in medicine easily gathers, but she keeps the innate accident so long as the thing or the body, in which that innate accident is, remains. You have an illustration in fire. In this the heat is an innate accident, and the nature of the three principles, which is evidently hot. Moreover, it cannot be otherwise but that the substance passes away together with the heat, if you wish to confer that heat on something else. Although that heat warms, still it is nothing more than a dead heat, nor does it heal disease or confer any other advantage, but is a certain superfluous heat added from without to the body. In this manner every innate accident puts forth and displays its power without any help to a sick person. Whatever, therefore, is adapted for the healing of disease should be prepared in the following manner : In the first place bring your medicine to him who separates the two essences, the one in the substance, the other in the vital spirit ; for wherever we wish to exhibit medicine, there it is necessary for the vital spirit to depart from the substance and to agree with the offending matter of the disease. Then the medicine will appear alone in its own body, and this in proportion to the three principles. The external elementated accidents go to that part where the disease lurks. And so I gather that in the universal nature of things a two-fold accident exists, an innate and an external. The innate confers little benefit on the health, but only the external. In fire there is no external accident, and therefore I assert that it is an imperfect work of nature.

CHAPTER III

In the beginning, when Nature brings forth in its proper element, the Archeus* prepares the same according to the proportion and nature of its peculiar Iliaster, so that the Ares consists altogether of three things, and generates in the same thing the substance of the body. This generation, *per se*, is, for the sake of the body alone, that it may appear the same with the relolleum. But what is this to the sick man ? For the fire is equally a relolleum accident, as is also snow. But they do not heal sicknesses, nor have any power in them for doing so, because they are a relolleum *per se*. Moreover, the external elements make up the cherio of Nature, which, also, you must bring to the relolleum, because, although you take this together with the cherio, it is the cherio that heals all sickness. Remark in this place concerning the cherio that the cherio is nothing but the heat or cold of these things which leaves the body and goes away into Nature. You have an illustration of this in camphor. This has its frigidity from the cherio, and so is a

---

* *The Archeus of the Metals.*—Ares contains within itself the first matter of all the metals, but with regard to the manner in which it distributes that matter over the globe, it must be held that it expels all matters not excocted into metals along a trinal line into the Yliadus, and separates them in division. Thus in one place there are branches of copper, in another branches of tin, and so of the other metals. Further, if thus they are brought from Ares along the triple line, out of some of them there is ejected a metal, such as tin, lead, iron, or copper, etc., before any of the marcasite, bismuth, cachimia, or zinc have been previously purged, or collected into fæces, but while they are all present, and according to their smaller or larger proportion an excellent or base metal is generated. For it is endowed with hardness in the triple line, when Archeus has extracted it out of Ares. For then they are found in Yliadus according to various modes and forms. By that preparation of the Archeus various colours are produced, no one colour being repeated, for just as from all fruit trees no apple or pear is exactly like another, so also these are not alike, as they philosophize concerning Thisma. But silver and gold are frequently found solid and pure, for this reason, that the marcasite, bismuth, and other metallic matters, have been properly separated from the metals, and are sent back along the triple line. Accordingly, when the metal has been made pure, gold and silver are produced. The other extraneous metals have already been expelled ; the rest, therefore, are found pure as Archeus has ordained them. Sometimes, also, spumes are found on the surface of the rocks which look like plates of silver, sometimes, again, in meadows like flames of gold. Then also in Yliadus there are many other forms. They are most frequently found in waters, because pure gold of this kind is compelled by the force of the waves, together with grains of sand to assume the shape of a bolt. It is afterwards deposited in grains, as takes place by the Rhine and other rivers. Cataracts of water, if they pass over the triple line where this kind of gold remains, then the water ejects it. The larger quantity is washed out by violent inundations, etc., on to the beach or coast. It also happens that two, three, or more metals are found mixed, as gold and silver are found in copper. The cause is that by the expulsive operation of Archeus in Ares, two or three are sent in company into Yliadus. This occurs chiefly in the case of cognate metals. While, therefore, they are mixed, and, being mixed, are coagulated, they cannot be separated again, but remain joined together.—*De Elemento Aquæ*, Tract III., c. 10.

most opportune remedy in case of inflations; but in the substance of its first elements it remains warm, as sulphur and the spirit of salt, together with the mercurial. Such, also, are gems and herbs. Moreover, whatever Nature produces has its cherio, that is, its external elemental accident. Wherefore, at this point, on the subject of degrees, I assert that there is a greater portion of cherionic heat, or cherionic cold, in one body than in another. Thus has the Archeus disposed all things, and that for the sake of the microcosm.

CHAPTER IV

But in order that our council concerning the compositions of recipes may be more clearly known, it must be noticed that, as I have before mentioned concerning relolleum and cherio, so, in this place, it is necessary that you again understand this with reference to the body, namely, that those sicknesses which are only of a cherionic nature lurk in the body and descend into the body without involving the destruction of the first three principles of the body wherein they exist. For, just as the Iliaster in the four elements, like a mother, produces the relolleum and cherio, so man exists in the four elements and receives, as it were, hereditarily, the sicknesses which forthwith germinate in the body, so that, eventually, they burst forth into external elementated ailments. Wherefore, for arranging cherionic recipes, it is necessary that the external elementated things should leave their own bodies, together with their substances, and should converge on the vital spirit. Thus, the sick person is set free. It should, therefore, be noticed that death is not cherionic, but relollaceous; although it arises hence that in no direction can death occur. For who can separate what is individual from that in which it lies hid? Here, however, we are speaking of cold and heat in cherionic not in relollaceous matters. The remaining desiderata on this topic you can read in the treatise on the Origin of Diseases.

CHAPTER V

As in the former book, I have conveyed a knowledge of the nature of things, with regard to cold and heat, together with many and various rules, so in this place the present rule must be observed with regard to herbs. Most of them are cold and dry, and these put forth a certain obscure greenness. These, though they are pointed out as hot, are in fact cold, as the verbena and the shepherd's purse. Some are reputed to be cold when they are hot, as the bugloss and the anise. The reason is that the coagulated moisture produces by its congelation great aridity, and the resolved dryness does not become dissolved without some little moisture on account of its cherionic nature. For it is certain that in no other way can anything be produced from the element of earth but it must be hot, nor from the element of water without being cold. This is the rule of Nature. But the reason why nothing of this kind takes place is that the external elementated condition corrupts and breaks through the former nature. Wherefore, it must be dealt with according to its cherionic nature, by the guidance of experience. Moreover, since the same nature, whether it be cold or hot, does not form the body under which it lies hid, there is no need that you should labour for the body, but you may spend all your experimentation on the aforesaid three natures, as we have prescribed in the first book.

CHAPTER VI

Lastly, the physician will have to observe the bodies of those things which lack sensation. For all those bodies in which these things lurk are nothing but a liquid under which is hidden that which is cherionic. But the liquid is congealed like its own element, just as the Iliaster produced it. Wherefore, the separations of Nature once again resolve that which Nature has congealed, and in this resolution the two above-mentioned natures are separated. Hence it is clear that the external elementated things of Nature are the relolleum—accident of nature, and exist apart without any virtue. So, likewise, it is clear that another nature is fully and perfectly present while the innate property and the accidental property remain each in its own separation. Hence it is gathered that nothing which is cold or hot is congenitally so, and more that whatever is inborn can do neither good nor harm to any person. But there is in addition a certain other nature which does induce heat or cold, and by which we judge the heat or the cold, that is to say, by the cherionic indication. When this interposes, all sickness can be healed. For that same coldness or heat, from the moment of its entrance, turns to the ailment—a thing which the innate property never does. All these matters are contained in the book on the Conjunctions of Things in the properties of the two natures, according to the three principles, and that according to the prescription of philosophy. Moreover, in the following chapters you will see the order of the degrees according to the reason and nature of their elements.

CHAPTER VII

Those things which come forth from the earth have a warm nature in the first degree of heat, and among these are the following :—

Dittany	Gentian	Clary
Lion's-foot	Elecampane	Filla
Anthos, or	Cypress	Calamus
Rosemary flowers	Great Sparge	Hirundinaria
Lacca	Gallingall	Peony
Dodder of Thyme	Philipendula	Ginger
Fig	Bloodwort	Flammula
Broom	Laudanum	Herb of Paradise
Costus	Cloves	Lavender
Pennyroyal	Monk's Rhubarb	Mustard

Humulus	Macropiper	Galbanum
Lencopiper	Fennel	Gamandrea
Hartwort	Grains of Paradise	Liquorice
Cretamus	Citonia	Succory
Scammony	Balm	Cubebs
Teazels	Chamæpitheos	Cardamoms
Basil	Bdellium	Marjoram
Horehound	Fumitory	Mother of Thyme
Sagapin	Thistle	Opopanax
Agrimony	Cheiry	Ammoniacum
	Mellilot	

Things which belong to the air are in the second degree of heat. These are :—

Tereniabin	Clouds	Chaos	Heat

Things which proceed from water are in the third degree of heat, as :—

Vitriol	Granate	Realgar
Sulphur	Red Marcasite	Cachimia of Sulphur
Golden Talc	Congealed Salt	Chimolæa Calcis
Copper	Sal Gemmæ	Jacinth
Topaz	Gold	Chrysolith
Carniola	Smaragdine	Ogorum
Red and White Arsenic	Copperas	Alumen Plumosum
Cachimia of Salt	Liquid Salt	Ruby
	Quicksilver	

Things which come forth from the fire are in the fourth degree of heat, and are these, namely :—

Hot Lightning	Hot Hail	All Ætnean Fires

CHAPTER VIII

The things which are here enumerated are of a cold nature.

Among these those which are produced out of the earth are cold in the first degree.

Dodder	Chestnuts	Pisa
Strawberries	Water Lily	The four greater cold
Comfrey	Lentils	seeds
Branca ursina	Eyebright	Flowers of Mulberry
Mandrake	Bitter Vetch	Ribes
Rose	Mallows	Dates
Acetum	Herb Mercury	Beans
Ciconidion	Pomegranate	Galls
Gourd	Henbane	Crispula
Sanders of all species	Purslane	Ash
Tragacanth	Citron	Darnel

Nightshade	Mirobolanes of all species	Lily of the Valley
High Taper	Ripe Apples	Cucumber
Lettuce	The four lesser cold seeds	Greater Arrow-head
Endive	Melon	Fleawort
Gladwin	Snapdragon	Poppies of all species
Bread Flour or Corn		

Things which are produced from the air are cold in the second degree, as Nebulgea.

Things which are produced from the water are cold in the third degree. They include :—

Lead	Antimony	Silver
Camphor	Hematites	Alumen Entali
White Cachimia	The three kinds of Tin	White Talc
Electrum terræ	Alumen de Glacie	The three kinds of Coral
Thalena alterrea	Silver marcasite	Lotho
Thalena frigida	Iron	Aqua Glariona

Things which proceed from the fire are cold in the fourth degree.

Crystal	Cold Lightning	Citrinula
Arles	Citrinæus	Snow
Beryl	Cold Hail	Ice

CHAPTER IX

It is, therefore, to be observed that the law which rules the procedure of each thing from a particular element, rules also that it should possess the same degree. The development of sensitive things from the elements is shewn in the following table.

Those which proceed from the earth occupy the first degree of heat, as:—

Men	The Lion	Rams
Boys	The Horse	The Wolf
The Goat	Oxen	Cocks
The Leopard	The Bear	Foxes, etc.

Those which inhabit the air belong to the second degree of heat.

| The Eagle | The Phœnix | The Sparrow |
| The Ostrich | The Swallow | The Heron |

And generally all winged animals which are not referable to water.

Those which relate to the water occupy the third degree of heat, as the Beaver.

Those which inhabit the fire belong to the fourth degree of heat, as the Salamander.

The following are of a cold nature, and, among these, those which proceed from the earth occupy the first degree of cold :—

| Women | Cows | All Species of Sperm |
| Girls | Menstruum | |

Those which belong to the air are in the second degree of cold, as doves storks, etc. Those which are referable to water occupy the third degree of cold, as fishes, worms, tortoises, frogs, etc. To the fourth degree of cold are referred those igneous creatures known as Gnavi, or Gnani, and Zonnetti.

CHAPTER X

Moreover, there are certain other simples which, by the intervention of composition, attain to the second grade. These, although they do not acquire their grade altogether according to the manner and nature of the elements, yet such as are in the first grade acquire the second; those which are in the second attain the third; while those of the third, in like manner, acquire the fourth grade, as shewn in the ensuing table.

Simples

Rose	Nenuphar	Flowers of the Centaury
Violet	Camomile	Flowers of the Bullace
Solatrum	Flowers of Tapsus	
Anthera	Flowers of Hypericon	

Addition of Compositions

Oil	Vinum Ardens
Crude Vinegar	And all fatty substances
Distilled Vinegar	

Further, although Nature by herself is not so frigid, yet composition effects such a reduction that, by means of addition, there results a certain grade of cold or heat, as is obvious in the case of the oil of roses, the vinegar of roses, and other matters of this kind. There are others which, properly belonging to the third grade, attain the fourth, as camphorated vinegar, oil of lead, etc. Moreover, there are grades which, by means of separation, ascend from the first into the fourth, as also from the third into the fourth grade, as will be seen in the third section of the Grades of the Spagyrists. Again, there are those things which are not intensified at all, of which kinds are snow and ice, by reason of their relolleous nature. Then there are those things which do not manifest their nature unless prepared, as is the case with the sperm of grass, the crystal, and sulphur. There are also those which are reduced from a hot grade into a cold, as gems, and others from a cold into a hot, as camphor, corals, etc. Lastly, there are those which lose their grade in preparation, as those which are congealed or resolved. Item. There are certain things which do not operate in the substance of their body, as oil of Jupiter, and the like. Experience will point out those matters which are omitted in this place.

CONCERNING
DEGREES AND COMPOSITIONS IN ALCHEMY

———

BOOK THE THIRD

CHAPTER I

AT the beginning of this third book, you are to observe that, besides those essences which I have already mentioned, there is another essence and nature which is called Quintessence, or, as the philosophers say, the Elemental Accident, or again, as ancient physics term it, the Specific Form. It is called quintessence for this reason, that in the first three essences there are four hidden, which in this place is called the quintessence, and is neither warm nor cold, without any complexion in itself. But to make the matter clearer by an example, the quintessence alone infuses robust health, just as the strength or robust health which is in man, without any complexion, is brought to its end. Thus virtue lurks in Nature, for whatever rejects diseases is nothing else than a certain confortative, even as, relying upon your strength, you repel a foe. It is part of the nature of things that there is nothing in the nature of things which is lacking in virtue, unless it be of a laxative quality. The same is the case with quintessence, because this is without complexion. But although coldness elsewhere relaxes, as also sometimes heat, yet it is beyond Nature, and from the virtue of a relollaceous nature. Whatever operates according to Nature possesses a quintessence, for its virtue is so ordered that it removes superfluities from the body, just as incarnatives for curing ulcers in such a manner promote the growth of new flesh, that by the intervention of their virtue the offensive matter is removed. These three are of a triple essence, but there is one virtue, which is justly termed the quintessence.

CHAPTER II

In order to become acquainted with the grades which exist of the quintessence, and specially of those things which are confortative, there are four points to be observed at the outset : firstly, whatsoever is of the earth holds the first grade of health ; secondly, whatsoever is from the air is referable to the second grade ; thirdly, whatsoever is of water belongs to the third grade ; fourthly, whatsoever is produced out of fire holds the fourth grade. But, further, it is labour in vain to seek a quintessence out of earthly things, equal to that which is extracted out of air. In like manner, that which is from air can never be compared with that which derives from water. Judge as follows of the fourth element. For example : To extract the quintessence of chelidonia is nothing else than toiling after the quintessence of the phœnix by that quintessence. Similarly, by the quintessence of the phœnix, the quintessence of gold ; by the quintessence of gold, the quintessence of fire ; but although in chelidonia, in melissa, and in valerian, there is more of the arcanum than in the rest, yet the grade excels so that by this superiority that arcanum is by far surpassing. Thus in every grade one thing is higher than another. Wherefore, with regard to earthly things, notice whether chelidonia is superior to melissa, as melissa to valerian. Judge in the same way concerning the three other elements.

CHAPTER III

Whatsoever has been dealt with in the former chapters has been with a view to proceeding subsequently to the following signs of the grades, so as to elucidate after what manner the grades stand in the elements. Platearius, Dioscorides, Serapio, and others, their followers, who have written much, but falsely, of the quintessence, do not signally differ from us herein. Yet do you, whoever you be, seek a knowledge of the quintessence from experience, for thus you will understand the grades in their division. That the manner in which diseases are repelled by the quintessence may become clear, we must first diligently notice the concordance of things in diseases. For some virtues contend only in synochia, others in mania, others in aclitis, and yet others in lethargic complaints, as is the case with concordances. In this place I think it worth while to know that which lies hidden in Nature, as in gelutta (carlinum) and melissa, which renew and remove disease without any virtue of grades, namely, in the restoration and repair of youth. The manner and the efficacy by which these things are done are indicated in the treatise, *De Vita Longa*, as certain peculiar mysteries which exist in the nature of things besides arcana. Wherefore, I think proper to pass them by here, and at length continue what I have begun concerning the four grades of the elements. Hence, although there be many and various virtues which cure maladies, some through their aperient nature, others through their narcotic nature, others again by other means, I leave such matters to those who devote their attention to theorems.

CHAPTER IV

Everything which strengthens is tempered. No hindrance will arise from the substance which, although it be cold or hot, will, however, not incommode the Quintessence in its body (*others read*, in its work). Moreover, every specific is a quintessence, without any corruption of its body. Furthermore, nothing is tempered except the Quintessence ; all bodies are elementated in nature and their accident.

GRADE OF HEALTH

Those things which proceed out of the earth hold the first grade of health.

Herbs
Seeds
Roots
Sponges
Animals > of all kinds.
Flowers
Barks
Fruits

Things which proceed from the air hold the second grade.

These are all kinds of winged creatures.

Things which proceed out of the water hold the third grade, as :—

Metals
Marchasites
Cachimiæ
Salts
Minerals > of all kinds.
Resins of Sulphur
Fishes
Gems
Stones

Things which proceed out of fire hold the fourth grade, as the

Tincture and the Philosophical Stone.

However, there are certain other virtues to be noted which are concealed in herbs, but not in winged things, nor in metals, as ursina and white thistle indicate, which, beyond their grade, admit foreign virtues. Among these there is also the emerald, which admits a foreign efficacy into itself, yet such in no wise conduce to health, for they are only external virtues which have no internal effect whatever.

CHAPTER V

Enough having been said of confortatives, we will now turn to laxatives and their grades. Accordingly, first observe that we shall not here make use of that classification whereby the laxatives are divided into four natures. They are described in this fashion according to ancient custom. The coloquintida and scammony purge cholera ; turbith and hellebore purge phlegm ; manna and capillus veneris purge the blood ; while lapis lazuli and black hellebore purge melancholy.

Moreover, there are others also which ward off *cholera vitellina*, others which ward off the rust-coloured and yet others the citrine-coloured water of dropsical subjects, with things of like kind, as elsewhere has been sufficiently described.

CHAPTER VI

As in the former chapter I made mention of the grades of laxatixes, so in this place, to impress it more deeply on the mind, I repeat the same—namely, that laxatives in no wise follow the four grades of the elements, but they have their grades mixed without any respect to the same. Wherefore, more diligent attention must be paid to the nature of the disease, lest you should carelessly misuse the confortatives designed for its cure. These should rather be accommodated so that they may agree with the nature of the disease. The grade and the disease should also be invariably compared. But lest with unwashed hands, as the saying is, we should rush in upon this question of purgations, it is needful to proceed as follows, namely, observing that the functions are sometimes unequal in the same operation in the fourth grade, as hellebore sometimes removes that which sea-lettuce cannot, and in like manner cataputia where both the above would fail. At one time precipitate, at another esula, and at yet another cassia, will prevail in the removal of fistula. Moreover, if we speak of fevers, such a laxative as centaury will occasionally purge febrile humours, and hellebore, another laxative, will be of use in an epileptic complaint. So, also, agaric, and things of this kind, will prevail in the case of worms. The reason is to be sought in Nature, not in the humours, for Nature has been equipped to remove whatever is melancholic, choleric, or phlegmatic, or, indeed, anything which could be mentioned here.

CHAPTER VII

Note the following things for very vehement and very gentle purging.

I	II	
		Lazuli
Polypodium	Mountain Osier	Scammony
Locusta Botim	Cyclamen	Centaury
Hairs of Venus	Turbith	IV
Turpentine	Azarabachara	Either Hellebore
Locusta Sambuci	Hermodactylus	Coloquintida
Senna	III	Sea-lettuce
Gamandrea	Rhubarb	Serapinum
Stomachiolum	Esula	Cataputia
Locusta Ebuli	Vitriol	Præcipitate
Succory	Diagridium	
Serum of Milk	Agaric	

CHAPTER VIII

Observe the following things concerning incarnatives and consolidatives. They contain in themselves the four grades, while the consolidatives, in the same manner as the laxatives, exclude the elements. In the first place, therefore, we must observe the manner wherein the ailments which we desire to heal stand in their grades. For out of these proceeds the grade of natural things. Now, some heal fractures of bones, others cure wounds, others ordinary ulcers, others eating ulcers (others fleshy). Hence arise four grades in the following fashion.

I

Broken bones are cured by Alchimilla, Periwinkle, Perfoliata, Diapensia, Aristolochia rotunda, Consolida, Serpentina.

II

Wounds are healed by Natural Balsam, Artificial Balsam; the oils of Hypericon, Bullace, Turpentine, Laterinum, Centaury, Anise, Benedictus; apostolic plasters and unguents; apostolic powders; potion for wounds.

III

Imposthumes and common ulcers are cured by Emplastra Gummata, Emplastra Mumiata, Emplastra Apostolica, and Unguenta Apostolica.

IV

Cancrous and eating ulcers are healed by composition of Mercury, of Brassatella, and of Realgar.

CHAPTER IX

There are, moreover, others besides the above which equally possess their own grades, of which kind are poisons, wherein the grades should indeed be specially observed. First, by reason of their elementated nature, they should be admitted into the composition of recipes. At the outset, therefore, have regard to the quantity of the poison; the weight must then be prepared, and that in the following way.

POISONS IN THEIR GRADES

I

Simples by themselves : Colcothar and Alum.

II

Reverberated : Spirit of Jupiter and Spirit of Saturn.

III

Calcinated : Tartar and Scissum.

IV

Sublimated : Arsenic and Mercury.

The other species of poison, as those of the spider, toad, scorpion, lizard, and serpent, as also the small dragon, among many, I pass over because they are not ingredients, except the Tyrian poison, which might be named. There are, moreover, those which provoke the courses in women, which also, being specially adapted for this purpose, may be placed among other recipes, according to the manner of their grades. There are others which suppress tumours, some which provoke the flow of urine; all these and their like are to be sought from experience and concordance. Now for the Grades of the Spagyrists.

CHAPTER X

Out of the spagyric industry four grades precede in the same manner with the four elements, and so surpass the other grades in their quantity. Further, wherever the last grade comes to an end, there the first point in the spagyric grades begins, and after this manner.

I

Oil derived by distillation from all herbs, roots, seeds, resins, gums, fruits, fungi, and tree mosses.

II

Oil of the vulture, the dove, the heron, the crow, and the magpie.

III

Water of vitriol, liquor, mercurial water, oil of quicksilver, viridity of salt, aluminous waters, calcined oils, oils of metals, liquors of gems, potable gold, essence of antimony.

IV

Oil of crystal, oil of beryl. tincture, stone of the philosophers.

All these are hot, for the grades remove that which is elementated, and over that which is element they advance their own grades. Therefore, to become acquainted with those grades there is needed full and perfect experience, so that you may see the preparation of these things which proceed out of the elementated, where they surpass the elementated.

Things which proceed out of the earth occupy the first grade of the Spagyrists, as, for example, out of the seeds of Anise, Juniper, Cardamum, Clove Tree; out of the roots of Jusquiam, Repontic, Angelica, Masterwort; out of the woods of Ebony, Juniper, Sandal.

Things which proceed out of the air occupy the second grade, as, for example, out of the fruits of Nuba, Ilech, Tereniabin; out of winged creatures, as the Phœnix, the Eagle, and the Dove.

Things which proceed out of the water occupy the third grade, as, for example, out of metals, as Gold, Mercury, Silver, Copper, Lothon, Iron, Lead, Tin, Electrum, Sapphire, Smaragdum, Granate; out of gems, as the Topaz, the Ruby, the Hyacinth, the Amethyst, and Corals; out of minerals, as Marcasite, Cachimia, Talk, Realgar, and Vitriol; out of salts and alums.

Things which have their origin from fire hold the fourth grade, as out of the Beryl, the Crystal, and Arles.

And the things which descend from the above-mentioned four elementated substances, as out of the earth, Water of Life, Distilled Balsam, Circulated Waters, Distilled Liquors; out of the air, Distilled Birds, also Tereniabin, Cloud, Ilech, distilled by retort; out of the water, Potable Gold, Sublimates, Resolutes, Liquor of Silver, Calcinates, Congelates, Resolution of Mercury, Reverberates; out of the fire, Liquor of Crystal, Liquor of Beryl, Liquor of Arles.

CONCERNING
DEGREES AND COMPOSITIONS IN ALCHEMY

BOOK THE FOURTH

CHAPTER I

THOSE herbs which are of a frigid nature and from the earth, are not altogether adapted to all diseases which are of a warm nature, nor, again, are those herbs which are of a warm quality adapted in all cases to diseases of a frigid nature. Hence, seven genera of diseases and seven genera of heats and colds are distinguished, and among them those of the heart as well as other members. This difference, therefore, should be very carefully observed, that those things which are wanting to the liver, whether it be warm or cold, may be sought from the same herbs. So, also, those things wherein the cerebrum is deficient require their special herbs. However, although herbs in general are either cold or hot, yet those which are for the spleen are in no wise appropriate for diseases of the reins. Wherefore, after an aquaintance with the grades there is required that of the difference between herbs in the manner following.

CHAPTER II

In the first place, herbs are divided into seven species, together with the rest of the elements, and this on account of the nature of the star, which, equally with these, is divided into seven species. Further, in the same way as they admit of a sevenfold division, the body also is subject to the same classification, and they correspond one to another, so that those things which are under the sun are appropriate to the heart, and these are twofold; while those under the moon are, in like manner, appropriate to the brain, and that in either grade. Those things which are under Venus are heating to the reins; those things which are under Saturn strengthen the spleen; those which are under Mercury defend the liver; those under Jupiter have regard to the lungs; finally, those subject to Mars are considered wholly adapted to the gall. Moreover, though herbs are not regulated together with simples of the planets, nor the planets regulated by them, there does certainly exist a singular supremacy in every element without mixture of another.

CHAPTER III

In order to become acquainted with those elements which pertain to the heart, we must, in the first place, observe that whatever regenerates is akin to the heart, as gold, balm, nuba, etc. Moreover, whatsoever removes phlegm, of which kind are the rose, camphor, musk, amber, etc., are brought to the brain through the medium of their native fragrance. In like manner, whatsoever heats or cools the blood is serviceable to the liver. What provokes the flow of urine or increases the semen benefits the reins. That which preserves long life benefits the spleen; that which purges, the stomach. Experience gives acquaintance with these things, but it is rather the experience which is derived from philosophy than from medicine, from regeneration rather than from disease, even that which is produced from transmutation. For when both medical and philosophical experiments concur there is derived a genuine diathesis of everything.

CHAPTER IV

When, therefore, you have become acquainted with the transmutation which indicates seven species, both of the hot and cold, you must observe that whatever regenerates or expels an evil growth, and purifies or restores the matter to wholeness, and thus to an incorrupt state, comes to be included under the same species, whether it arises from the heat or cold of the elements. Moreover, everything in transmutation consumes superfluous humours, as salt removes leprosy of the moon, whence it is a most speedy remedy for the cerebrum. In this place you may observe that the herbs are not to be administered in this fashion, because they are Lunar, but because they reduce and compel Lunar things into their power. By silver or the Moon the brain is in no wise healed, but by those things which are opposed to these. Moreover, whatsoever prevents rubefaction and putrefaction, and conserves into an essence, as fixed things which obtain in the transmutation of metals, in the same manner preserves the spleen incorrupt. Similarly whatsoever resolves a substance and a body into liquor strengthens the liver and expels that which is repugnant thereto. But whatever dissolves to such an extent that the contraries are separated from one another, is beneficial to the stomach. Of this kind are the alkalies in tin. Finally, whatever prepares things and renders them suitable for the augments of transmutations—of which kind are the conjunctions of arcana—is to be used above all. Seek an experience of these things from the transmutation of Nature, but waste not your whole life in those miserable grades, nor in the profitless catalogues of herbs, which are found in senseless codices. For these are inimical rather than beneficial to the stomach.

CHAPTER V

The following table will indicate the manner in which the seven aforesaid species are distinguished in the four elements, namely, which have their origin from earth, which from air, fire, and water.   Hence you may judge concerning the method of composing recipes, as follows :—

Those which come from the earth, and are of a hot nature :—

THE CEREBRUM, *Viriditas Salis*, Liquor of Vitriol, Liquor of Lunaria. THE HEART, Essence of Melissa, Quintessence of Gold.   THE REINS, Correction of Sibeta, Essence of Satyrion.   THE LIVER, Liquor of Brassatella, Liquor of Manna, Xylo aloes.   THE SPLEEN, Mystery of Black Hellebore, Mystery of Valerian, Mystery of Verbena.   THE CHEST, *Extractio de Pulmone*, Extract of Tree Moss.   THE GALL, Quintessence of Chelidony.

Those which come from the earth, and are of a cold nature :—

THE CEREBRUM, Essence of Geloen, Essence of Anther.   THE HEART, Matter of Laudanum, Matter of Pearls, Matter of Sapphires.   THE REINS, *Materia Sintocorum*, Matter of Lettuce Seed.   THE LIVER, Liquor of Senna, Quintessence of Blood, Quintessence of Gamandrea and Cichorea.   THE SPLEEN, *Compositio Candi*, Confection of Dubel Coleph, or Dubelteleph. THE CHEST, Matter of Dew, Matter of Sulphur, Matter of Ologan.   THE GALL, Composition of Agresta (verjuice), Composition of Pomegranate Flowers.

CHAPTER VI

Those which come from the air, and are of a hot nature :—

BRAIN AND HEART, Nuba, Symona.   REINS AND LIVER, Ilech, Hallereon. SPLEEN, CHEST, AND GALL, Tereniabin.

Those which come from the air, and are of a cold nature :—

BRAIN AND HEART, Halcyon.   REINS AND LIVER, Crude Ilech.   SPLEEN, CHEST, AND GALL, Crude Arles.

CHAPTER VII

Those which come from the water, and are of a hot nature :—

BRAIN, Oil of Mercury of the Moon, Essence of Silver, Essence of the Sixth, *i.e.*, of Venus.   THE HEART, Potable Gold, Liquor of the Sun, Oil of the Seventh, *i.e.*, of Saturn.   THE REINS, Essence of Vitriol, Quintessence of Sulphur, Flower of Venus.   THE LIVER, Mystery of Mercury, Mystery of Antimony.   THE SPLEEN, Mystery of Asphalt, *Rubedo de Nigro*.   THE CHEST, Flower of Jupiter, Extract of Tin, Resolved Talc.   THE GALL, Crocus of Mars, Topaz from Iron.

Those which come from the water, and are of a cold nature :—

THE BRAIN, Juice of Amethyst, Liquor of Granates, Composition of Gems.   THE HEART, Composition of both Marcasites, Composition of White Talc.   THE REINS, The Tincture, The Physical Stone.   THE LIVER, Spirit of Saturn, Essence of Lead.   THE SPLEEN, Mystery of Coagulated Mercury. THE CHEST, Flower of Crude Jupiter.   THE GALL, Dust of the fifth metal.

CHAPTER VIII

Things which come from fire, and are of a warm nature :—
Warm Nostoch.
Things from the same element, but of a cold nature :—
Arcana of Crystal, Mastery of the Beryl, Citron Liquors.

CHAPTER IX

When you have become acquainted with the grades and their species, you must then advance to the composition of recipes, according to the direction of the rule following. For as there are four elements, so four recipes are to be prepared. There are some kinds of diseases which require earthy remedies, more require atmospheric, others aqueous, and yet others igneous. In the first place, therefore, you must notice the diseases in the seven members, among which the elements are distributed. Hence simples are to be extracted of which you may prepare a composition according to the nature both of the grades and the species thereof. Accordingly, with elementated diseases, as, for example, terrene, a composite is not to be prepared higher than its own grade, but is to be left in that same grade. Similarly, with regard to the atmospheric, nothing out of a foreign element is to be introduced. Judge in this same manner concerning the other elements. This, then, is the crucial point, to accommodate each disease to its proper element, for hence arises that common error which is continually recrudescent in the case of gout, paralytic diseases, and others of this kind, by reason of the preposterous healing method which is adopted by unskilled men. Take epilepsy as an instance ; a species hereof is subject to the element of water, wherefore it is to be healed by means of minerals.

CHAPTER X

Take general rules for the composition of recipes as follows. All those which are prepared for elemental diseases consist of six things—two of which are from the planets, two from the elements, and two from narcotics. For although they can be composed of three things, one out of each being taken, yet, these are too weak for healing purposes. Now, there are two which derive from the planets, because they conciliate and correct medicine ; two derive from the elements, in order that the grade of the disease may be overcome. Lastly, two are from the narcotics, because the four parts already mentioned are too weak of themselves to expel a disease before the crisis. Observe, then, concerning composition, to forestall the critical day. Recipes prepared in this manner are very helpful for diseases in all degrees of acuteness.

CHAPTER XI

Lastly, concerning weights, observe the following rule. Observe the grade, whether it be surpassed by the medicine, or whether the medicine agree with the grade. But in order that the three species may not corrupt one another, dispose the weights as follows: The proportions from the planets should be as four, from the elements as three, and from the narcotics as one.

CONCERNING GRADES AND COMPOSITIONS*

BOOK THE FIFTH

CHAPTER I

IN the prescription of recipes divide the disease into four species, and distribute these among the elements, taking the grade which occurs, and proceeding in the following manner: If the disease passes from one grade into another, take the same grade, for thus are healed diseases of the first grade, which are of the earth; of the second grade, which are of the air; of the third grade, namely, of the water; while that which attains to the fourth grade must be cured by the Tincture alone, for otherwise there is nothing which which can be accommodated to this case. Moreover, although in the fourth book I have prescribed that the recipe should be constituted of six parts, it could, notwithstanding, be confected of three, or the six might be doubled. Again, they might be distributed as follows: Of those which are of the planets take four drachms; of those which are of the elementated take three drachms; of those which are of the narcotics take one drachm. Thus the matter consists in the weight, not the number of the simples.

Again, the strength, and what is more, the effect of those things which are admitted into the description of this recipe, are referred neither to the weight nor to the recipe, but to the dose, so that in those things which are of the planets you obtain greater efficacy than with the elementated things. But this is of the dose, not the weight or the recipe. Wherefore the above method is to be observed. Finally, signal skill in medical matters is herein required, to avoid premature application of the healing process; you must rather proceed so as to purge where there is need of purgation, heal where healing is required, and consolidate where consolidation is necessary.

CHAPTER II

You must know that everything which comes to be tested by Nature pertains to that subject which the physician makes his province. That only which is of ocular demonstration is to be considered in doses. Every dose,

---

* The treatise *Concerning Grades and Compositions in Alchemy* consists of seven books, but of these the last three are so much outside the purpose of this translation that they have been compressed into a very small space. All the information which the Hermetic student is likely to require on the subject of complexions and qualities is embodied in the first four books.

according to its proper arete, is either hot or cold. With regard to the composition of recipes, neither the humid nor the dry is to be considered in doses. As there are only two complexions, so there are only two doses. Whoever is acquainted with the grade of hot or cold, will know that there is joined thereto not only dry and humid, but also a dry resolute and a humid coagulate. No arcanum or aniadus resides in the warm or dry, inasmuch as no disease occurs which only seeks one of these. The chief point is in the hot or cold, for that diathesis dominates either in the hot or the cold. The sole inclination of every disease is that the physician should simply observe whether it be hot or cold. Every grade is the dose of its disease, and from every grade should the dose be taken, as may be understood by the comparison of fire, which has only one grade, and yet it is abundantly sufficient to consume its contrary, which is, indeed, according to heat. So every disease has its own grade, neither more nor less. The dose has a like relation to each and every disease. But natural things are not graded equally as to the disease in the matter of the dose ; each has a grade equal to its disease, and that is the grade of the dose.

CHAPTER III

Since there is only one grade, and nothing is graded higher in warm or in cold, equality is the chief help to ascertain the dose. There is one grade in disease and in things of Nature. No disease becomes worse because the grade of its medicament is higher. It becomes worse only according to the capacity of its nature. The extent of the disease regulates the amount of the dose. Wherefore the physician must know with what weight the disease is loaded, for the dose demands for it the same weight of medicine. The weight, therefore, and not the grade, is to be administered. Herein consists the chief principle of administering any dose : this ought to proceed from the number, not the body of those things. The end is that the ares of the microcosmus, and not the medicament, should effect the cure. As soon as the disease has been brought to an equality, Nature herself cures that which is contrary to her. The quantity of the dose must therefore not exceed the number which is taken from the disease. There are twenty-four numbers in Nature, and within this number the medicine should also be confined. In the anatomy of Nature there are twenty-four *minuta* of diseases. Medicine, therefore, has twenty-four lotones. The physician should administer his medicines with reference to these two series, so as to produce the same number of each in the microcosmus. When this is done Nature will cure the sick. The absence of such an equilibrium will sometimes cause death, when the disease itself has run its course.

CHAPTER IV

While any disease in itself is one, it has, as we have indicated, twenty-four numbers, degrees, or minims, and the lotones of the medicine must correspond. The proportion of the dose to the disease in any particular case cannot be learnt from theory, but is gathered from experience only. The anatomy of the dose must correspond to the anatomy of the disease. It must not exceed the number twenty-four. Indeed, the object is to restore health, both to the nature of the Microcosm and to that of the external elements, when these agree in the body. The conjunction is the same as zinobrium, which is graded by minium. In that elementated exaltation, they afford their own exaltations to the virtues of the Microcosm, and so the grain passes into the scruple, the drachm, and loton, some, indeed, into the pound, some also into the kist, and others into the talent.

HERE ENDS THE BOOK CONCERNING DEGREES AND COMPOSITIONS

CONCERNING
PREPARATIONS IN ALCHEMICAL MEDICINE*

TREATISE I

CONCERNING ANTIMONY AND MARCASITE OF SILVER.
CONCERNING WHITE AND RED CACHIMIA.
CONCERNING TALC, FLUIDIC AND SOLID.
CONCERNING THUTIA, CALAMINE, AND LITHARGE.

CONCERNING ANTIMONY

The virtues of Antimony obtain in Morphew, Leprosy, Elephantis, Wounds, and Ulcers.

HERE FOLLOWS THE PREPARATION OF ANTIMONY FOR THE SEVERAL SPECIES OF LEPROSY

℞   Of the best pounded Antimony, lb.j.
Of highly distilled (*sic*), lb.iiij.
Of crude white Tartar, lb.ss.

Reduce to a fine powder in a phial, distil by retort, and a red oil will result.

The preparations of Antimony vary with the diseases for which it is administered. That which is used for wounds differs from that which is applied in the case of leprosy. And so of the rest. To take the same preparation of Antimony both in wounds and in leprosy would be a serious error.

ELEPHANTIS

The preparation of Antimony for Elephantis is, however, the same as for leprosy.

THE PREPARATION OF ANTIMONY FOR MORPHEW.

℞   Of the best pounded Antimony, lb.ss.
Of calcined Tartar ⎱
Of Alum   ⎰ an equal quantity:

Arrange in alternate layers; reduce in a fire of reverberation of the fourth grade; then distil, and a thick red oil will come over. By alternate layers

---

* It is in every way highly desirable that this important collection of Hermetic prescriptions should find place among the Hermetic Medicine of Paracelsus. It is of very considerable value as evidence of the extent to which mineral, and especially metallic, substances were applied by Theophrastus in all varieties of disease. The author's intention seems to have been the compilation of a whole alchemical pharmacy, and there is a small fragment extant of a second book, under the title *De Nascentibus ex Terra*.

understand one layer of the Tartar and Alum, afterwards a layer of the Antimony, and so forward.

THE PREPARATION OF ANTIMONY FOR WOUNDS

℞   Of Antimony } an. lb.ss.
           Of calcined Tartar }
           Of Alcool of Wine, 1 kist.

Mix ; distil by the alembic till the matter is resolved.

℞   Of the substance dissolved as above, ʒj.
           Of Alcool of Wine, ʒiiij.

Dry by coagulation, and reduce into oil on a marble slab.

There is no greater cure for wounds than that which is obtained from Antimony, except in wounds of the head. The Antimony should be distilled upwards till what is below becomes aqueous.

THE PREPARATION OF ANTIMONY FOR ULCERS

℞   Of Antimony }
           Of Colcothar } any equal proportion.
           Of Flos Aeris }

Reduce S.S.S. to the grade of reverberation. Make an extract with red wine, and reduce into alkali.

The said alkali made into an unguent with olive oil and laid over ulcers is of great healing virtue.

ADDITIONS OF ANTIMONY FOR LEPROSY

℞   Of the said Antimony, ʒj.
      Of Oil of Wine Fæces, ʒj.
      Of Oil of Bitter Almonds, to the weight of both the above.

Mix. If there be no hoarseness of the voice, anoint once or twice every seven days. If the voice be hoarse, it is useless.

ADDITION IN MORPHEW

℞   Of the said Antimony, ʒj., with Kist, *i.e.*, Alcool of Wine.
      Of Tragagantum, ʒij.
      Of consolidated Royal Mucilage }
      Of Psyllus Seed } each ʒij.
      Of Gum Arabic }

Make an unguent. The process is the same for Morphew and Alopecia. The unguent is to be applied warm, once or twice in every seven days. By this means the scab will rise up. If it peels off, the ulcer may be healed with the following unguent.

℞   Of Spermiola }
           Of Camphor } each ʒj.ss.
           Of Oil of Ceruse to the weight of both.

Make an unguent. After the scab has come off the place must be anointed with this unguent every eight days.

ADDITION FOR WOUNDS

℞   Of the said prepared Antimony, ℈vij.
        Of the juice of White Tartar ⎱ each ℈v.
        Of Oil of Myrtles   ⎰

Mix. Apply once every other day, and no mischance need be feared. Note.—
In summer, add camphor at pleasure.

ADDITION FOR ULCERS

℞   Of the said prepared Antimony, ℥iij.
        Of Oil of Colcothar, ℥ss.
        Of Oil of Mastic (*Oleum Lentiscinum*) to the weight of both.

Make an unguent. Anoint around the ulcer. This does not cure cancer,
elephantis, or esthiomena. *Oleum Lentiscinum* is oil from the bark of the
mountain osier.

CONCERNING LITHARGE

The virtues of Litharge obtain in Cancer and Fistulas, in Tentigo Prava,
Esthiomensis, Red Jaundice, Persian Fire, and Wounds.

HERE FOLLOWS THE PREPARATION OF LITHARGE FOR CANCER

℞   Of pounded Litharge, lb.ss.
        Of Salt Water ⎱ each lb.j.
        Of Alum   ⎰
        Of White Vinegar, lb.iiij.

Reduce over hot coals till their moisture is consumed. Take of the said
Litharge with an equal quantity of spring water: reduce *ad colores* for a
night, and dry.

The same preparation of Litharge prevails in fistulas.

PREPARATION FOR ESTHIOMENSIS

℞   Of Litharge, lb.j.
        Of Calcined Tartar, lb.ss.
        Of Spring Water, or *Aqua Fuliginis*, a sufficient quantity.
        Of common melted Salt ⎱ each ℥vj.
        Of Rock Alum   ⎰

Reduce according to the fourth grade of reverberation with the aforesaid
water into an alkali.

Aqua Fuliginis is water from sooty roofs obtained during rain.

PREPARATION FOR RED JAUNDICE

℞   Of Myrrh   ⎱ each ℥j.
        Of Frankincense ⎰
        Of Litharge, ℥iiij.
        Of very strong Vinegar, lb.ss.

Reduce into a decoction.

PREPARATION FOR WOUNDS

R̟   Of Litharge, with four times Whitened Vinegar, lb.ss.

Of the juice of the herb Pellitory   ⎫
Of the lesser Comfrey   ⎬ an equal quantity.
Of the round Aristolochy (Birthwort)   ⎭

Make a compound with *mucilago lumbricata*.

PREPARATION OF LITHARGE IN TENTIGO PRAVA

R̟   Of washed Litharge, 1 lb.

Of Rock Alum, lb.jss.

Mix. Pound well, place in the fourth grade of reverberation for four hours, then extract the alkali with spring water, together with the remaining litharge and rock alum in equal quantity. Pound as above till all the litharge has been used.

The process should be as follows : When the litharge has been placed for four hours in rock alum, take of this distilled alkali, of fountain water, and of soot, each half a pound, and mix together.

PREPARATION OF LITHARGE FOR PERSIAN FIRE

R̟   Of Litharge, lb.j.

Of Red Realgar, ℥ij.

Of Sal Ammoniac, ℥ss.

Mix, and place in a sublimatory. This must be done ten or twelve times. Then pour on warm water, and let the litharge be separated.

ADDITION IN TENTIGO PRAVA

R̟   Of the said Litharge, ℥j.

Of Common Realgar, ℨj.

Of the juice or water of Chelidony, a sufficient quantity.

Make into an unguent, anoint very thinly, and apply four or five times. The skin turns red, and the rank smell goes away. Then use the following recipe :—

R̟   Of the said Litharge, ℥jss.

Of mucilage of Fœnugrek   ⎫
Of Lumbrici Nitri   ⎬ an equal sufficient quantity.

Make into an unguent. Lumbrici Nitri are worms found in dung.

ADDITION IN WOUNDS

R̟   Of the said Litharge, ℥iiij.

Of Oil of Camphor, Əj.

Of Crocus of Mars, Əiiij.

Make into an unguent.

Apply to the wound once or twice daily, and rub in well.

ADDITION IN ESTHIOMENSIS

R̟   Of the said Litharge, ℥iiij.

Of Powder of Chelidony ⎫
Of Oak Apples   ⎬ each ℥ij.

Reduce to a powder.

℞   Of the said Litharge, ℥iij.
       Of Mucilage of Consolida ⎫
       Of Lumbrici Nitri   ⎬ each a sufficient quantity.
       Of Oil of Myrrh   ⎭

Make into an unguent.   The disease is cured thereby.

ADDITION IN CANCER

℞   Of the juice of Marrubius ⎫
       Of Persicary   ⎬ each ℥j.
       Of prepared Litharge, ℥ij
       Of Oil of the Yolk of Eggs, a sufficient quantity.

Compose an unguent.

ADDITION IN RED JAUNDICE

℞   Of prepared Litharge, ℥ss.
       Of Rock Alum, ii. oz.
       Of Salt Water, ℥jss.

ADDITION IN PERSIAN FIRE

℞   Of Elect Vitriol, ℥iiij.
       Of Oak Apples, ℥ss.
       Of Frankincense, ℥j.
       Of prepared Litharge, to the weight of all.
       Of Wine and Vinegar, as may be wanted.

When it boils (? ferments) then it is to be used, and the more it boils the better it is.

CONCERNING MARCASITE

Gold or silver Marcasite has angles like tiles.   The virtues of Marcasite are in Restriction of the Blood, the Menstrua, and Hemorrhoids.

PREPARATION OF MARCASITE

℞   Of Marcasite, ℥iiij.
       Of Pitch   ⎫
       Of Colophony ⎬ each ℥vi.
       Of Resin of the Fir, to the weight of all.

Reduce to calx.

To reduce into calx is to place in a brick kiln and burn till the resin flows out twice or thrice until it glows.

PREPARATION IN RESTRICTION

℞   Of best pounded Marcasite, ℥ij.
       Of Oil of Flax, ℥vij.

When these two are conjoined and set on fire, the true matter will remain.

PREPARATION FOR HEMORRHOIDS

℞   Of Marcasite, ℥jss.
       Of best dried Alcool of Wine, lb.j.

Mix.

ADDITION IN RESTRICTION OF THE BLOOD

℞　　Of the said Marcasite, ℥j.
　　　Of Corals, ℥ss.
　　　Of Plantain Seed, ℥ss.

Reduce to a fine powder.

The powder is to be sprinkled upon the wounds, or mixed with vinegar, and be bound up below the wound; thus it will hold it together. Those who, by reason of an accident, bring up blood, should drink it.

ADDITION FOR MENSTRUA

℞　℥ss. of this Marcasite, and as much as suffices of Oil of Sandarach. Make into an ointment.

If the menstruum flow to excess, let the umbilicus be anointed twice or thrice.

ADDITION FOR PILES

℞　　Of the said Marcasite, ℥iij.
　　　Of Sal Gemmæ ⎫
　　　Of Mumia　　　⎬ ℥j. each.

Make into a powder.

The swollen piles must be cut and then anointed.

CONCERNING CACHIMIA

The virtues of Cachimia obtain in Dysentery, Diarrhœa, and Lienteria.

PREPARATION FOR DYSENTERY

℞　　Of Cachimia well ground, ℥vj.
　　　Of Rust of Iron, ℥ss.

Reduce by the second grade of fire for six or seven hours. Afterwards take out and reduce into alkali.

PREPARATION FOR DIARRHŒA

℞　Of prepared Cachimia as above, and of Nutmeg what is sufficient for incorporation. Reduce to the second grade in the form of a bolus.

PREPARATION FOR LIENTERIA

℞　Of Cachimia prepared as above, and of Gum Arabic dissolved in plantain water. Make a bolus. Reduce to second grade.

ADDITION FOR DYSENTERY

℞　　Of the said Cachimia, ℥ss.
　　　Of Boiled Dove, a sufficient quantity.

ADDITION FOR DIARRHŒA

℞　　Of the said Cachimia, ℥j.
　　　Of Theriaca, ℥iij.
　　　Of Sealed Earth, ℥ss.

Make a bolus, the dose containing from ℥j. even to ℥ij.ss. Administer morning, noon, and evening. Abstain for three days. Afterwards repeat it. Do this thrice.

ADDITION FOR LIENTERIA

℞ Of prepared Cachimia, ʒj.
  Of Crocus of Mars, ʒij.
  Of Red Corals, ʒss.
  Of Theriac, as required.

Make a bolus; the dose is from ʒij. even to iij. or iiij. morning and evening. Let a portion of this be administered daily.

CONCERNING THUTIA

The virtues of Thutia are for spots in the eyes.

PREPARATION FOR SPOTS IN THE EYES

℞ Of Thutia, ℥j.
  Of White Vitriol ⎱
  Of Juice of Eyebright ⎰ equal quantities.

Make into a bolus with Gum Arabic, and bring to the second grade of fire. It becomes an unguent beneficial to the eyes.

PREPARATION FOR WHITE SPOTS IN THE EYES

The Thutia must be extinguished in milk and placed for the night in rose water. This water removes the white speck when applied to it.

FOR WENS

℞ Of Thutia, ℥iiij.
  Of Fused Salt ⎱
  Of Live Calx ⎰ each ℥vj.
   S.S.S. ℥vj.

Arrange S.S.S. Apply fourth grade of fire; reduce into alcali.

ADDITION FOR THE SPOTS OF THE EYES

℞ Of the said Vitriol, ʒss.
  Of Spawn of Frogs, ʒij.
  Of Laterine Oil, Əss.

Make an eye salve. If yellow spots appear in the eye and sparkle, they return.

ADDITION FOR WENS

℞ Of the said Thutia, ʒj.
  Of White Vitriol, ʒvij.
  Of pounded Camphor, Əjss.

Make a mixture with water of roses or fennel. This disease attacks all pedal animals. The wen assails the eyes of goats, other animals, and men. In the case of human beings camphor must be administered with it, lest inflammation supervene.

ADDITION FOR WENS

℞ Of the said Thutia, ℥j.
  Of Sal Anatron, or Gall of Glass ⎱
  Of Fused Salt   ⎰ ℥j. each.
  Of distilled urine, lb.ss. and ℥iiij.

Mix. The process consists in administering this potion to such as are afflicted with wens, in the morning and in the evening, for three or four weeks. This medicine removes all wens, except *grisonum*.

CONCERNING TALC

The virtues of Talc obtain in ulcers and humid wounds.

PREPARATION FOR WOUNDS.

℞ lb.j. of Talc and an equal weight each of Cinder of Beans and of Oats. Reduce at the fourth fire for a day and night ; cleanse and dry. Talc dries the bottom or base of the wound, so that it does not change into a fistula. It also powerfully desiccates ulcers. It must not be used beyond the third day.

ADDITION FOR WOUNDS AND ULCERS

℞ Of the said Talc, ℥j.
Of Liquor of Mumia
Of Washed Turpentine } each q.s. for an unguent.

Cures eating, cancerous, and other suppurating ulcers.

CONCERNING CALAMINE

The virtues of Calamine are suitable for plasters, eye salves, and the Persian fire. Add Calamine both for ulcers and for wounds, for plasters where a growth of fresh flesh is required, also for eye salves where neither albugo or spots of the eyes are present, as in the case of red eyes, where the greatest experience is requisite.

PREPARATION OF CALAMINE FOR PLASTERS

℞ Of Washed (that is pure) Calamine, ℥j.
Of Colcothar
Of Live Sulphur } each ℥j.ss.

Arrange in layers and apply the fourth fire for a day and night. Reduce by the second ablution.

PREPARATION FOR UNGUENT

℞ Of the said prepared Calamine, ℥iij.
Of the Oil of the Yolk of Eggs, ℥j.

Make a bolus with gum arabic ; reduce at a fire of the second grade for four hours, then reduce by ablution.

FOR THE EYES

℞ Of the said prepared Calamine, ℥j.
Of Distilled Vinegar, ℥vj.

Make an extraction ; then dry it.

PREPARATION FOR PERSIAN FIRE

℞ Of Crude Calamine, lb.ss.
Of Water of Water Lily, ℥vj.
Of Alumen Plumosum, ℥ss.

Reduce these by digestion in a glass for the space of a week, and distil.

lt is a recipe of Geber. Petrus de Archilata errs in the process : for the medicine is to be used for pandaricia, but not for combustions.

    ℞   Of Apostolic Plaster, ʒvj.

       Of the said Calamine, ℥ss.

       Of Camphor, ʒj.

Make a plaster.

For Unguents

  ℞   Of Agrippine Unguent, ℥iiij.

      Of Unguent of the Flower of Copper, ℥ss.

      Of the said Calamine, ʒx.

Make a mixture. Most excellent for ulcers, itch, and scab.

Addition for Eye Salve

   ℞   Of the said Calamine, ℥j.

       Of Water of Eyebright ⎫

       Of Water of Fennel ⎬ ʒiij.

       Of Water of Roses ⎭

To be applied at night.

Addition for Persian Fire

   ℞   Of the said Calamine, ʒvj

       Of Waters of Vitriol and Oak Apples

Some use cobblers' atrament for Persian fire and red jaundice.

TREATISE II

CONCERNING BLOODSTONE, ARSENIC, SULPHUR, SAXIFRAGE, ORPIMENT

CONCERNING BLOODSTONE

The virtues or chief arcana of Bloodstone are for bloody ulcers, resolved menstrua, premature profluvia of the matrix, lax dysentery, diarrhœa.

PREPARATION FOR BLOODY ULCERS

℞ ʒiij. of Bloodstone, and ʒiij. each of *lutum Lephanteum* (that is, clay from which small cucurbits are made), and of Bolus Armenus. Make a bolus with traganth dissolved in vinegar : Reduce by the fourth grade of reverberation ; then extract the alkali.

In the case of wounds, of lupus, and of herpata, bloodstone proves extremely beneficial. It binds the veins so that the flow of blood ceases. Let it be sprinkled upon the parts.

PREPARATION OF BLOODSTONE FOR MENSTRUA

℞ Of Bloodstone, ʒiiij.
Of Mastic dissolved, ʒx.
Of Carabe, ʒjss.

Make a mixture with a decoction of water of alum ; reduce by ablution.

The flow of menstrua should be checked when it causes pallor in the face. The use of this preparation is safe, and effects a complete cure. After decoction for seven hours lute is produced from bloodstone, out of which trochisks are made for menstrua.

THE PREPARATION FOR IMMATURE MENSTRUA IS AS FOLLOWS

℞ Of Bloodstone, ʒj.
Of Oil of Nutmeg ⎫
Of Oil of Grains of Actis ⎬ ʒiij.
Of Petroleum ⎭

Reduce into a composition. The dose is Əj. It ought to be administered with water of roses, decocted with roots of plantain, or with water of plantain.

Bloodstone stops profluvium *sine torsura*. But if there are gripings, it is the generation of the stone.

PREPARATION FOR LOOSE DYSENTERY

℞ Of Bloodstone ⎫
Of Red Corals ⎬ v.ss. each.
Of Spodium ⎭
Of Tanacetum, to the weight of all.

Make trochisks with mucilage of the glue of *botin*. The dose is ʒss.

PREPARATION FOR DIARRHŒA

℞   Of Ice Alum ⎫

Of Bloodstone ⎬ equal quantities.

Of Crocus of Mars ⎭

Make trochisks of gum arabic dissolved in plantain water. The dose is from ℥j. to ℥j.ss.

Water of plantain is to be extracted from the roots and herbs. The cornelian, if carried in the hand, stops blood, but not so bloodstone.

ADDITION FOR BLOODY ULCERS

℞   Of Prepared Bloodstone, ℥ss. (al. ℥j.)

Of Oak Apples, ℈ss.

Of Seraphinus, ℈j.ss.

Of Oil of Kerua (Keyri) from Violets, sufficient for incorporation.

Let an unguent be made. In cases of acute ulcers, add in place of Oil of Keyri a proportion of the Liquor of Mumia, as in the case of herpeta, and in eating and cancerous ulcers.

ADDITION FOR MENSTRUA

℞   Of Bloodstone, ℥j.

Of Long Pepper ⎫

Of Nutmeg ⎬ each ℨss.

Of Cinder of Frogs' Follicles, ℈iiij.

Make trochisks with Mint Water. The dose if from ℈ss. even to ℈j.ss.

ADDITION FOR DYSENTERY.

Let the prepared Bloodstone be given in red wine. Therein let iron be extinguished or let it be given with Tyriaca.

ADDITION FOR DIARRHŒA

℞   Of the said Bloodstone, ℥iij.

Of Pearls, ℈ss.

Of Liquefied Mumia, to the weight of all.

Mix. The dose is from ℥iij. to iij. or iiij.

CONCERNING SAXIFRAGE

By Saxifrage understand any stone which removes growths like tartars, mosses, sand, frost, and hail.

Saxifrage is properly a pale crystal, called also Citrine Stone or Citrinole Stone. Citrine Stone stands between crystal and yellow beryl. Let the liquor be produced after the mode of an alkali. The dose is ℈ss. in good wine.

FIRST PREPARATION FOR SAND, MOSS, FROST, HAIL

℞   Of Saxifrage, ℥j.

Of Borax, ℨij.

Of Salgemmæ, ℨvj.

Of Fused Salt, ℥j.

Reduce S.S.S. at a fire of reverberation through the fourth grade, from the setting of the sun till morning. Reduce into alcali. The dose is ℥ss. in white wine.

SECOND PREPARATION

Take of the said Saxifrage ℥j. Reduce it by itself to the fourth grade of reverberation. Also take :—

℞. Of Reverberated Saxifrage, ℥j.

   Of Cinder of the root of larger Radish, ℥j.

   Of Alkali of the roots of Petroselinon, Əj.

Make a mixture by itself. The dose is from Əj. to Əiij. or iiij.

FIRST ADDITION

℞. Of the said Saxifrage, ℥j.

   Of Millet of the Sun, ℥ij.

   Of White Wine, ℥x.

The dose is from ℥iiij. to vj.

SECOND ADDITION

℞. Of the Saxifrage, ℥j.ss.

   Of Seed of Parsley ⎞

   Of Rocket   ⎭ each ℥j.

   Of Clarified *aqua mulsa*, ℥x.

The dose is from ℥iiij. to vj. or vij.

It is necessary to continue this prescription as long as the tartarised urine issues.

CONCERNING ARSENIC

The virtues of Arsenic are for ulcers, wounds, and other openings.

Arsenic is a soot out of metals, and especially from lead, and it is realgar or fulgurr (or soot) out of metals.

THE FIRST PREPARATION IS THE REDUCTION OF ARSENIC INTO MUMIA.

During the preparation the venom must be removed. Nothing cures ulcers and wounds more perfectly than prepared arsenic. It also cures *syrones* and all ulcers, gangrenes, and fistulas. The arsenic from lead is the best. Next, that which exudes from iron and resembles copper.

THE SECOND PREPARATION OF ARSENIC IS THE REDUCTION OF ARSENIC INTO BALM

THE THIRD IS THE REDUCTION INTO LIQUOR

Arsenic has three preparations, into Mumia, Balsam, and Liquor.

℞. Of White Arsenic, ℥vj.

   Of Fused Salt, ⎞

   Of Colcothar, ⎭ ℥j.ss.

Mix and reduce to the second grade of reverberation for three or four hours. Take out.

It must be removed from the top five or six times, pounded, and again prepared as above. This must be repeated five or six times.

PREPARATION OF BALSAM

R Of White Arsenic, ʒx.
   Of Talc, ʒiiij.
   Of Live Calx, ʒxv.

Make a subtle mixture. Reduce to the fourth grade of reverberation for 24 hours. It is like glass and the venom sticks at the bottom of the calx. That which is removed from the top is to be pounded and placed in a glass vessel. Let it be set in a cellar, whereupon an oil or balsam will come forth.

PREPARATION OF LIQUOR

R Of Crude or White Arsenic, lb.ss.
   Of Saltnitre, lb.j.
   Of Salgemmæ, ʒss.

Pound. Reduce in an open reverberatory for twenty-four hours.

If these being united are placed over a fire of reverberation, the Arsenic glows for three hours; afterwards it liquefies. In this condition it is poured into water, and coagulated after the manner of an alkali.

ADDITION FOR BALSAM

R Of the said Balsam, ʒiij.
   Of Oil of Yolk of Eggs, ʒx.
   Of Distilled Turpentine, ʒj.

Mix. Like Mumia, the balsam is to be applied for the space of twelve hours.

ADDITION FOR LIQUOR

R Of the said Liquor, ʒxv.
   Of Skins of Pomegranate, ʒvj.
   Of the Bark of the Frankincense Tree, ʒij.
   Of Mucilage of Botin, to the weight of all. Mix.

CONCERNING ORPIMENT

Orpiment is a Yellow Minera like gold.
The virtues of Orpiment obtain in fistulas, cancers, and eating ulcers.

PREPARATION FOR FISTULA

R Of Orpiment, ʒj.
   Of Calcined Tartar, ʒiij.

Arrange in alternate layers.

Reduce by fourth grade of reverberation for a day and night, that is, for twenty-four hours.

It melts when thus decocted. Let it be removed again, pounded, and poured into water, whereupon a white powder will settle at the bottom, which is the prepared Orpiment.

If it be put into a glass, an oil results, which must be injected into the fistula, or it may be applied by means of a rag. But let ulcers be sprinkled with the powder.

PREPARATION FOR CANCER

℞ Of Orpiment, ʒv.
   Of Fuligo, ʒss.
   Of Sal Ammoniac, ʒiij.

Reduce by the fourth grade of reverberation a day and a night. Reduce into alcali. Alcali is the chief arcanum for cancer.

PREPARATION FOR ESTHIOMENIS

℞ Of Orpiment, ʒiij.
   Of Calcined Alum, ʒvj.

Reduce by the fire as above with extraction of alcali.

ADDITION FOR FISTULAS

℞ Of the said prepared Orpiment, ʒss.
   Of Resin of Pine, ʒj.
   Of Wax, to the weight of all.

Make into a cerotum. It is to be applied for fistulas.

ADDITION FOR CANCER

℞ Of prepared Orpiment, ʒv.
   Of Cinders of Pigeons' Dung ⎱
   Of Oil of Yolk of Eggs   ⎰ q.s. for unguent.

This is used for cancer.

ADDITION FOR EATING ULCERS

℞ Of Orpiment, ʒv.
   Of Liquor of Mumia, ʒij.
   Of Oil of Roses, ʒj.ss.
   Of Mucilage of the Seed of Fleawort, to the weight of all.

Make an unguent or cataplasm. Should the sick person complain of heat, the ulcer must be anointed with oil of camphor previous to application of the remedy.

CONCERNING SULPHUR

The virtues of Sulphur apply in cases of very acute imposthumes and asthma; they serve to maintain health. Imposthumes include pleurisy, pest, and the like.

THE FOLLOWING IS THE PREPARATION FOR EXTREMELY ACUTE ULCERS

℞ Of Live Sulphur, lb.j.
   Of Colcothar ⎱
   Of Fused Salt ⎰ each lb.ss.

Make a fine powder; reduce by sublimation. As soon as it has been sublimated, make an addition again, and sublimate thrice as above. Live Sulphur coheres in fragments, and is not yet dissolved.

PREPARATION FOR ASTHMA

℞   Of Fused Sulphur, lb.j.
Of Shavings of Red Sandal ⎫
Of Cypress Shavings   ⎬ each to the weight of the Sulphur.
Of Pine Shavings   ⎭

Arrange in layers.   Reduce by fire of reverberation, finally into alcali.

℞   Of this Alkali, ʒx.
Of Myrrh, ʒv.

Sublimate

PREPARATION TO CONSERVE HEALTH

℞   Of Sulphur, ʒiiij.
Of Oriental Crocus ⎫
Of *Myrobalani*   ⎪
Of *Chebuli*   ⎬ each ʒj.
Of *Bellirici*   ⎭
Of Oil of Juniper Seeds, sufficient for incorporation.

Sublime by a very slow fire.

ADDITION FOR VERY ACUTE IMPOSTHUMES

℞   Of this prepared Sulphur, ʒss.
Of Oil of Nutmeg, ʒj.
Of prepared Aqua Veronica, to the weight of all.

Make a potion.

ADDITION FOR ASTHMA

℞   Of the said Sulphur, ʒss.
Of corrected Thebanus, ʒiij.
Of Tyriaca, q.s.

Make a bolus.   The dose is from Əj. to two or three.

ADDITION FOR CONSERVATION OF HEALTH

℞   Of the said Sulphur, ʒss.
Of Red Myrrh   ⎫
Of Oriental Crocus ⎬ ʒss. each and Əj.
Of Aloepaticus, to the weight of all.

CONCERNING GEMS, TRANSPARENT AND OTHERWISE; OF CORALS, THE MAGNET, THE CRYSTAL, RUBIES, GARNETS, SAPPHIRES, EMERALDS, HYACINTHS, &c. EVERY STONE POSSESSED OF MEDICAL VIRTUES IS A GEM

CONCERNING CORALS

The virtues of Coral are for menstruum and profluvium; poison taken internally; thunderings or rumbling of the stomach; charms, if any be enchanted; obsession, if any be mad; nervousness, if any be timorous; melancholy, for those who appear wise in their own eyes but are foolish. The virtue and the substance is one and the same. Virtue is a thing by itself. Coral simples restrict urine and evacuation, and after a long time the menstrua. When prepared their operation is sudden and safe.

PREPARATION OF CORALS TO RESTRAIN MENSTRUUM AND PROFLUVIUM

R̵ Of Corals, ℥ss.
Of Oil of Myrtles, ʒj. (*al.* ℥ss.)
Of Olibanum, ʒj ss.
Of Fused Salt, ℥ij.

Make a mixture; reduce by calcination through the fourth grade of reverberation for twelve hours or more; afterwards reduce by ablution with water of plantain. Corals restrict urine but not menstruum.

PREPARATION OF CORALS AGAINST INTERNAL POISONING

R̵ Of Corals, thoroughly pounded, ℥ij.
Of Waters of Ligusticum, lb.ss.
Of Sal Gemmæ ⎱ each ℥ij.
Of White Vitriol ⎰

Reduce by digestion in the second grade of fire for a month; take out the red and coagulate. For poisons the medicine must be without a body; for poison is also without a body. The red which settles at the bottom is a good remedy against poison.

PREPARATION FOR RUMBLINGS IN THE STOMACH

R̵ Of Corals, ʒvi.
Of Cinders ⎱
Of Roman Cummin ⎰ each ʒiii.
Of Beans ⎰
Of prepared Alum to the weight of each.

Mix these. Reduce by digestion with lb.ss. of desiccated alcool; desiccate. It becomes yellow all over. This preparation of corals ought to be used for *diacymimum*. The colour becomes red when the corals have been prepared. Separate and desiccate.

PREPARATION OF CORALS AGAINST CHARMS, OBSESSION, NERVOUSNESS, AND MELANCHOLY

℞   Of Corals, ʒiij.

Of Glue of Oak, ʒiiij.

Of St. John's Wort, ʒiiij.

Of *Storax Calamita* ⎫
Of Laudanum   ⎬ each ʒj.ss.
Of Gum   ⎭

Of distilled wine add lb.ij.   Reduce by decoction in a closed alembic for a day and a night.   Distil and pour over again as above.

Corals, if prepared in this manner, become red and exceedingly hard. Consequently, they must first be pounded.

ADDITION FOR MENSTRUUM

℞   Of prepared Corals, ʒj.ss.

Of Tanacetum ⎫
Of Plantain ⎬ each ϑj.ss.

Of Long Pepper ⎫
Of Nutmeg   ⎬ each ϑss.

Make a powder.   The dose is ϑj. in a tempered egg.

When salt is injected into the egg, it must also be eaten, otherwise it does no good.

ADDITION FOR CORALS AGAINST POISON

℞   Of prepared Corals, ϑv.

Of Theriac, ʒss.

Of Root of Larger Lapathius, to the weight of all.

Of Alcool of desiccated Wine, lb.ss.

Reduce by digestion for a week.   The dose is from gr.xv. to ϑij.

One who has drunk poison should have administered to him ʒij. of water of bullace or of roses.   Let it be repeated several times, so as to produce perspiration, until the evil be felt no longer.

ADDITION FOR RUMBLINGS

℞   Of the said Corals, ʒij.

Of Species of Diacymini, ʒiiij.

Of prepared Blood of the Goat, to the weight of all.

Make a physical powder with Saccharum ; the dose acts as a sedative.

The Goat of the Spagyrists is the castrated young of a coney.   It must be fed with aperient herbs, then it is good.

ADDITION FOR THE OTHER THREE SPECIES

℞   Of the said prepared Coral, ʒviiij.

Of Masterwort, ʒj.

Of Angelica, ʒv.

Of Glue of Oak, ʒj.ss.

Mix with water of St. John's Wort : the dose is from ʒss. to ʒvj.

This is the best medicament at the commencement of *tympanis.*

CONCERNING THE MAGNET

It has a virtue for wounds and ulcers *cum flaxis et ramentis.*

PREPARATION FOR THE ABOVE

℞   Of the Magnet, ℥j.

         Of Calx of Eggs, ℥ij.

Make S.S.S. in the fourth grade of the fire of reverberation for a day and a night.   Remove the calx of eggs.

ANOTHER PREPARATION

℞   Of Magnet, ℥j.

         Of Calx of Eggs, ℥vj.

Set in layers in a crucible.   Place in a fire of reverberation a day and a night. Extract, and it will be prepared.

Otherwise, if not prepared by pounding, it misses its true extractive efficacy.   But if previously prepared, then pounded and mixed, the oppodeltoch has an excellent effect.

ADDITION FOR WOUNDS CUM FLAXIS ET RAMENTIS

℞   Of the Magnet, ℥ss.

         Of Carabe, ℥ij.

Make a subtle powder with lb.ss. of the plaster oppodeltoch or plaster of apostolico.   Reduce to a wine by shaking.

This plaster when applied extracts splinters of bone, and bullets received from guns, out of ulcers and other wounds.   If the magnet be pounded in unprepared condition, it loses its efficacy, but if you pound and mingle with *apostolico*, it extracts by itself.   Unprepared it effects nothing.

CONCERNING GEMS

The crystal has the property of producing an abundance of milk if administered internally to women.

PREPARATION OF GEMS BY DIAPHANITAS

The preparation of gems is fourfold.   First, by reverberation.   Secondly, by calcination.   Thirdly, by elevation, and in the fourth place by means of distillation.

A woman requires over lb.j. of crystal before she experiences an increase of milk.   Accordingly prepared crystal is necessary.

REVERBERATION OF CRYSTAL

℞   Of Crystal, lb.j.

         Of Water of Entali, lb ij.

Make a mixture by imbibition.   Reduce by reverberation for twenty-four hours.

Thus there is left from lb.j. a *verto* (that is, a kind of weight).   Dose ℥ij.

CALCINATION OF CRYSTAL

R   Of Crystal, ℥iiij.

Of Mastic

Of Colophony   } ℥ij. each.

Of Sulphur

Reduce in an athanor.   The dose is ℈j.

An athanor is a furnace in which things are burnt.

ELEVATION OF CRYSTAL

R   Of Crystal, ℥j.

Of Sal Ammoniac, ℥iij.

Reduce in a sublimatory to powder. The sublimation is to be performed five or six times and the crystal is always to be removed.

R   Of the said elevated Crystal, ℥ss.

Of common distilled Water, ℥iiij.

Reduce to an alkali.   Dose ℈ss.

DISTILLATION OF CRYSTAL

R   Of elevated Crystal, ℥j.

Of Water of Nitre, and

Of Alumen without distillation   } ℥ij. each.

Reduce into digestion for three or four days; then distil; coagulate that which is distilled, and dissolve the coagulate.

Coagulation must take place over a slow and small fire. This coagulate, if placed in a cellar, passes into water, which is the last preparation of the crystal. The dose is ℈j. With all other valuable gems the preparation is as with the crystal.

The chief virtue of rubies is for dysentery. The dose is ℥j. if crude, but if reverberated then the dose is ℈ij. After calcination the dose is ℈j. After elevation, ℈ss. After distillation, ℈j.

Also garnets, thus distilled, constitute a most powerful salve for spots of the eyes.

Emeralds, if prepared by means of distillation, are beneficial to those with bloodshot eyes.

Sapphires, being prepared to the third or fourth preparation, remove trembling of the heart if prepared into distillation. The dose is gr.v. Sapphires dispel *synthena* and palpitation.

The case is the same with the other gems. Bartholemew, the Englishman, has written voluminously concerning gems and precious stones.

ADDITIONS FOR THE GENERATION OF THE MILK OF CRYSTALS

R   Of the said Crystal prepared, ℈ij.

Of Spermaceti

Of Seed of Lettuce   } ℥ iiij. and ss.

Make a powder with administration of water of almonds.

ADDITION OF GARNETS FOR TREMBLING OF THE HEART AND BLOODSHOT EYES

℞   Of Garnets, ℥ss.

Of Aloë Epaticus, ʒiii.

Of Prepared Sulphur, ʒj.ss.

Mix with clarified Zuccarum. The dose is from ʒij. to ʒij. This medicine must be diligently administered and continued for five days, although the trembling of the heart may disappear previously.

ADDITION OF SAPPHIRE

℞   Of Sapphire, ʒiij.

Of Dissoved Amber, ʒj.ss.

Of Storax Calamita, Əj.

Make a mixture. The dose is from Əj. even to Əj.ss.

The Emerald strengthens women in labour, and is the sovereign arcanum for their ailments if prepared by distillation, as crystal.

℞   Of the said Emerald prepared, Əj.

Of Liquor of Melissa, ʒj.

Of Southernwood, ʒij.

Mix. The dose is from three to six drops.

ADDITION FOR PREPARED JACINTH

℞   Of Prepared Jacinth, Əj.ss.

Of Laudanum, that is, gum, Əj.ss.

Mix. This is a chief arcanum for fevers arising from putrefaction of air or of water. Should fevers of this kind be usual with any persons, let them drink, every new moon, four or five drops; thus they will be safe from being attacked a second time, and will be absolutely secure during the new moon.

TREATISE IV

CONCERNING SALTS

SAL GEMMÆ, SAL ENTALI, SAL PEREGRINORUM, ALUMINOUS SALT, SAL ALCALI,
SAL NITRI, SAL ANATRON, SAL TERRÆ, SALT FROM VITRIOL
All Salts are from the element of water, as also are all alums.

CONCERNING VITRIOL

The virtues of Vitriol obtain in falling sickness, suffocations of the Matrix, Siphita Stricta, or Noctambulones, Gutta, and Obesity.

The varieties of falling sickness are Analentia, Catalentia, Epilentia, etc. The administration of Vitriol is the same in all.

℞ Of cuprine Vitriol, 1lb. Reduce by separation from the phlegmatic part. Reduce the said phlegmatic part over its colcothar. Distil. Reiterate in the fourth grade of fire. The dose is from Ɔss. to Ɔj. before and after the paroxysm.

If the disease arises from the element of Vitriol, it is cured by Vitriol. The disease of falling sickness is occasioned by salt of Vitriol. The medicine is to be administered during the paroxysm and on the day when it is expected. Epilepsy is a mineral disease; its cure is also mineral, that is, by the salts of Vitriol, and by the spirits before and after the paroxysm. Before the paroxysm the body is in great agitation; after the paroxysm the patients sleep. The medicine should be given after the sleep, while the body is still under the excitement thereof. When the body is healthy it should not be administered.

PREPARATION OF VITRIOL FOR SUFFOCATION OF THE MATRIX
℞ Of Vitriol, purged from Phlegma and Colcothar, ʒij.
Of Pennywort, ʒiij.
Of Alcool of Wine, ℥ss.

Reduce by distillation. The dose is from Ɔss. to Ɔj. This is the most efficacious medicine for the complaint in question.

PREPARATION FOR GUTTA AND SIPHITA STRICTA
In Gutta—
℞ Of the said prepared Vitriol, ʒij.
Of Alcool of Wine, ʒij.
Of Jamen Alum, ℥ss.

Reduce into liquids by the fourth grade of fire; applied externally, the dose is ʒss. Applied internally, the dose is from six to nine grains.

Jamen alum is white, like that of Crete, and sweetish.

The outside application is on the place of the Syntheoma, and this is where the disease begins, that is, *in pulsu*, which is the Syntheoma thereof. But if the patient still walks, the medicament is then to be bound about the pulse of the wrist and the neck. For siphita parva castigation is an effectual medicine, but it is of no use for siphita stricta. In gutta the medicament should be placed on the tip of the tongue. The paralysis being arrested, apply to mouth and tongue. It is the best medicament for the complaint.

ADDITION IN EPILENTIA

℞   Of the said Vitriol, ʒj.
Of the Viscous Liquor of the Oak with Orisons, each ℈ss. and gran. iij.
Mix.

The Syntheoma of Caducus is in the nape of the neck. In the case of young persons to anoint the nape of the neck with castor oil after the paroxysm is an excellent method.

ADDITION IN SIPHITA STRICTA

℞   Of the said prepared Vitriol, ʒj.
Of Seed of St. John's Wort, ℥ss.
Of Amber, gr.vj.

Mix. Seed of St. John's Wort takes away Siphita Stricta.

ADDITION IN SUFFOCATION OF THE MATRIX

℞   Of the said Liquor of Vitriol, gr.vij.
Of Grains of Actis, ʒj.
Of Alcool of Wine, to the weight of both.

Make into a composition.

Unless the spot is on the umbilicus let it be applied thereto. Should there be suffocation attended with vomiting, the medicaments previously mentioned are to be taken internally. The most important preparation of vitriol is to separate it from colcothar. Then add an equal quantity of alcool of wine. This done, place burnt bread (namely, bread made *ex furfure scalino*, which is dried so that it can be pounded in a mortar) in liquor of vitriol. Next set it for a month in horse-dung. Then prepare alcool, by means of a bath of the first grade, from vitriol. If the vitriol has lost its acetivity, its strength is gone.

CONCERNING WHITE VITRIOL

There are external species, as scotomia and spots of the eyes.

The virtues of White Vitriol obtain in affections of the external parts of the eyes, and in Neutha. Sometimes a cuticle covers the eyes or the ears of children at their birth. Neutha are pellicles growing anywhere from time of birth, as on the face, on the mouth, the eyes, ears, etc. White vitriol is a great cure in such cases, and also for exterior complaints of the eyes.

PREPARATION FOR EXTERNAL DISEASES OF THE EYES

℞   Of White Vitriol, ʒv.

Of Oil of Siligo, ʒss.

Of Oil of Camphor, ʒij.

Putrefy for a month by means of horse-dung, and distil by descension.

Oil is produced out of Siligo. The Siligo is placed on a red hot iron plate, when it becomes encircled by a kind of grease, which is the oil in question. If, after birth, a pellicle covers the neighbourhood of the eyes, it must be most carefully treated with water of eyebright, of roses, or of fennel.

PREPARATION OF VITRIOL IN NEUTHA

℞   Of White Vitriol, ʒj.

Of Oil of Tartar, ʒvi.

Of Laterine Oil, ʒv.

Distil together.

Neutha (Teutha) should never be cauterized.

ADDITION IN EXTERNAL COMPLAINTS OF THE EYES

℞   Of prepared Vitriol, Эj.

Of Liquor of Eyebright, Эij.

Of Red Poppy, ʒj.

Make an eye salve.

ROCK ALUM

The virtues of Rock Alum obtain in open ulcers, scab, itch, eating ulcers, putrid, lascivious, and humid ulcers.

FIRST PREPARATION

℞   Of Rock Alum, lb.ij.

Of White Vinegar, lb.ss.

Of Burnt Salt, one verto.

Mix till it passes from ebullition to coagulation. Then distil.

If open (cavernous) ulcers are washed with this water, the result is wonderful. If they are not thus cured nothing else will avail.

ANOTHER PREPARATION

℞   White (? Alum), lb.x.

Of the Juice of Chelidony ⎞
Of the Juice of Plantain ⎟ each lb.j.

Of Pellitory, lb.ss.

Distil. Take of this water lb.j., and of common water lb.x   Make a lixivium. Foment warm in the case of alopecia, tinea, and ulcers.

ALUMEN PLUMOSUM

The virtues of Alumen Plumosum obtain in paralysis, lethargy, and benumbed limbs.

PREPARATION FOR PARALYSIS

℞   Of Alumen Plumosum, ʒvi.

Of resolved Colcothar, ʒiiij.ss.

Of Sal Ammoniac, ʒiiij.

Resolve. Alumen Plumosum confers strength imperceptibly. Hence it is the best medicament for paralysis.

ADDITION FOR LETHARGY AND BENUMBED LIMBS

℞   Of the said prepared Alum, ʒj.

Of Dragon's Blood, ʒiij.

Of Liquor of Mummy, ʒvij.

Make an unguent. The seat of the disease is in the occiput and in the nape of the neck.

CONCERNING ENTALI

The virtues of Entali are in profluvium and hæmorrhoids.

PREPARATION

℞   Of common Tartar, ⎫

Of Entali,   ⎬ each ʒij.

Of Karabe, Əj.

Of Mastic, ʒij.

Reduce by reverberation to the second grade, and afterwards into alkali.

In profluvium the seat is in the umbilicus ; in hæmorrhoids it is in the spine.

ADDITION IN PROFLUVIUM

℞   Of the said prepared Entali, ʒj.

Of burnt Bolus, ʒiij.

Of corrected Hematite, ʒj.ss.

Mix.

ADDITION IN HÆMORRHOIDS

℞   Of the said prepared Entali, ʒiij.

Of prepared Corals, Əiiij.

Of Oil of Nutmeg, as required.

Make an unguent.

CONCERNING ANACHTHRON

Anachthron is a salt growing in rocks, and is like a moss in appearance. When the said moss is decocted a salt results, that is, glass gall. Its virtues obtain in fistulas, cincilla or cintilla (diarrhœa), and scropulas. Cintilla is from the diaphragm, and it is cured by pure anatron.

℞   Of Anatron, ʒvj. (al. ʒj.)

Of Bean Ashes, ʒij.

Reduce by the fourth grade of reverberation for twelve hours ; extract the alkali.

Anathron with deer grease is good for cincilla. Anathron possesses a volatile Mercury, which must be corrected. It is then an addition in fistulas, cincillas, and scrofulas.

> ℞   Of the said Anatron, ʒij.
> Of crude Butter, ʒiiij.
> Of the Fat of Marmots, ʒiij.

Make an unguent. The said unguent is the best for fistulas, cintillas, and scrofulas.

CONCERNING SAL GEMMÆ

Sal Gemmæ is called Sal Granatum by the Spagyrists and Sal Lucidum.

It is a laxative of intense salt, *i.e.*, of cholera, and is like colocinthis. It cures jaundice, yellow dropsy, and sufferings arising from corrupt blood.

PREPARATION IN DROPSY AND JAUNDICE

> ℞   Sal Gemmæ   } each ʒj.
> Of Tithymal, *i.e*, Esula Major (al. Minor) }
> Of Gum of Cherries, to the weight of both.

Make a bolus. Reduce by the third grade of reverberation for two hours; extract the alkali. The dose is from eight to twelve grains.

You may use it in place of diagridium, and add trochisks of alhandal.

ADDITION

> ℞   Of Sal Gemmæ, Ɔss.
> Of Rebotium (*i.e.* true Mumia), } each Ɔiiij.
> Of Liquor of Centaury,   }

Make a compost. The dose is from four or five to ten or twelve grains in an egg.

PREPARATION IN OTHER DISEASES

> ℞   Of the said Sal Gemmæ, ʒj.
> Of the Juice of Cataputia, ʒij.
> Of Ground Flour, to the weight of all.

Make a roasted loaf. The dose is from ʒj to ʒij.

CONCERNING SAL PEREGRINORUM

The virtues of Sal Peregrinorum obtain in fortifying the digestion, also against infection of the air, and against future imposthumes.

PRESCRIPTION OF HERMES

> ℞   Of Burnt Salt Nitre, } each ʒj.
> Of Sal Gemmæ,   }
> Of Galanga,   }
> Of Mace,   } each Ɔj.
> Of Cubebæ,   }

Make a powder. The dose is three grains in the morning. It prevails against seasickness, and confers long life on old persons.

PREPARATION OF SAL PEREGRINORUM

℞   Of the said Salt, ʒiij.

Of dried Alcool of Wine, lb.ss.

Extract the alkali.

℞   Of the said Alkali, ʒj.

Of the Liquor of Juniper Seeds, one kist.

Make a compost.   The dose is one grain.

CONCERNING SAL NITRI

Sal Nitri obtains in pleurisy and open ulcers.

PREPARATION IN PLEURISY

℞   Of Sal Nitri, lb.ss.

Of Crude Tartar, lb.j.

Distil in *sextum alembicum*.   The dose is from Ɔj. to Ɔj.ss. in spring water or good wine, in the morning, at evening, and at midnight.   Administer often.   It purges through the urine.

PREPARATION IN OPEN ULCERS

℞   Of Alumen,

Of Nitre,   } each lb.ss.

Of Spring Water, lb.ij.

Distil into water.

ADDITION AGAINST PLEURISY

℞   Of the said Nitre, Ɔij.

Of Aqua Regis, Ɔss.

Of dried Alcool of Wine, ʒv.

Mix.   The dose is ʒss. or ʒi.ss.

ADDITION FOR OPEN WOUNDS

℞   Of Plantain Water,

Of Chelidony,   } each lb.j.ss.

Of Oak Leaves,

Use for ulcers of the legs.

TREATISE V

CONCERNING METALS

CONCERNING GOLD, SILVER, TIN, COPPER, IRON, LEAD, MERCURY

CONCERNING GOLD

The virtues of Gold obtain in Paralysis, Synthena, Fevers, Palpitation of the Heart, complaints of the Matrix, Ethica, Peri-pneumonia, and in acute diseases generally.

PREPARATION FOR PARALYSIS, PALPITATION, AND SYNTHENA

R̥   Of pure Gold, purged from its alloys, ℥ij.

         Of the water of Sal Gemmæ, ℥vj.

Reduce into one by separation with alcool of wine. Then—

R̥   Of the Crocus, ℥ij.

         Of corrected Alcool, ℥vi.

Mix. The dose is from three or four to six grains.

PREPARATION IN FEVERS AND ACUTE MALADIES

R̥   Of melted Leaves of Gold from the Water of Honey, ℥j.

        Of Alcool of Wine, ℥ij.

Reduce by separation from the honey. The dose is from Əss. to Əj.

PREPARATION FOR COMPLAINTS OF THE MATRIX, ETHICA, AND PERIPNEUMONIA.

R̥   Of Gold extinguished in Chelidony Water, ℥xiij.

         Of Indian Myrobalani, ⎫
         Of Chebuli,   ⎬ each Əj.

Reduce to digestion for a week by separating the superfluous aquosity. The dose is from Əj. to ℥j.

ADDITION FOR PARALYSIS, PALPITATION, AND SYNTHENA

R̥ Of the said prepared Gold, Əj.

    Of Lavender Water, with corrected Alcool of Wine, and Spicula, each ℥i.
Dose, Əj.

ADDITION IN FEVERS AND ACUTE STAGES

R̥   Of the said prepared Gold, Əiiij.

        Of the juice of Centaury, ⎫
        Of the juice of Sage,   ⎬ ℥ij.

Dose, from Əss. to Əi.ss.

ADDITION IN COMPLAINTS OF THE MATRIX, ETHICA, AND PERIPNEUMONIA

R̥   Of the Oil of Nutmeg, ℥ss.

        Of the Oil of Cloves, ℥j.

        Of the said prepared Gold, Əj.

Dose, from Əss. to Əj.

PROCESS FOR WATER OF SAL GEMMÆ

℞   Of Sal Gemmæ, lb.ss.

Of Rain Water, lb.j.

Distil by retort till the whole substance of the salt is perfected.

PURGATION OF GOLD

℞   Of Gold, ʒss.

Of Antimony, ʒij. or ʒiij.

Melt into a regulus. By this means the antimony takes up the impure part, and the gold remains at the bottom.

CONCERNING SILVER

The virtues of Silver obtain in complaints of the cerebrum, the spleen, the liver, and in the retention of the profluvium.

PREPARATION IN COMPLAINTS OF THE CEREBRUM, SPLEEN, AND LIVER

℞   Of Laminated Silver, ʒiij.

Of Sal Gemmæ, ʒvi.

Arrange in layers. Reduce to the fourth grade of reverberation for twenty-four hours, and extract the alkali. The said alkali is placed for three or four days in sublimated wine, when the silver becomes like the wine itself. Let stand. Evaporate The alkali sinks to the bottom, and, being received into a glass, liquefies in a cold place. The dose is from five or six to twelve grains.

PREPARATION IN PROFLUVIUM

℞   Of filings of Silver, ʒj.

Reduce into calx by means of Aqua Regis.

℞   Of the said Calx, ʒij.

Of crude Tartar, ʒiiij.

Reduce to the fourth grade of reverberation with extraction of the alkali.

PROCESS FOR AQUA REGIS

℞   Of Alumen, ⎫

Of Vitriol, ⎬ lb.ss.

Of Nitre, ⎭

Distil into sweet water.

METHOD OF THE EXTRACTION OF ALKALI

℞   Of the said Silver, as required.

Of Alcool of Wine, ⎫ each ʒx.

Of Water of Chelidony, ⎭

Reduce as above. The dose is from ℈j. to ℈j.ss. If the red flows forth with the profluvium it is a sign that it is going to be restrained. ʒj. of the said water should then be taken.

PREPARATION OF SILVER FOR ALL THE ABOVE COMPLAINTS

℞   Of Laminated Silver, ʒj.

Of Purged Sulphur, ʒiiij.

Of Pine Resin, ʒij.

Make a bolus, set alight, and reduce to preparation with spring water. The dose is from Əj. to Əj.ss. When thus prepared it is good in all the cases, but the first is more efficacious.

CONCERNING TIN

The virtues of Tin obtain in jaundice, asclitis, and worms.

PREPARATION FOR JAUNDICE

℞   Of Calcined Tin, lb.j.

Of Salt, ʒv.

Of Bean Ashes, lb.ss.

Reduce into Litharge by a fire of reverberation.

℞   Of the said Litharge, ʒx.

Of Alcool of Wine, lb.ss.

After resolution reduce into Alkali. The dose is from six to ten or twelve grains.

PREPARATION FOR ASCLITIS

℞   Of purged Tin, ʒj.

Of Antimony, ʒij.

Of Filings of Cinetus, to the weight of all.

Reduce into calx by reverberation for twenty-four hours. Then

℞   Of the said Calcined Matter, lb.j.

Of Alcool of Wine, lb.j.ss.

Reduce into alkali. The dose is from ʒj to ʒj.ss.

PREPARATION FOR WORMS

℞   Of Tin, ʒiij.

Of Common Salt, ʒiiij.

Of Asphalt, ʒj.

Make into a powder by burning. The dose is from ʒss. to ʒiij.

ADDITION IN JAUNDICE

℞   Of the said prepared Tin, Əiiij.

Of Alipta Muscata, Əj.

Of Bdellium, Əij.

The dose is from Əj. to Əij.ss.

ADDITION IN ASCLITIS

℞   Of the said prepared Tin, ʒss.

Of Dragon's Blood, ʒij.

Of Liquor of Tapsus, ʒj.

Mix. The dose is ʒss.

ADDITION FOR WORMS

℞   Of the said prepared Tin, ʒj.

Of Colocinth seed, ⎫
                     ⎬ each ʒvi.
Of Plantain seed,   ⎭

Make into a powder. The dose is from ʒj. to ʒj.ss.

CONCERNING COPPER

The virtues of Copper obtain in ulcers, wounds, worms, and ulcers of the mouth.

PREPARATION FOR ULCERS

℞   Of Copper, lb.j.

Of Immature Botrum, lb.v.

Of Vinegar, lb.j.

Of Sal Ammoniac, ℥ss.

Digest for a month in a closed vessel, afterwards reduce by ablution, and convert into a salt of alkali, *i.e.*, verdigris in ulcers (*sic*).

PREPARATION FOR WOUNDS

℞   Of Copper, lb.ss.

Of Distilled Turpentine, lb.j.

Of Common Salt, ʒj.

Of Vitriol, ℥ij.

Mix in a closed glass vessel for three months. If plates of Copper are taken and thus prepared, the best Balsam results. Afterwards take ℥j. of Flos Æris, and ℥j. of common oil.

PREPARATION FOR WORMS

℞   Of Calcined Venus, ℥j.

Of Water of St. John's Wort, and of Centaury, each ℥vj.

Of Plantain Water,   ⎫

Of Sour Wine,   ⎬ each ℥iiij.

Digest for seven or eight days. Reduce into Alkali. The dose is from ℈j. to ℈iiij. or v.

PREPARATIONS FOR ULCERS OF THE MOUTH

℞   Of Venus, laminated or cemented, ℥ij.

Of Burnt White and Rock Alum, each ʒvj.

Of Distilled Vinegar, lb.j.

Extract the Alkali for a day and night.

ADDITION FOR ULCERS

℞   Of the said Flos Æris, ℥j.

Of Aggripine Ointment, ℥j.

Of Earth Worms, ʒiij.

Make an unguent.

ANOTHER ADDITION IN COMMON ULCERS

℞   Of the said Flos Æris, ℥v.

Of Alum Water, ℥xv.

Make a mixture after the manner of a lotion.

ADDITION FOR WOUNDS

℞   Of the said prepared Flos Æris, ℥j.

Of Oil of Anise, ℥iij.

Of Oil of the Yolks of Eggs, ℥ (imperfect quantity).

Compose an oil.

ANOTHER

℞   Of the said prepared Flos Æris, ʒss.

Of the Hepatic Aloe, ʒj.

Of Liquor Consolida, ʒiiij.

Make a Gum.

ADDITION FOR WORMS

℞   Of the said Flos Æris, ʒj.

Of Zuccarum Taberzet, ⎫

Of Liquorice Juice, ⎰ each ʒij.ss.

Reduce to a powder. The dose is from ℈ss. to ℈j.

ADDITION FOR ULCERS OF THE MOUTH

℞   Of the said Prepared Flos, ʒj.

Of Chelidony Water, ʒiij.

Of Alum Water, ʒj.

Make a gargle or mouth wash.

CONCERNING IRON

Iron has styptic, constrictive, and drying qualities.

PREPARATION FOR STYPTIC QUALITY

℞   Of Iron Filings, lb.j.

Of Common Salt, lb.v.

Of Spring Water, sufficient for incorporation.

Reduce for the space of a month, and afterwards reverberate into a powder. Incorporation is treatment with water till a pulp is formed.

CONSTRICTIVE PREPARATION

℞   Of Iron Filings, lb.ss.

Of Alum Water, lb.j.ss.

Of Distilled Vinegar, lb.ss.

Digest for a month. Reduce by ablution and afterwards by reverberation to a crocus.

DRYING PREPARATION

℞   Of Iron Filings, lb.ij.

Of Water of Vitriol, lb.ss.

Digest for a month and reverberate into a powder.

Styptic virtue is closing and drying to fistulas and cancers. Constrictive obtains in lienteria, dysentery, and diarrhœa. Exsiccative is for phlegmatic complaints.

ADDITION FOR A STYPTIC

℞   Of the said Crocus, ʒj.

Of Burnt Bolus, ʒiij.

Of Sealed Earth, ʒv.

Reduce to a powder. It is an incarnative for ulcers and wounds. If taken internally the dose is ʒj.

ADDITION FOR A CONSTRICTIVE

℞   Of the said Crocus of Mars, ʒj.

Of Myrrh, ʒss.

Of Oriental Crocus, ϶j.

Reduce to powder. The dose is from ϶ij. or iij. to iiij.

ADDITION FOR AN EXSICCATIVE

℞   Of the said Crocus, ʒiij.

Of Pomegranates, ʒj.

Of the Sap of Acacias, to the weight of all.

Make an electuary.

CONCERNING SATURN

Saturn has an incarnative virtue.

PREPARATION

℞   Of Lead Ashes.

Decoct with vinegar for three or four hours. This is the first preparation; it cures wounds, and grows solid flesh. Ceruse is also made from lead if washed with water in the sun. Minium is decocted from cerussa in a kettle. All medicaments for wounds and ulcers should be prepared from metals.

CONCERNING MERCURY

Mercury has an incarnative and laxative virtue.

PREPARATION AS AN INCARNATIVE

℞   Of Prepared Mercury, powdered, ʒij.

Of Aqua Regalis, ʒx.

Reduce by distillation in a bath several times a day, and convert into an oil. It is a most speedy incarnative for wounds and ulcers. It has two objections: it salivates and produces cerussa. Hence it is generally rejected and disliked. Otherwise it consolidates well and quickly.

PREPARATION AS A LAXATIVE

℞   Of Mercury coagulated by the Albumen of Eggs, ʒj

Of Alum Water, ʒvi.

Distil through ashes and make into a powder. The dose is from three to four or five grains. It is a potent purgative for diseases which originate from leprous humidity, such as paralysis, pustules, the varieties of gutta, and humid dropsy.

HERE ENDS THE BOOK CONCERNING PREPARATIONS IN ALCHEMICAL MEDICINE

THE ALCHEMICAL PROCESS AND PREPARATION OF THE SPIRIT OF VITRIOL

By which the Four Diseases are cured, namely, Epilepsy, Dropsy, Small Pox,* and Gout

To abolish those errors which are usually committed by Philosophers, Artists, and Physicians

———

THE spirit is extracted out of vitriol by means of colcothar, which is useless and of no efficacy. That which they call a phlegm is the most noble of all spirits, and all virtues should be ascribed to it.

But although the oil of colcothar is indeed of great efficacy in gravel and stone, as also in alopecia, yet it is of no use for the aforesaid diseases, to which, however, it is commonly applied.

Hippocrates, with whom almost all others agree, hands down certain stages and symptoms, which supervening epilepsy and gout must be reckoned as incurable. But since they had no acquaintance whatever with the spirit of vitriol, let their opinions go to the winds.

In the first place, the extraction of the spirit from vitriol must be effected by means of a powerful fire in an upright cucurbit, so that it may be driven into a fresh alembic, and, remaining in the athanor four days and nights, may be most skilfully passed through the reverberatory. And thus have you prepared this spirit of vitriol.

Afterwards colcothar must be distilled through a phial placed in the athanor for the space of three days over a very fierce fire of wood and coal, until there shall appear in the receiver, out of 1lb. of colcothar, ʒvj., which is tinctured with a scarlet colour.

This having been accomplished, the alkali is to be extracted from the *caput mortuum*, and the same having been resolved four or five times, is to be then coagulated. Thus the three things which exist in vitriol shall be extracted and separated.

PROCESS

Proceed as follows in an epileptic disease. After each paroxysm let ℈j. be given to the sufferer. Let a dose of four grains of the oil of colcothar be

---

* The term *pustula* also signifies St. Anthony's Fire.

administered morning and evening in peony water. This order is to be followed up to the fifteenth paroxysm. If the paroxysms become less frequent, half the dose is to be taken for the next thirty days.

In gout a daily dose must be taken for a period of thirty days. Afterwards the afflicted part must be anointed with the spirit of vitriol till the pain is removed. If it be gout of long standing, add a fourth part of the liquor of mumia* to the said spirit.

In dropsy half a scruple of the spirit of vitriol must be administered in liquor of Serapinus, the dose being repeated three or four times, according to the stage of the disease. In the absence of the liquor of Serapinus, the liquor of crude tar must be substituted.

In the case of small pox observe this order and method. The whole seat of the disease and the part of the skin affected is to be anointed with the spirit of vitriol for nine days. But if the skin be ulcerated, let the oil of colcothar be applied in combination with its alkali, according to chirurgic method. The bandages are to be loosened after the sixth day.

The regimen and diet must be adapted to the condition of the patient, for the whole circle of disease lies in medicine and not in diet. The medicine, therefore, must be administered sedulously. So are the four diseases here dealt with fundamentally and radically cured.

AN ADDENDUM ON VITRIOL

Alchemy has produced many excellent arts for physicians, whereby admirable cures of various diseases are effected. For this reason, therefore, in the commencement of medicine, alchemical doctors did always labour that it might become the mother and parent of many advantages. These two faculties were long cultivated together as companions, until there arose the triflers and sophistic humourists, who mingled poison with medicine, and rendered it meretricious, which medicine will still continue to remain so long as the humourists survive. I preface this that you may pay more diligent attention to the point, by reason of its great medical utility. But this is to be passed over. Wherever unskilful men rush into any art there they corrupt and defile everything, and out of a pearl make a fetid marsh. A like thing happens with regard to vitriol. At first the spirit of vitriol is taken, and is

---

* A PROCESS FOR MUMIA, OPPOSED TO THE ERRORS OF THOSE WHO ADMINISTER IT FOR POISONS. - Many have laboured in experiments, compositions, and recipes, whereby they might draw forth the poison into the universe, yet have they accomplished nothing. For among all things, both in experiments and recipes, it is only Mumia which brings an immediate remedy against all kinds of poisons. The method of dealing with Mumia is as follows. In the first place, cause the Mumia to putrefy in olive oil, and that for four weeks. Then separate in a retort. To each pound which has proceeded from the separation add one drachm of Alexandrine musk, and of Alexandrine theriac six ounces. Lastly, dissolve the mixture in the Bath of Mary for a whole month. You will then have theriac of Mumia. As regards its administration to the sick, give at first one ounce in oil of almonds, by way of a drink. Next see that the poisoned individual takes to a bed wherein he can sweat well, and the medicine may take effect. By this method any animal or mineral poison is expelled. Moreover, such is the virtue of Mumia, that if it have been administered before the reception of the poison, the latter will work no harm. A single dose (ℨ) taken in the morning will obviate the possibility of poisoning during the whole day. In cases of poisoned ulcers, plague, carbuncle, anthrax, and pleurisy, give ℨ j., and repeat the dose at the end of six hours, when, if the sufferer has survived to take the second quantity, his recovery is assured. Lastly, there are innumerable other diseases which by this theriac are completely and perfectly cured.—*De Mumia Libellus.*

usually raised to the highest grade. By this exalted they cure epilepsy, whether it be of recent or long standing, in men and women of whatsoever condition. But here the unskilled workmen rushing in, and about to enter on a better way, have attempted to apply the virtues of vitriol to another purpose, and thus departing from the first method and arcanum, they have suffered it to expire, and then have sought oil in colcothar, which can in no wise be usefully done. For whatsoever is to take away epilepsy must have a subtle, sharp, and penetrating spirit.

For therein consists the faculty of pervading the whole body, and passing over nothing. By such prevading or penetration the disease is attacked on its own ground, for it is certain and beyond all doubt that its seat, centre, or sphere cannot be known ; hence it is inferred by the physician that there is need of those remedies which penetrate the whole body. And this is the reason why the mercenary humourists cure none, but prostitute all their learning and profession. I therefore freely affirm that in that oil which these workmen have sought there is no penetrating spirit whatever. They supply, so to speak, a mere earthiness which does not penetrate far, but where it falls there it remains. It is therefore to be regretted that, owing to their ignorance, the true process is prejudiced, and a false one is substituted. For I am persuaded that the devil has devised these things in order that health may not be restored to the sick, and that the sect of the humourists may shortly come into still greater power. To return, however, to the beginning, and to the manner in which the spirit of vitriol was invented. They first distilled the humid spirit of vitriol by itself from colcothar, then they intensified its grade by distilling it and circulating it to the highest point, as the process teaches. In this manner, the water comes to be used for various external and internal diseases, as also for the falling sickness. Thus a marvellous cure is obtained. But in the extraction they were much more diligent, for they took the spirit of vitriol, corrected as above, and distilled it from colcothar eight or ten times over a very strong fire. Thus the dry spirits were completely mingled with the humid. They continued their work until the dry spirits departed, by reason of the uninterrupted and vehement extraction. Afterwards they graded each spirit, both the humid and the dry, received in a phial together, to their terminus. They regarded this medicine as of great efficacy against diseases, and were so successful therewith that they completely confounded the humourists. But there is added unto it a certain correction by the artists by means of sublimated wine, and therefore of greater penetrating power, but it has not attained a higher grade.

But I will communicate to you my process, which I recommend to all physicians, especially for epilepsy, which has its cure in vitriol alone. Wherefore charity towards our neighbour demands that we should take greater care in case of this disease. My process is that the spirit of wine be imbibed by vitriol, and afterwards distilled, as I have said, from dry and humid spirits.

This done, I discovered that the following addition was very useful. Let the spirit of corrected tartar be mingled with the third part of vitriol, and let there be added the spirit of the theriacal water of lavender in the proportion of one-fifth in respect of the vitriolated spirits. Then let it be administered to the sick person before the paroxysm, or several times in the day. This medicine possesses a signal efficacy against the said disease, so that it is not lawful to expect a better one from Nature. Accordingly, the first process was invented and retained by the ancients with the said correction. For thus the heart and the whole virtue of Nature is attained.

But I hope that I shall not be reproved by all good persons who think of the terrible nature of this disease, which ought to move the stones to pity, for whether the vehemence and atrocity of that disease be so great or not, it would be permissible for any one to say, Cursed be all the physicians who, passing by the sick, give them no aid, like the priests and Levites in Jericho, who, deserting the wounded man, left him to be treated by the Samaritan! For they were worthy of the fire of Gehenna, from which there is no redemption, and who will admit anything else than that all these physicians, without exception, look at the disease, and yet pass by the sufferer? Who can say anything else but that they will be judged at the last day? For scarcely would they spend one penny to secure a more certain foundation for the cure of this disease. Did they strive to imitate the Samaritan, God would not then judge them, but, in consideration of their faithfulness, would manifest to them all the secrets of Nature, whereby they might assist the sick, and, although the required properties did not exist in Nature, He would create them afresh. Wherefore I testify to you, men both of high and low degree, that all the doctors have shamefully strayed; whatsoever the seducers Galen or Avicenna have concocted, they adhere to it, and weary themselves with lies. To such an extent are they obsessed by the devil that they cannot exercise charity towards their neighbours, and therefore make of themselves children of condemnation. First of all the Kingdom of God is to be sought, yet not with the Levite or the priest, but with the Samaritan.

If we are merciful, and follow the example of the Samaritan, God is with us, and He will immediately confer upon Nature a remedy not hitherto created. While men have levitical or sacerdotal dealings with the sick, God puts off the medicine, and keeps it to Himself. The sick flee to the Kingdom of God, but the physicians to the abyss of hell. The same place is prepared for both doctors and Levites. Therefore, open your eyes, there are two paths —one taken by the Levites, the other, which leads to heaven, along which the Samaritan proceeds.

That vitriolated extraction is not only excellent for falling sickness, but in the same way for cognate diseases, as swoon and trance, also for constipation and internal imposthumes, etc., and for strangulation and precipitation of the matrix. But far other than the aforesaid virtues would be discovered by diligent inquiries. The devil, whom the false physicians serve, however, obsesses them, and he incites them, so that they cannot endure a lover of the truth. So there is an end to the health of the good.

Further, it is to be known that the aforesaid recipes for making a humid spirit of vitriol cannot be more clearly described. An artist is required to understand it, but sordid cooks can by no means grasp a matter of so much moment. It is from artists, therefore, from alchemists, and from experimentalists, that you are to expect sufficient information on all points. Similarly, we shall, by the same, be more fully instructed in the correction of the spirits of the wine. The doctors of the academies are so ignorant that they can scarcely distinguish between agaric and manna. The art and virtue of all vitriol consists in this, that the spirit of the vitriol should be properly extracted, raised to the highest grade, and by addition should be made potent to enter the penetralia where the centre, root, and seed of the disease can be found. For it is impossible to discover these places so exactly as those doctors assume when discoursing of their humours. The fundamental principle has not yet been discovered as to what makes the disease, or where it is situated, or what it is that throws a man into such a severe paroxysm. Therefore the whole operation must be committed to the arcanum alone which Nature has appointed for the disease. That arcanum will find out the disease just as the sun penetrates all the corners of this world. In short, whoever desires to act as a true physician should first of all study to be a Samaritan, not a priest or Levite. If he be a Samaritan all things of which there is need will be given him, nor will anything be concealed from him.

OF THE OIL OF RED VITRIOL

You must know how a most blood-red and vinegar-like oil is prepared from colcothar by distillation in a retort after the alchemistical fashion. This oil the operators have regarded as more efficacious in the aforesaid diseases than the spirit itself, but erroneously. The process of preparation is well known, and need not be here described. The most important part consists in the manual work, in diligent inspection, and, finally, in suitable instruments. Also you should know concerning its virtues, that, in the first place, it is an acid matter, surpassing all acidity, so that there is nothing more acid. Next, it possesses a corrosive nature, whence it follows that it must be used with caution, as also not by itself, but diluted in such a manner as will be in harmony with the nature of the case. By reason of its acidity it is beneficial to a stomach which is free from cholera and ulcers, but not otherwise, for acidity is aggravating to ulcers. If cholera be present, a continual conflict ensues, even as between aquafortis and tartar. It will also conduce to health in the case of all fevers and loss of appetite, the same method being observed, for many virtues are ascribed to this oil, though few are confirmed by experience. Those, indeed, who have boasted of effecting marvellous cures thereby, have been proved by use to have lied disgracefully. It is useful in stone and gravel, though I know no case which has been cured by it.

In all instances it effects something, but it does not get to the root of any disease.

But with regard to surgery, understand concerning this oil that it creates severe pains, but, nevertheless, brings immediate health even in the most desperate diseases. In certain complaints it is better to distil the red oil of vitriol into a spirit, and thus a minimum quantity becomes sufficient to effect a cure. These things have been taught me by experience ; the other decoctions of vitriol are of no importance.

The White and Green Oil of Vitriol

Out of raw vitriol there can be distilled, by descent, an oil, sometimes white and sometimes green, according to the conditions of the vitriol. The same calls for special praise. Because it is prepared out of crude vitriol, it therefore also contains the spirit of the same. It is excellent and advisable for internal diseases. The green oil is better if it be circulated and mingled with a commixture drawn from the spirit of vitriol. Let no one who possesses this despair that he has a certain and undoubted remedy for the falling sickness, and all its varieties. It should be brought to the highest grade, that it may be separated from its earthiness and its fæces by the bath of Mary, and afterwards by fire. Thus, in the bath of Mary the phlegm will be removed, the earthiness will be removed by fire, and the spirit of oil must be then collected alone that it may circulate by itself. Afterwards you may take an addition of the spirit of wine, nor is there need that many things be added. It is taken in peony water before the commencement of the paroxysm. When the spirit of the oil has searched out the centre of the disease, the paroxysm abates, creating a certain vertigo at first, but soon after the patient subsides into gentle sleep, and experiences relief. Yet it is necessary all the same to persevere with the administration of the medicine.

THE ALCHEMIST OF NATURE *

BEING THE SPAGYRIC DOCTRINE CONCERNING THE ENTITY OF POISON

CHAPTER I

CONCERNING the nature and the essence of poison by which our bodies are affected, we would thus establish the foundation and the truth. It is agreed among all parties that our bodies stand in need of conservation, that is, a certain vehicle, by the aid of which they flourish and are nourished. Wheresoever this is wanting, there life itself departs. But this, however, is equally to be borne in mind, namely, that He who built up or created our bodies, the same, in like manner, procreated the foods thereof, and that with the same facility, though not indeed in an equal perfection. I wish the matter to be understood thus : We are endowed with a body which is devoid of poison. But that which we administer by way of nourishment to our body has poison combined therewith. Thus our body is created perfect, but not also the other. Hence, observe that the other animals and fruits are for us designed as food, and so, also, as poison. They are not in themselves either foods or poisons, but, as regards themselves, and inasmuch as they are creatures, they share their perfection equally with us. When they are taken by us as food they are thus poison to us. Thus a thing becomes poison to us which in itself is by no means a poison.

CHAPTER II

To consider the matter further, everything in itself is perfect, is made good in relation to itself, and according to its own law. But if we have regard to its external uses, it has been formed both good and bad. Understand this as follows : The ox which feeds on grass receives both health and poison, for the grass contains in itself both nourishment and medicament. In the grain itself there is no poison. Thus, also, whatever man eats or drinks is at the

---

* Perhaps the chief utility of this treatise will be the illustration which it affords of the extremely wide sense in which the terms Alchemist and Alchemy were applied by Paracelsus. The little work itself is derived from that portion of the *Paramirum* which is entitled *Textus Parenthesis super Entia Quinque*, and, in addition to the entity of poison, is concerned with the astral entity, the entity of seed, the entity of virtue and quality, the natural entity, and the spiritual entity. Finally, there is the entity of God. The whole constitutes a kind of general introduction to the body of exoteric medicine for which we are indebted to Paracelsus. It is to the several treatises dealing with these subjects that reference is intended in the fifth chapter.

same time both venomous and healthy.   Take this statement in two ways, one concerning man himself, excluding the nature of animals and plants, the other concerning the assumption.   To impress this more plainly on the mind—the one in man is the great nature, the other is the poison inserted into the nature ; and, in order that we may conclude this matter, remember that God has formed all things perfect, in so far as regards their utility to themselves, but imperfect to others.   Herein rests the foundation of the entity of poison.

God, indeed, has appointed for man or for creatures no Alchemist for His own sake, but He has destined the Alchemist as one to whom we may betake ourselves if any of those things whereof we have need be imperfect.   And He has done so for this purpose, that we may eat the poison which we take under the appearance and in place of healthful food, not as poison, but that we may separate and divide it from such healthful food.

What we tell you about this Alchemist do you regard with the most careful attention.

CHAPTER III

When therefore, anything, which is, in other respects, perfect, assumes at one time the form of poison and at another that of healthful sustenance, we say, proceeding with our subject, that God has appointed an Alchemist for him who eats and uses anything which, when taken, tends as much to destruction as to health ; and this Alchemist is such an artist that he can only separate these two elements one from the other by banishing the poison to its own place, and by introducing the food into the body.   In this way, as we have said above, we would have our fundamental principle understood and accepted by you. Take an illustration of a different kind.   Whoever is a lord or prince is, so far as he himself goes, perfect, as a prince should be.   But a prince cannot be without servants who shall minister to him in his princely character.   Those servants too, are, so far as concerns themselves, perfect, but not as they stand related to the prince.   To him they are as a poison, as a loss; they receive pay from him.   So understand the natural Alchemist.   God has granted that science shall exist bountifully in him, as in a prince.   He teaches him how to separate the poison from his ministrants and to accept the good among them.   Here you will find the fundamental principle of our present subject, even though the illustration may not recommend itself to you at first.   Its teaching is according to the doctrine of the wise man, where the whole is unfolded. The matter stands thus.   Man must eat and drink.   For the body of man, which is the temporary abode of his life, absolutely requires food and drink.   Man, therefore, is compelled to take into him poison, diseases, and death itself, by means of his food and drink.   So then, this argument might be used against Him who endowed us with a body and then added food in order to slay us.   Learn, however, that the Creator takes nothing away from the creature, but leaves to each his own perfection.   And, although one thing is poison to one and another thing to another, the Creator is not to be accused or blamed for this.

CHAPTER IV

But in this way you will track out the Creator. If all things are perfect in themselves, and this from the settled arrangement of the Creator, according to which one serves for the conservation of the other, as when the grass nourishes the cattle and the cattle nourish man; and if thus the perfection of one thing be to another which partakes of it now an evil, now a benefit, and thus imperfect; then we must assume that the Creator, for the sake of enlarging His creation, arranged matters so for this reason: He determined all things should be so created that in whatever is necessary for some other thing such virtue and efficacy should be latent, that by means thereof the poison should be separated from the good for the health of the body and supply of nutriment, and that this mutual arrangement should be preserved. For example: the peacock devours the snake, the lizard, and the newt. These creatures, so far as they themselves are concerned, are perfect and wholesome, but with reference to other animals, all those mentioned are mere poison, except in the case of the peacock. And hear the reason of this difference. Its Alchemist is so subtle that the Alchemist of no other animal can come up to him. He so thoroughly and purely separates the poison from the good that this diet is innoxious for the peacock. So is it true in other respects also that to every animal is assigned that particular food which is adapted for its preservation, and besides this a special Alchemist is given who separates the nutriment. To the ostrich such an Alchemist is given who separates the iron, that is, the dung, from the nutriment, and it is not possible that this should be done for any other creature. To the salamander is given fire for food, or rather a body of fire. For this purpose it has an Alchemist appointed. The swine feeds on dung, although it is poison; and for that reason it is extruded from the human body by the Alchemist of Nature. Nevertheless, it serves as aliment for the swine, since the Alchemist of swine is much more subtle than the Alchemist of man; for the Alchemist of the swine separates that aliment from dung which the Alchemist of man cannot so segregate. On this account, too, the dung of swine is not eaten by any other animal. There is no other Alchemist more subtle, or who can separate aliment more cleverly, than the Alchemist of the swine. And so of others, which we advisedly omit, to make our discourse the shorter, it must be understood in like manner.

CHAPTER V

We have already said something about the Alchemist; and you must believe him to be appointed by God solely for this reason, that he may separate from the good that which differs from it in the body when, by the divine arrangement, it takes something into itself for the support of life. Recur now to the data elsewhere supplied, namely, that there are five things which have power over man, and whereto man is subject. These are the Astral Entity, which we have dealt with, and next the Entity of Poison. Now even though a man may be in no way affected by the stars, still he is not equally safe and secure from the Entity of Poison, but there is reason to fear lest he may go wrong thereby. These we leave as we described them in the prefaces.* But in order that you may more easily embrace all, observe the initial principle, so that you may more clearly understand in what way the poison can or actually does hurt you. Since, then, we have within us an Alchemist placed by God the Creator in our body for this purpose, that he may separate the poison from the good, and that so we may suffer no harm, it is necessary that we should next discuss about this, namely, what is the principle and what the mode according to which all diseases issue from the Entity of Poison as well as from the other sources. In this disquisition we will pass over all that which brings no injury but some advantage to bodies, as we shall in due course make clear.

CHAPTER VI

But first you must know that in this matter astronomers are deceived, for while pointing out the maladies of our body, they make the body fortunate and the body healthy. But if this be not the case, this one cause remains, that the remaining entities, of which there are still four, do injury to the body, and by no means the stars of themselves. Hence we rightly ridicule and explode their writings, wherein they make such lavish promises of health, and do not at the same time consider this, that there exist four other entities, of equal power with the stars. But we ought to have a game with them, for what is the good of a cat without a mouse, and what is the good of a prince without a fool? Clearly, the physiomantist has also concocted a similar history, yet he will never excite our tears. He promises health, nor does he think of the four entities of which he is ignorant. For he augurs from the sole natural entity, and says nothing about all the rest, which tickles us not a little. It is the part of a well-informed man to declare of many things those which depend upon a course, for of motions or courses there are five, of which man is only one. He who omits some of these motions and proceeds with the rest is truly a vain prophet. To divide and to speak according to division, each according as he has learnt, according to his judgment and opinion, this is extremely praiseworthy. Thus, accordingly, the pyromantic Entista delivers his judgment concerning spirits ; similarly, the physiognomical Entista prophesies of the nature of man ; the theological Entista of the course of God ; the astronomical Entista of the stars. Each by himself is a liar, but they are true and just if they unite in one. We tell you this lest you should proceed to prophesy before you have learned the five entities of the entities. Then indeed we shall repress our laughter.

---

* The *Libellus Prologorum* prefixed to the *Textus Paramiri* observes that there are altogether five modes of cure, which is as much as to say that there are five medicines, or five arts, or five faculties, or five physicians. Each of these faculties, taken separately, is sufficient for the cure of all diseases. The professors of the first are termed natural physicians, for this reason, that they effect cures by the administration of contraries. Those of the second are termed specific physicians ; they cure all diseases by means of the *ens specificum*. The professors of the third are called magical physicians ; they effect the cure of diseases by the use of magical words and characters. The spiritual physicians are those who understand the nature of the spirits of roots and herbs, and have them under their control. The last are the physicians who heal by faith.

CHAPTER VII.

In order that you may have a fundamental knowledge of the Alchemist, know now that God has dispensed to each creature His own substance and all things which are necessary for this, not for His own regulation, but for the use of those who need those things which are conjoined with poison. That creature has within him, in his own body, one who separates the poison from whatever is applied to the body. This is in very truth an Alchemist, because in his mode of action he makes use of chemical art. He separates the evil from the good. He transmutes the good into tincture. He tinges the body for the sake of its life. He arranges and disposes all that is subject to Nature in it, and tinges it so that it turns into blood and flesh. This Alchemist dwells in the bowel, as in his instrument, with which he decocts and where he operates. Understand the matter thus. Whatever flesh man eats has in it both poison and good. In the act of eating, all things are regarded as healthful and good. Under the good, indeed, poison is latent; but under the bad there is nothing good. As soon as ever food, that is to say, flesh, is taken into the stomach, the Alchemist, immediately fastening upon it, effects a separation. Whatever does not tend to the health of the body he puts aside in its special places; whatever he finds good he also sequestrates into its proper abodes. This is the Divine ordinance. In this way the body is preserved so that it shall not be killed by the poison of what it eats. Now, this separation is made by means of the Alchemist without anything being done on the part of man. And thus it is with the virtue and power of the Alchemist in man.

CHAPTER VIII

Understand, moreover, after the following manner, how, in every single thing which man takes for his use there is a poison hidden under what is good. The essence is that which sustains the man. The poison, on the other hand, is that which destroys him and brings diseases upon him. And this is true of every alimentary substance, without exception, in respect of that animal which uses it. Now, physicians, attend to this! If it be thus with the aliment of the body and the body cannot do without it, but is altogether dependent upon it, then the body simply takes the aliment, such as it is, under the twofold species of good and ill, and delegates to the Alchemist the duty of separation. Now, if the Alchemist be weak, so that with all his care he avails not to separate the poison from that part which is evil, then, from the poison and the good there arises a combined putrefaction and, eventually, a kind of digestion, and this it is which inflicts diseases of humanity upon us. For every disease in man begotten of the Entity of Poison emanates from a putrefied digestion, which ought to take place so gently that the Alchemist should perceive in it no measure of excess. But when digestion is interrupted, then the Alchemist cannot remain perfect in his instrument. So, then, corruption necessarily ensues, and this, in its turn, is the mother of all diseases. This is what physicians ought most carefully to watch, and it should not be involved in any of your intricacies. Corruption defiles the body, and it is produced in this way. Water which is clear and limpid can at pleasure be tinged with any colour. The body is like such water : corruption is the colouring matter ; and there is no such colour which does not derive its origin from decay. It is at once the signal and proof of poison.

CHAPTER IX

Learn this with the view of more fully following up the subject, that corruption is produced in two ways, locally and emunctorially, according to the following method : If, as we have said, there be an Alchemist present in digestion, and if in the process of separation, he succumbs to the fault of defective digestion, then in place of him there is generated putridity, which is poison. Everything putrid is poisonous for that place where it is detained, and so becomes the mother of certain and deadly poison. For putridity corrupts that which is good, and if the good be stripped of its virtue, then the evil triumphs over the good, and this good no longer appears otherwise than under the false appearance of that good which is really subject to putridity. And so this becomes the source of diseases which, in their turn, are subject to it. But know that which is produced emunctorially occurs through the failure of the expulsive nature, in the following manner: If the Alchemist expels the poison, he does so in every case by the proper emunctories ; the white sulphur by the nostrils, arsenic by the ears, dung through the anus, and so other poisons according as each has its own special emunctory. Now, if one of these poisons he hindered by the weakness of Nature, or by itself and in other ways, it then becomes the mother of the diseases which are subject to it. So, universally in all diseases two sources are patent. We will not say more on this subject.

CHAPTER X

Moreover, as has been said above on the subject of natural Alchemy, that it is situated in every animal, on account of that separation which must take place in the bowel, listen to the following doctrine as to how, in the aforesaid manner, all other diseases also can be investigated and searched out: If the man be well and strong in respect of all entities—if, for example, he has a suitable Alchemist who separates well with appropriate instruments, reservoirs, and emunctories, then it is necessary, in addition to the instruments, to regard many other matters, especially to see that the stars are favourable, and that all the other entities are well disposed. And yet these general entities do not greatly affect us ; for assuming that they are all good and effectual, still many accidents happen to the body which either break or spoil, befoul or impede, the reservoirs or emunctories. Fire is contrary to the Nature and to the body. For this in its quality, nature, ardour, dryness, and other force, can so corrupt us that by its presence the instruments of the Alchemist are violated, and so he appears to be of feeble powers afterwards. So, too, water itself is adverse in its nature to our body and its reservoirs, so much so that the instruments are either stopped up or perverted by it, or in some other respect altered. The same is the case with the air and with other necessary things, and also with external accidents, which are of universal power, so that they break asunder, change, and render useless the instruments and emunctories. Then, too, the Alchemist, being weak and dead, proves unequal to the accomplishment of his work.

CHAPTER XI

But it should not escape your notice that the reservoirs, instruments, and emunctories can be corrupted through the mouth by the air, food, drink, and other things of this kind, in the following way : The air which we breathe is not without its poison, and to this we are specially subject. Concerning the quantity of food and drink, and its bad quality, which disagrees with the organism of the body, the truth is that by these means the organs are thrown out of order to an excessive degree, and that, in this way, the Alchemist is clearly disturbed in his operations. Hence ensue digestion, putrefaction, and corruption. And whatever be the properties of the poison which man takes, such a nature the bowel assumes, and, together with the bowel, all the rest of the body. Thereupon this becomes the mother of diseases in that self-same body. Hence, you physicians ought to understand that one poison and not more produces the mother of diseases. Thus, if you eat flesh, herbs, pulse, spices, and from the consumption of these poison is generated in the belly, then it is not all these foods which are the cause, but only one of them, as, for instance, the poison of the herb, or of the pulse, or of the spices. This you should consider a great secret. For if you know this thoroughly, what poison is the mother of diseases, then we will allow you with justice to be called physicians. In this way you will have discovered what remedy you ought to use ; otherwise, you will attempt this in vain. This becomes for you the foundation of the mother of all diseases, whereof indefinite numbers are reckoned up.

CHAPTER XII

We will now communicate to you some brief information on the subject of poisons that you may understand to what and to what kind of poisons we allude. We have pointed out to you that poison exists in all foods. From food, therefore, is educed a certain entity which has power over our bodies. We afterwards explained the Alchemist who is in our bodies, who for the well-being of the body separates the poison from the food by means of his instruments and reservoirs. When this has been done, the essence passes off into a tincture of the body. The poison goes from the body through the emunctories. Whilst each operation proceeds in this way man is sound and healthy by means of the entity itself. At the same time, also, we mentioned the hostile accidents which might impinge upon the entity itself so as to destroy it. In this way it becomes the mother of diseases.   Having repeated so much, now let us speak about the different kinds of poisons.   I think you now understand what the emunctories are, and how many of them there are.   Reasoning from these you can get at a knowledge of poisons.   Whatever exudes substantially through the pores of the skin is resolved Mercury.   What is excreted through the nostrils is white sulphur.   Arsenic is ejected through the ears ; sulphur through the eyes.   Through the bladder there is a resolution of salt ; through the anus putrefied sulphur.   Possibly your reason may seek to know under what form and appearance each of these can be recognised ; but that is a matter which our present parenthetical treatise does not include.   You will gather from our book on the Human Construction of Philosophy the fundamental principles which it is necessary for a physician to know.   There, too, are given at some length the appropriate remedies in many cases, such, for example, as in putrefactions, and it will be well for you to read these.   In the same treatise, too, you will learn in what way poison is latent in food.*

CHAPTER XIII

We will give you an illustration from which you may briefly learn how poison lurks under aliment ; and how it is that the condition of a substance which is perfect in itself becomes vicious and poisonous in respect of men and animals who use it.   The ox is created with all appendages sufficient for its own use, its skin adapted to the accidents of its flesh, its emunctories ready to use for the Alchemist.   But this illustration does not seem altogether to square with our purpose.   We will give another.   The ox is created, with a view to its own requirements, in that form wherein we find it ; not for man by way of nutriment in food.   Mark, then ; half of that ox is a poison to man. If the ox had been created for man's sake alone, it would not have needed horns, bones, or hoofs.   There is no nutriment in these ; and what is made from them does not come under the category of necessities.   You see, therefore, that in respect of itself, the ox is created altogether good, nor is there in itself anything which it could do without or which it has in excess.   But now, if that ox is taken for human food then man eats at the same time that which is hurtful to himself, which is, in fact, poison, but which to that ox had not been poison.   This, then, must be separated from the nature of man, and the work of separation is undertaken by the Alchemist, in cases where several poisons are generated without any provision for carrying them off.   By the operation of the Alchemist each poison is draughted off to its emunctories, and these are filled with such poisons.   But every Alchemist among men can perform the same office as the Alchemist in the body ; no art is lacking to him.   This should be an example to every one of them that as the Alchemist of Nature works so he himself should endeavour to

---

* This treatise cannot be identified from its title among the extant writings of Paracelsus.   The matters referred to are the subject of frequent instruction throughout the medical books,

work. And if poisons are so separated that poison no longer appears, think how even from a caul is produced the most beautiful golden oil, which, nevertheless, is of all oils the most detestable. The mucus of the nostrils, too, is not reckoned a poison, but it is, nevertheless, a most accursed poison, from which arise all the diseases of distillations, and it can be easily recognised from these diseases.

PARTICLE I

We seem to have sufficiently explained the entity of poison, namely, that it proceeds solely from that which we use as food and drink. Hence, further note that digestion is the same as corruption, if it is corrupted. Next, further, know, how every poison is generated in its place; and how in the process of time either diseases or deaths are occasioned from that poison.

PARTICLE II

Notwithstanding, in dealing with the entity in question, we shall not explain to you the manner in which every disease originates from the above-mentioned poisons of food, which poisons are removed to their emunctories. However, to avoid mistakes, pass it over in this parenthesis, and seek it in the Books of the Origin of Diseases. We shall there clearly explain it to you according to this fundamental principle. So ye shall at once understand what are the diseases of arsenic, salt, sulphur, and mercury, according to the distribution of every form and species, just as it is fitted to itself and to the generation of diseases. Thus we desire to conclude these matters with this entity. We wish them to be understood as an introduction to all our other books.

HERE ENDS THE ALCHEMIST OF NATURE

PART III.

HERMETIC PHILOSOPHY.

THE PHILOSOPHY ADDRESSED TO THE ATHENIANS*

BOOK THE FIRST

TEXT I

OF all created things the condition whereof is transitory and frail, there is only one single principle. Included herein were latent all created things which the æther embraces in its scope. This is as much as to say that all created things proceeded from one matter, not each one separately from its own peculiar matter. This common matter of all things is the Great Mystery. Its comprehension could not be prefigured or shaped by any certain essence or idea, neither could it incline to any properties, seeing that it was free at once from colour and from elementary nature. Wherever the æther is diffused, there also the orb of the Great Mystery lies extended. This Great Mystery is the mother of all the elements, and at the same time the spleen of all the stars, trees, and carnal creatures. As children come forth from the mother, so from the Great Mystery are generated all created things, both those endowed with sense and those which are destitute thereof, all things uniformly. So, then, the Great Mystery is the only mother of all ephemeral things, from which these are born and derived, not in order of succession or continuation, but they came forth at one and the same time, in one creation, matter, form, essence, nature, and inclination.

---

* The whole literature of alchemy, so far, at least, as regards the Western world, appeals to the cosmological philosophy contained in the writings attributed to Hermes Trismegistus as to its source and fountain-head, and a right understanding of its mysteries is regarded by its authors as impossible in the absence of a right understanding of that philosophy. At the same time, the cosmology of the Hermetic books has at first sight nothing to do with alchemy. In the same way, the cosmological philosophy of Paracelsus does not at first sight seem to have any distinct bearing upon the alchemy of Paracelsus, and yet it is a complement of his alchemy, and is indispensable to students thereof. The editors of the Geneva folio regarded it in this light, and collected his chemical and philosophical works into a distinct volume. It is therefore advisable that readers of the present translation should have an opportunity of judging after what manner the physician of Hohenheim was accustomed to philosophise hermetically. The two most compendious treatises are here given. The Philosophy to the Athenians, that is, to the followers of Aristotle, deals cosmologically with that separation of the elements which is the subject of such frequent reference in other writings of Paracelsus. The *Interpretatio Alia Astronomiæ* is concerned with man and the sciences in relation to the greater world. It illustrates many matters which have been the subject of reference in the sections devoted to Hermetic Alchemy and Hermetic Medicine. Paracelsus is never characterised by extreme lucidity, and if his alchemy, his medicine, and his philosophy are not illustrated by each other, seeing that he intimately connected them together, their individual difficulties are likely to be increased in proportion. At the same time, it is requisite, for many reasons, that this section should be kept within somewhat narrow limits.

TEXT II

This mystery was such that none other like it ever appeared to any creature ; this was the first matter from which all transitory things sprang ; and it cannot be better understood than by considering the urine of man. This is produced from water, air, earth, and fire. Of these, no one is like another, and yet all the elements proceed from thence to another generation, and so on to a third generation. And yet, as the urine is only a creature, there may be some difference between this and that. The Great Mystery is uncreated, and was prepared by the Great Artificer Himself. No other will ever be produced like it ; neither does it return or is it brought back to itself. For as the cheese does not again become milk, so neither does the generation hereof return to its own primal matter. For though all things may be reduced to their pristine nature and condition, still they do not go back to the Mystery. That which is consumed cannot be brought back again. But it can return to what it was before the Mystery.

TEXT III

Moreover, although the Great Mystery appertains to all created things, sensible and insensible, still, neither growing things, nor animals, nor the like, were created therein ; but the truth about it is this, that it left and assigned to all, that is, to men and animals, general mysteries ; and to those of each sort it gave the mystery of self-propagation according to their own form, to each its own essence. And so, by a similar example, it conferred on each of the rest alike the special mystery to produce its own shape by itself. From the same origin spring also those mysteries out of which another mystery can be produced, which, also, the primal mystery arranged. For a star (*alias* stercus) is the mystery of scarabæi, flies, and gnats. Milk is the mystery of cheese, butter, and other substances which belong to this class. Cheese is the mystery of worms that grow in it. So, again, in turn, worms are the mystery of their fæces. In this way, therefore, twofold mysteries exist ; one the Great Mystery, which is the mystery uncreated. The rest, as if springing out of it, are called special mysteries.

TEXT IV

Since, therefore, it is certain that all perishable things sprang from, and were produced by, the Uncreated Mystery, it should be known that no one thing was created sooner or later than another, or this or that by itself separately, but all were produced at one and the same time, and together. For the supreme arcanum, that is, the goodness of the Creator, created or brought together all things into the uncreated, not, indeed, formally, not essentially, not qualitatively ; but each one was latent in the uncreated, as an image or a statue in a block of wood. For as this statue is not seen until the rest of the superfluous wood is cut away, but when this is done, the statue is recognised, so with the Uncreated Mystery ; all that is fleshly, whether sensible or insensible, arrived straightway at its form and species by a deliberately planned process of separation. Here no section was lost or passed away, but all found their way to their own form and essence and to their own likes. Nowhere, in any age, could sculptor be found so careful and industrious in the work of separation who by like art could utilise the smallest and most shortlived thing, and shape it into a living being.

TEXT V

In this way it should be understood, not that a house was built out of the Great Mystery, or that all animals were brought together, piled up, and then perfected, and that the same was done with other growing things ; not so, but as a physician makes up some compound, of many virtues, though it be but a single matter, in which appear none of those virtues, which lie concealed under that particular species. So must we suppose that creatures of all kinds, which are comprised in the ether, were brought together into the Great Mystery and set in order, not, indeed, perfectly, according to their substance, form, or essence, but according to another subtle standard of perfection, which is hidden from us mortals, and according to which all things are included in one. For all of us are sprung from what is perishable and mortal, and are born just as if by the procreation of Saturn, who in his separation puts forth all sorts of forms and colours, of which not a single one shews visibly in himself. Since, then, the mysteries of Saturn exhibit procreations of this kind, much more surely will the Great Mystery have this miracle in itself; in the separation whereof all foreign and superfluous matter has been cut away ; yet nothing has been found so empty and unserviceable as not to produce some growth or useful matter from itself.

TEXT VI

Know, too, that in the cutting or carving of the Great Mystery different fragments fell down, and some went to flesh, of which the species and forms are infinite in number ; some to monsters of the sea, also marvellous in their variety ; others to herbs ; not a few to wood ; more still to stones and metals. As to how Almighty God carved these things, there are at least two methods of art in answering such inquiries. The first is that He constantly arranged for life and increase. The second, that it was not a single and everywhere similar matter which fell down. If a statue is carved out of wood, all the chips cut from that block are wooden too. Here, however, this was not the case ; but a separate form and motion were given to each.

TEXT VII

In this way, distribution ensued upon the working of the Great Mystery, and the things separated from the superfluous ones shone forth. At the same time, from these same superfluities which were cut off, other and diverse things were produced. For the Great Mystery was not elementary, though the elements themselves were latent therein. Nor was it carnal, though all the races of men were comprehended in it. Neither was it wood or stone ; but the matter was such that, whilst it embraced every mortal thing in its undivided essence, it afterwards, in the process of separation, conferred upon each one by itself its own special essence and form. Something of the same kind takes place with regard to food. When a man eats it, flesh is generated therefrom, though the food itself in no wise resembled flesh. If it be allowed to putrefy, then grass grows up from it, though there was nothing in the flesh that was like grass. And this is much more the case with the Great Mystery. For in the Mysteries it is abundantly clear that one has gone off into stones, another into flesh, another into herbs, and so on in different and infinitely varied forms.

TEXT VIII

The separation, then, having been now made, and everything being reduced to its peculiar shape and property, so that each shall subsist by itself, then, at length, the substantial matter can be distinguished. What was fit for compaction has been compacted, the rest (so far as its substance is concerned) remaining empty and thin. For when the compaction first took place, the whole could not be equally compacted, but the greater part remained void. This is clearly shewn in the case of water. If this be coagulated, the mass or quantity of that which is compacted is small. The same takes place in the separation of the elements. The whole compaction took place—stones, metals, wood, flesh, and the like. The rest remained more rare and void, each single thing according to its own nature and the properties of the planets. And so the Great Mystery in its compaction was just like smoke which is diffused far and wide. But it does not contain in itself much substance, save a little soot. The rest of the space which the smoke occupies is pure, clear air, as can be seen in the separation of the smoke from the soot.

TEXT IX

The principle, mother, and begetter of all generation was Separation. It is true that men ought not to philosophise about these things beyond the grasp of human reason ; but the following is the method of learning about such things, how they come to pass. If vinegar be mixed with warm milk, there begins a separation of the heterogeneous matters in many ways. The truphat of the minerals brings each metal to its own nature. So was it in the Mystery. Like macerated tincture of silver, so the Great Mystery, by penetrating, reduced every single thing to its own special essence. With wonderful skill it divided and separated everything, so that each substance was assigned to its due form. In truth, that magic which had such an entrance was a special miracle. If it were divinely brought about by Deity, we shall in vain strive to compass it in our philosophy. God has not disclosed Himself to us by means of this. But if that were natural magic, it certainly was very wonderful, marked by intensest penetration, and most rapid separation, the like whereof Nature can never again give or express. For whilst that operation went on, part of the things was cleft asunder to the elements, part went to other invisible things, part went to produce vegetables, and this is deservedly held to be a singular and supreme marvel.

TEXT X

So when the Great Mystery was filled with such essence and deity, with the addition of eternal power, before all creatures were made the work of separation began.* When this had commenced, afterwards every creature emerged and shone forth with its free will ; in which state all will afterwards flourish up to the end of all things, that is, until that great harvest in which everything shall be pregnant with its fruits, and those fruits shall be reaped and carried into the barn ; for the harvest is the end of its fruit, and signifies nothing else than the corporeal destruction of all things. The number of those fruits is, indeed, almost infinite ; but the harvest is one wherein all the fruits of creation shall be cut down and gathered into the barn. No less marvellous will be this harvest, the end of all things, than was stupendous at the beginning that Great Mystery. Although the free will of all things is the cause of their mutual affection and destruction, nothing exists without friendship and enmity ; and free will exists and flourishes only in virtues ; but it is friendly or adverse in its effects. These belong in no way to separation. This is the great divider, which gives to everything its form and its essence.

TEXT XI

But in the beginning of the Great Mystery of the separation of all things, there went forth first the separation of the elements, so that before all else those elements broke out into action, and each in its own essence. Fire became heaven, and the chest of the firmament. The air was made mere emptiness, where nothing appears or is visible, occupying that place where no substance or corporeal matter was located. This is the chest or cloister or the invisible Fates. The water went off into liquidity, and found a seat for itself around the channels and cavities of the centre, within the other elements and the æther. This is the chest of the nymphs and sea-monsters. The land was coagulated into the earth, which is not sustained by the other elements, but propped up by the columns of Archialtis, which are the mighty marvels of God. The earth is the chest of growing things, which are nourished by the earth. Such separation was the beginning of all creatures and the first distribution both of these and of others.

---

* The Book of the Philosophy of the Celestial Firmament, otherwise the *Philosophia Sagax*, thus explains the *rationale* of that primal creation by which the heaven, the earth, and all creatures were made : The first body ; after the body the Rector or Moderator ; then the sensible body ; after this the King governing it ; lastly, the King ruling men. The first body is that of the upper and lower sphere; the Rector is the living spirit which informs it, and this is the motive power. The Rector of man is the divine reason with which he is endowed,

TEXT XII

The elements having been thus produced according to their essence, and separated from one another, so that each should subsist in its own place, and no one encroach upon another, a second separation ensued upon the first, and this emanated from the elements themselves. Thus, all that was latent in the fire became transformed into the heavens, one part being, as it were, the ark or cloister, while the other developed from it as a flower from its stalk. In this way the stars, the planets, and all that the firmament contains, were produced. But these were begotten from the element, not as a stalk with its flowers grows out of the earth. These, indeed, grow from the earth itself, but the stars are produced from the heavens by separation alone, as the flowers of silver ascend and separate themselves. Thus all the firmaments were separated from the fire. Before the firmament was separated from the fire, this fire had existed as the one universal element. For as a tree in winter exists only as a tree, but when summer comes on, the same tree, if those leaves which have to be separated are removed, still puts forth its flowers and fruits (for this is the time of gathering and of separation), so must be understood as like in all respects that ingathering in the separation of the Great Mystery, which could not any longer restrain itself or be delayed.

TEXT XIII

To the separation of the elements there succeeded another separation from the air, made at the same time with the fire. The whole air was predestinated to all the elements. But it is not in the other elements in the way and manner of mixture. It lays hold of all kinds of things in all the elements, and seizes upon these. Nor does it occupy that which was possessed before. No mixture of the elements remained joined or united, but each separate element withdrew according to its own pleasure, in no way united or conjoined with the rest. Now, after the element had in this way withdrawn from the Great Mystery, there were forthwith distributed from it fates, impressions, incantations, superstitions, evil deeds, dreams, divinations, lots, visions, apparitions, fatacesti, melosiniæ, spirits, diemeæ, durdales, and neufareni. From this separation of the aforesaid, which had now been accomplished, there was assigned to each its own prearranged seat, and its own peculiar essence was predestinated. Thus it befell that things in themselves invisible became to us objects of our perception. No element was created by the supreme arcanum more subtle than the air. The diemeæ dwell in hard rocks. They were so created, together with the air, in vacuity. The durdales withdrew to the trees. A separation of them was made into a substance of this kind. The neufareni dwell in the air of the earth or in the pores of the earth. The melosiniæ took up their abode in human blood. The separation of them was made out of the air into bodies and flesh. The spirits were distributed into the air, which is in chaos. All the rest are, and abide in, special places of the air, each occupying its own definite position, and being separated from the element of air, but still so that it of necessity dwells therein, and cannot change that position.

TEXT XIV

By the separation of the elements the water was segregated to the place predestined for it. In this way, everything that was latent in its elementary virtue and property was more fully segregated by another separation, and the water was divided into many special mysteries, all of which had been moulded from the element of water. A certain part, by means of that separation, was cut off to form fishes of manifold forms and kinds; a certain part went to form fleshy animals; some went to salt, and no small portion to marine plants, as corals, trina, and citrones. Many also went to form marine monsters, contrary to the manner and course of all the elements. Not a few went to nymphs, syrens, dramas, lorinds, and nesder. Some went to rational creatures bearing in their bodies something eternal and propagating their kind; some which finally die out altogether, and some which, in course of time, are at length separated. For the perfect separation of the elements has not yet been fully made. As the great harvest overwhelms us, or draws nigh year by year, new growths may emerge in the element of water. And, indeed, that separation was made at the same moment as the separation of the elements, in one day, and by the motion of sequestration. And, so, by that means everything spending its existence in the water, simultaneously and in a single moment, was created and revealed by the process of separation.

TEXT XV

In like manner, also, when the element of earth had been separated from the rest, a terrestrial separation was made, that is to say, of all things which are, or have been, born from the earth. Four elements, in all respects alike, were latent in the Great Mystery at the time of the first creation. The same, in like manner, were also divided at one and the same time. Moreover, these were also, in like manner, divided from one another in the second separation, which is called elementary. By such elementary separation there were divided out of the element of earth, severally, sensible and insensible things, eternal and non-eternal, each being allotted its own essence and free will. Whatever was of a wooden nature therein was made wood. A second went off into metallic minerals, a third into marcasite, talc, bismuth, granate, metallic cobalt, pyrites, and many other substances; a fourth into gems of manifold forms and kinds, and to stones, sands, and chalk; a fifth into fruits, flowers, herbs, and seeds; a sixth into sensible animals, of whom one part are partakers of eternity, as men, the others cut off from it, as cattle and the rest. Very many species and differences can be enumerated; for in the terrestrial element far more species have been separated than in any other of the rest. For by means of the semen, and of the congress of two, the father and mother, to wit, all things were propagated, and this was not arranged and predestinated after the same fashion in the other elements. Here are the gnomes and the sylvesters and the lemures, of which the one are destined for the mountains, others for the woods, and the rest only for the night. Moreover, the giants were separated to the third generation. Great essences were also distributed, forming, as it were, stupendous miracles among men, cattle, and growing things. This is a matter hard to be understood by any philosophy and so is esteemed to be brought about contrary to the serial order and the customary method of Nature.

TEXT XVI

After that, as has been said, the four elements of things were in the beginning severally separated from one single matter, in which, however, their complexion and essence were not present—those complexions and natures emerged by that process of separation. The warm and the dry withdrew to the heavens and the firmament, each falling according to its own property. The warm and moist withdrew to the air, whereupon the warm and the moist were invisibly separated. The cold and moist was cleft asunder to the sea, and places bordering thereupon. The cold and dry degenerated to the earth and all terrestrial things. But contrarieties were originated from this separation of the elements, which appear in no respect like their elements. Among these is reckoned lime, which in respect of its nature is not fire, but originates in fire. The cause of this is, that in the course of the separation of the element, its dissolution departed too far from the fiery nature. Fire contains in itself both coldness and humidity. In fact, fire is fourfold. So, again, the colours which proceed from fire are not always alike. One fire makes a white and a lazurium colour. A dry fire makes red and green. A humid fire makes ashen and black. A cold fire makes saffron and red. For this reason one procreation is warmer than another, because the one fire was more or less graduated than the other. For fire was not only one simple thing, but some hundred kinds existed of which no one had precisely the same grade as another. There was a procreation of each one, as if it were made a certain pre-ordained mystery according to its definite subject.

TEXT XVII

Moreover, neither did the water itself obtain the complexion of a simple species. An infinite number of waters were latent in that element, yet all were veritable waters. It is not matter of discovery for a philosopher that the element of water is only cold and moist of itself. It is a hundred times colder, and not more moist, and, moreover, this is to be referred not so much to its warmth as to its coldness. The element of water does not live or flourish only in cold and moisture of one degree ; nay, indeed, it does not even exist of one degree. Some waters are springs, and these are manifold. Some are seas, which again are very numerous, and very different one from another. Some are streams and rivers, whereof no single one is precisely like any other. Some aqueous elements were appointed to stones, as in the case of the beryl, the crystal, the chalcedony, the amethyst. Some went to plants, as the coral and amber; some to chime, as the liquid of life. Not a few went to the earth, as the liquid of the earth. Such are the elements of water, manifold in species. Those things, for instance, which grow out of the ground from seed scattered there are also referred to the element of water. So, too, those which are fleshy, as the nymphs, are referred also to the element of water. For although, in this instance, the element of water must be understood to be transmuted into another complexion, it never lays aside or oversteps the nature of the element itself, from which it proceeds. Whatever is of water, that again becomes water. So, too, whatever comes from fire again becomes fire, what comes from earth becomes earth, and what from air becomes air.

TEXT XVIII

By parity of reasoning, too, it is plain that all things which are constituted from earth retain the nature of the same. Although mineral liquids are taken to be fire, yet they are not really so. Not even sulphur burns so much as to make it a fiery element. Indeed, the cold as well as the warm burns. That which in burning produces an ash is not the element of fire, but a fire of the earth, and that fire must not be esteemed an element. It is not an element, but only a consumption of the earth, or of its substance. Water itself can be quite as easily made to burn and to cause a conflagration. If this be so, then this is aqueous fire. Besides, the mere fact that the fire of earth burns and flames does not justify our considering it to be igneous, though it be true that it is in some respects like fire. That philosopher is at once simple and sensuous who names an element from what he perceives. The element is something widely different from such a fire as this. Why do we say so? Everything that moistens is not the element of water. The element of earth can be reduced to water, yet it still always remains earth. In the same way, whatever is in earth belongs to the element of earth. For it exists and is known by the properties of that from which it proceeded, or to which it appears like. The hard flint and the chalcedony alike emit fire from themselves. That fire, however, is not an elemental fire, but a strong expression in great hardness.

TEXT XIX

The element of air contains within itself a large number of procreations, which, nevertheless, are all merely air. Every philosopher should know that an element does not procreate anything else save of the same kind as it is *per se*. Like is constantly produced from like. So, therefore, since the air is invisible, it cannot beget any visible thing of itself. In the same way, since it is impalpable, it produces nothing which is palpable. Thus it melosiniates

—if I may so say. If the melosinia be from the air it is air, nothing else. Still, a conjunction takes place with some other element, which, in this case, is the earth. For here a conjunction can be made from the air to form a human being, just as in all evil deeds and incantations it happens by means of spirits. The truth is the same as with regard to nymphs, who, though they are ranked under the element of water and are nothing else, can still freely have commerce with terrestrials and generate with them. A similar compaction is made from the air. It is visible and tangible, not, however, as a procreation of the primal separation, but only a sequel of the same. As the scarabæus is produced from dung, so a monster can assume bodily shape from the aërial element with aërial speech, thoughts, and action through commixture with the terrestrial. But miracles and consequences of this kind return again to air, just as nymphs return to water, and as man by decay is reduced to earth and consumed because he was born from the earth.

TEXT XX

In this way, then, by means of the great separation, procreations are produced, the one following from the other. But from these procreations other generations have emerged, which have their own mystery in these procreations, not, indeed, as a separation in the exact form of those before mentioned, but as an error, or an abortion, or an excrescence. Thunder arises from the procreations of the firmament, and this itself exists from the element of fire. Thunder is, as it were, the harvest of a star, at that precise point of time in which the thunder has grown mature for acting according to its nature. Magical storms arise from the air and end in the air again, not because the element of air produces them, but rather the spirit of the air. Some are conceived corporeally from fire, as gnomes are from the earth. Just in the same way, dung is produced by men and cattle, not from the earth. The lorind emerges from the naturals of the water, and is not from the water itself. From that abundance, or that error, or that harvest, many other things, too, are begotten. By impressions are born deformed men, women, and other generations like these. From fatal storms arise infections of particular districts, pestilences, and dearness of the market (famine, etc.). From dung are born scarabæi, erucæ, and dalni. From the lorind is collected or understood prophecy of that particular region, and this is in a certain way a presage of future things, stupendous, rare, and never heard of before.

TEXT XXI

We have seen a threefold separation into threefold forms already made from the mystery. There remains, in like manner, a fourth and ultimate separation. It will be the last of all, none will come after it. Then all the others will perish, and not even the Mystery will remain. By that separation all things are reduced to their supreme principle, and that only remains which existed before the Great Mystery, and is eternal. That, however, should not be accepted in the sense that I am to be turned into *something*, or that, in the ultimate separation *something* will be produced out of me, except by death. I am reduced to nothing, as if being produced by reason, I came forth from nothing. And when the sum total of things shall return to their principle it is well that we should know how it takes place. When they are turned to nothing, then they exist in their prime. That prime must be sought in the beginning. What that means—the going into nothingness—must be accounted one of the secrets. The soul which dwells in me was made of *something*. Therefore it does not pass into nothingness, because it is built up of this *something*. Of nothing nothing comes—nothing is generated. A figure painted into a picture, when it is there, has been certainly made of *something*. But we are not thus constituted in the æther out of something—like a picture. Why? Why, because we came forth from the Great Mystery, not from anything procreated therefrom. So we return to nothing. If the figure be blotted out with a sponge, it leaves nothing behind it, and the picture returns to its former shape. So, assuredly, all creatures will be reduced to their primeval state, that is, to nothingness. And if we would know *why* all things must be reduced to nothingness, let us learn that it is on account of the eternity that stirs within rational corporeities. Such an ultimate separation is the ultimate matter. There will be centred all the numerous procreations, permixtions, conversions, transmutations, alterations, and the like ; to know the sum total whereof surpasses the scope of the human intelligence.

TEXT XXII

Moreover, as it is certain from philosophy that all those things which minister as auxiliaries to what is perishable and mortal are themselves equally mortal with that to which they subserve, and that what is divided cannot be again conjoined—as coagulated milk is not restored to its freshness—so we must philosophise that the Great Mystery does not return to that from which it proceeded. Hence it should be realised that all creatures are a picture of the supreme arcanum, and so nothing more than, as it were, the colouring spread over a wall. So we spend our time beneath the æther. One or another may be destroyed and turned to nothingness. As a picture is liable to destruction or conflagration, so is the Great Mystery, and we with it. All creatures, together with the Great Mystery itself, perish, are wiped out, and reduced like some wood which burns down to a small heap of ashes. But out of those ashes is made a little glass ; the glass is made into a small beryl, and the beryl passes away into wind. In the same way shall we be consumed, passing from one thing to another, until nothing more of us any longer survives. Such as was the beginning of creatures, such shall be their end. If from a little seed a cypress tree can be produced, surely it can be brought back again into as small a compass as that first little seed was. The seed and the beryl are alike ; and as the beginning is from the seed, so in the beryl will be the end. The separation having been thus made, and every single thing brought back to its nature or first principle, that is, *to nothing*, then within the æther there will be nothing that is not eternal, but all will be without end. That from which the non-eternal came into existence will flourish far more widely than before the beginning of creatures. It has no frailty in it, no mortality. And as glass cannot be consumed by a creature, so neither can that eternal essence ever be reduced to nothingness.

TEXT XXIII

When, therefore, the last separation shall be the dissolution of all created things, and one after another is consumed and perishes, from that circumstance the time of those things is recognised. For after the generation of created things there is in them no passing away. The seed of the old supplies the place of that which perishes. In this way, something eternal shews itself in mortals without any distinction, by the renewal of other seed. This the philosopher ignores. He neither admits nor takes count of any eternal seed. And yet he admits putrefaction ; in course of which that which is eternal is taken back again into the eternal. In this respect, man alone of all created things comprises in himself something eternal joined to that which is mortal. Since, therefore, on the aforesaid reasoning, the mortal and the eternal are combined, it should be known that the mortal prepares an essence in the stomach, and sustains the merit of the body. And this is so because the eternal part of man lives for ever, but the mortal part feebly dies out. And as with the body, so with that eternal part which proceeds from that body. This is the great marvel of philosophy, that the mortal part dominates and sways with its nod the eternal part itself, and this depends upon the man himself. He, therefore, in this way, is more the partner of his eternal portion than is that one from whom emanated both the mortal and the eternal parts of man. Hence it can be gathered that the mortal parts of all creatures dwell together—that is to say, the rational and the irrational parts—the one subserving the use of the other in succession, and that an eternal element is inserted into a mortal one, both dwelling together simultaneously. Whence philosophy teaches that all those things which dwell together without disagreement, without deceit or fraud, without good or evil, cannot be destroyed and consumed. But this would be the case if one were opposed to the other. In those, however, wherein the eternal does not dwell no indication is found. But those where the eternal is will not lack such indication. Now, if discord ensues in this way, the one eternal is compelled to give account to the other, and to pay what is due for injuries inflicted by the one upon the other. And when compensation regards the eternal it is not undertaken by that which is mortal ; for though bodies conciliate one another, still, if anything remains over and above here, that is eternal. So, then, is that which alone is eternal judged in us. And whilst one demands account of the other, all the mortal parts which bear the eternal within them are compelled to die out, so that the eternal may alone survive without the concurrence of the body. So is the judgment completed. For that alone is eternal, nor is there more of it ; in the final passing away all is mortal.* If, then, that which they embraced within themselves of an eternal nature thus perished, nothing remains but that it was *per se* eternal, and that it nourished and increased the mortal portion. That which profits not does not remain in the creature. All the rest is present only for the sake of the eternal. Hence it happens that, together with that which the mortal holds within itself of eternity, all those elements which sustained it perish and die out. So, then, it is clear, that the end of human affairs is in nothingness, whereto they all tend. From their essence they are separated into nothingness—that is, from something to nothing. In man, however, a perfect separation, that is, of the eternal from the mortal, is yet wanting. For here the judgment is clear which, in all things within the æther, denounces frailty. And if no reason could co-exist with frailty, there would be no passing away in creatures ; all would be eternal. There is one single cause for all this, namely, that we mortals do not dwell in justice, nor do we give just judgments, nay, we have not received the faculty of judging what is eternal. This power belongs to the eternal part. And in order that we may gain this power, it is necessary that we should all be collected and come together. And so the passing away of all things is desired.

TEXT XXIV

Since it has been said, then, that all things were created from the Great Primal Mystery, and that they pass away in like manner, it follows as an evident consequence that some such Great Mystery must exist. This is nothing more than saying that a house has been built by a word. This is an attribute of the eternal, just as it is possible for man to elicit fire where there is no fire, and from that which is not fire. The flint has no fire, though it emits fire from itself. So in the Great Mystery all Primal Mysteries were existent in a latent form and after a threefold manner, in respect of vegetative, elementary, and sensible things. The vegetative things were many hundreds, nay, many thousands. Every kind in the Great Mystery had its own specialty. With regard to the elements, there were only four. These four only had their principles. But men were innumerable. One kind were the loripedes, another the cyclopes, another the giants, another the mechili. In like manner, there are those who dwell in the earth, in the air, in the water, in the fire. So, also, every kind of growing thing had its own proper mystery in the Great Mystery, and from thence emerged all the multifarious created things. There are as many mysteries as there are trees or men. The eternal alone dominates in

---

* The text at this point is scarcely capable of an intelligible rendering.

man, and in the whole of his mystery, not in one more than in another.   In the Great Mystery was no kind which was not infinitely formed and coloured, one differently from another.   Now, all these things are to perish.   What then happens we forbear to say, but—A new Mysterium Magnum is not possible. That would be a greater miracle than we are able even to speculate about.

THE PHILOSOPHY ADDRESSED TO THE ATHENIANS

BOOK THE SECOND

TEXT I

SINCE, then, there was a something by which, in the course of separation, all created things issued forth, in the very beginning we must hold that there is some difference of gods, and that of the following kind : Since created things are divided into eternal and mortal, the reason of this is that there existed another creator of mysteries who was not the supreme or the most powerful. For that Supreme One should be the judge and the chastiser of all creatures, and should know how much had been conceded to them, seeing that they might do either good or evil, although not produced by himself. Moreover, created things are continually stirred up and instigated to evil rather than good; they are driven by the stars, by fate, and by the infernal power ; when, if created things had issued forth from the Supreme himself, it would be impossible that we should be forced to these properties of evil or of goodness, but we should enjoy free will in all respects without any such impulsion ; whereas the creature has not wisdom enough to know good or evil, or to distinguish between what is eternal and what is mortal. For many are foolish and destitute of mind, scarce one in a thousand is wise. The greater part are false prophets, teachers of lies, masters of ignorance. They are openly esteemed foremost persons, though they are in no sense such. The cause is obvious. We are creatures who do not receive what is good and perfect from our masters, but we are chiefly built up by the mortal gods, who in the Great Mystery had indeed some power, but, nevertheless, were placed by the Eternal for judgment, both to themselves and to us.

TEXT II

Now if, as we know by separation, the created universe consisted of four elements only, and proceeded therefrom, these four will be literally the matrices of all creatures, and are called the elements. And although every created thing is so far an element, or has a portion of an element, it is not really like an element, but like the spirit of an element. But then, too, nothing can subsist without an element. Moreover, even the elements themselves cannot coexist. Indeed, there is nothing which consists of four, three, or two elements. Each single element exists apart; and every created thing has only one element. It is mere blindness on the part of those who set down humidity for an element of water, or burning for an element of fire. An element is not to be defined according to body, or substance, or quality. What is visible to the eyes is only the subject or receptacle. But an element is spirit, and lives and flourishes in those things as the soul in the body. This primal matter of the elements is invisible and impalpable, but present in all. For the first matter of the elements is nothing else than life, which all created things possess. What is dead subsists no longer in any element but in ultimate matter wherein flourishes neither any taste, nor virtue, nor force.

TEXT III

All created things in the universe, then, were born of four mothers, that is to say, of four elements. These four elements, it should be remarked, exactly sufficed for the creation of all things. Neither more nor fewer were requisite. In mortal things more than four natures cannot subsist. In immortal natures, it is true, temperaments can subsist, not elements. Whatever " elementure " (if I may use that term) exists, is dissoluble; but, on the other hand, temperature does not involve in itself the idea of dissolution. Its condition is such that nothing can be added to it or taken away from it, nothing decays, nothing perishes. This being the mortal condition, therefore, as has been said, it can be understood that all things subsist in four natures, and each nature has the name of its element. The warm is the element of fire. The cold is the element of earth. The moist is the element of water. The dry is the element of air. The next thing to be considered is how each of the aforesaid natures is what it is peculiarly or separately. Fire is only warm, not dry, not moist. Earth is only cold, not dry, not moist. Water is only moist, and not warm, not cold. Air is only dry, and not warm, not cold. Thus it is they are called elements. They are of one single and simple nature, not of a twofold one. This declaration on the subject holds good for all created things, that an element is that which subsists with body and substance, and thus operates. The chief point to be understood on the subject of elements is that they each and all have only a single and simple nature, moist, dry, cold, or hot. This is the condition of spirits. Every spirit is simple, not double, in its nature; and this is the case with the elements.

TEXT IV

Now, if it should be that composites do exist in us mortals, this scarcely coincides with the opinions of the ancients. The colic comes from the element of fire. It is not compounded of heat and dryness, but it is only hot. In like manner with the other complexions. And so, if any disease be detected where heat and dryness are combined, one can only come to the conclusion that the two elements exist, one in the spleen and the other in other members. Two elements cannot be in the same member. It is certain that in every member dwells some one element which is peculiar to it ; a subject which we leave for physicians to define. This, however, cannot consistently be said, that two elements exist at the same time in the same place, or that one and the same element is both warm and moist. Such a composite is not forthcoming. The elements know nothing of a compound, for the reason already given. Where heat is there cold cannot be, nor dryness, nor humidity. So, in the same way, where cold is, there none of the others is present. The same is the case with moisture and dryness. Every element is *per se* simple and solitary, not twofold in its composition. Whatever philosophy may concede as to the possibility of conjunction of elements—as of heat with humidity—all this goes for nothing. No element of water bears heat. No heat can subsist in humidity. Every element stands by itself alone. Cold cannot of itself tolerate dryness. That subsists by itself uncontaminated. This may be said and understood of the special essence of the elements. All dryness is a dissolution of cold. Just as humidity and dryness cannot be conjoined, so much the less can coldness and dryness and humidity, or warmth or dryness meet together and co-exist. As heat and cold are contraries, so heat and cold stand in the relation of contrariety to moisture and dryness.

TEXT V

It would be erroneous if we wished to assert that because all things are made up of four elements, they therefore necessarily exist in conjunction. Every conjunction is a composition. Since they are compounded, they cannot be the Great Mystery. Every Mystery is simple and a single element. There is a similar difference between elements and compounds. An element, as likewise a mystery, can generate a divertallum. A compound can generate nothing except what is like itself, as men make men. But a mystery does not produce a mystery like itself. It produces something different as a divertallum. The element of fire is the producer of stars, planets, and the whole firmament, and yet none of these is formed or constituted like fire. The element of water has constituted water, which is altogether contrary to the element of water. The element of water *per se* is not at all moist. It is true the element itself has this kind of moisture, that it softens stones and hard metals. But this remarkable power of softening is taken from it by substantial water, so that its virtue is not perfect. The element of air is so dry that in one moment it dries all waters. But this power is taken away and destroyed by substantial air. The element of earth is so cold that it reduces all creatures to their ultimate matter, as, for example, water to crystal and * to Dualech, animals to marble, and trees to giants. The foundation of the knowledge of the elements is to understand that they are of such remarkable and prompt activity and efficacy that nothing else like them can be found, or even conceived. The things in which they reside are attracted and assumed by them as if it were by a fate which is corporeal, and without these has not the smallest amount of virtue.

TEXT VI

In order more fully to understand what an element is, it should be realised that an element is really neither more nor less than a soul. Not, indeed, that it is of precisely the same essence as a soul, but it corresponds with a certain degree of resemblance. There is a difference between the elemental soul and the eternal soul. The soul of the elements is the life of all created things. Fire which burns is not the element of fire, as we see it ; but its soul, invisible to us, is the element and the life of fire. The element of fire can be present in green wood no less than in fire. But the life itself is not equally present as it is in fire. There is again a difference between the soul and the life. Fire, if it lives, burns. But if it be in its soul, that is, in its element, it lacks all power of burning. It does not follow that a cold thing must have proceeded from a cold element. It may often proceed from a warm one. Many cold things issue forth from the element of fire. Whatever grows is of the element of fire, but in another shape. Whatever is fixed is from the element of earth. Whatever nourishes is from the element of air ; and whatever consumes is from the element of water. Growth belongs only to the element of fire. Where that element fails there is no increment. Except the element of earth supplied it there would be no end to growth. This fixes it ; that is to say, it supplies a terminus for the element of fire. So, also, unless the element of air were to act, no nutrition could be brought about. By the air alone are all things nourished. Again, nothing can be dissolved or consumed unless the element of water be the cause. By it all things are mortified and reduced to nothing.

TEXT VII

But although the elements are thus in other respects altogether invisible and impalpable, and are thus hidden from us, they have, nevertheless, the power of putting forth their own mysteries. Thus, the element of fire gives forth from itself the firmament ; not, indeed, by any corporeal method, but on the scale of the elementary essence. The sun has a body different from that which he has received from the element of fire. Nevertheless, in this same he exists essentially with heat. For his heat does not arise from his motion or his rotation ; it is from himself. Though the sun stood still and never moved, he would still put forth his heat and brightness. The crystal has given the sun from the element of fire, though it has no other body than that which it has received from the element of the fire.* For thence the elements are (if I may so say) incorporated. The sun and the other stars, in like manner,

---

* This literal rendering seems to be without a reasonable meaning.

derived their origin from the element of fire, but only in a red colour, in which no heat or glow, but only a kind of dead splendour, inheres. And although signs appear in the sky differing in form and species—which we do not enumerate here—still the same form is to be understood as with ourselves on earth. It is not one only but multifarious, whether known or unknown to us. When the mystery of fire was divided off, a certain something was produced, such as we see. So the stars are daughters of the element of fire, and the sky is nothing but chaos, that is, vapour exhaled from the firmament, and so fervid as to be, in this respect, indescribable. That fervour or heat it is which gives coruscations, and colours, and appearances; for in that region is the pure element of fire, as shall be pointed out at greater length in its proper place.

TEXT VIII

In the same way as the fire put forth its different species and essences, so in like manner did the element of air. And yet there is a certain difference in the four elements as to those things which are procreated from them. Each of them produces after his kind. The firmament is like neither of the other three. Fate is from the air, and is not like those other three. Those signed by the earth cannot be in any way compared to the other three. In the same way are marine monsters related to the rest. For each single element has procreated in itself both rational and irrational things. The sky has in its firmaments rational creatures just as much as the element of earth has. In the same way, too, the fate of the air is distinct as to its signature in the matter of rationality and brutality. And the same is true of earth and of water. Who is he, then, who shall assure us as to the truth about these signed elements? Who are they that have the true faith handed down and entrusted to them, and the right way of salvation, or who alone shall possess eternity—matters which we pass by for the present? It cannot be but that human beings dwell in all four of these just as in one of them, that is, the earth. And on the subject of fate, it should be made known that its generation from the element is manifold, and yet without any body or substance, according to the property of the air (which is incorporeal), together with its habitation. For some things have it corporeally, others incorporeally, as the thing is understood.

TEXT IX

It is perfectly well known that from one seed emerges a root, divided into many filaments. Afterwards rises a stem, then branches are suspended there; flower, fruit, and seed come forth. Exactly similar is the method of manifold procreations from the four elements. Thus, all those which are from one element cohere, as the herb is produced from one seed. Yet all these growths are not exactly like their own seed. Those things procreated from the water are partly men, partly animals, and partly what these feed upon. One element has left its signature as both its necessity and its sustentation ; then it also gives sign of its course and its advent, which are easily recognised by the stars, not because these rule or influence us, but only because they run concurrently with us, and imitate the inner movement of our body And no less in the element of water all those things are produced which are produced in the element of earth. The lorind is a commotion and change of this element of water. This moves itself ; and the lorind is like a comet. A monster in the water should be accepted in just the same way as an error of the firmament. So in the water a special world is to be recognised, together with its mystery, even to the end of the world. There is no beginning in these save that which is in the other elements ; nor is there any other end than is found in the other elements. The only difference is of forms, essence, and natures, accruing to them, with their signatures and elements. Thus we must understand the four worlds according to the four elements and their primary habitations. With regard to justice there is only one eternal, which is to be recognised alike in all four.

TEXT X

In the element of earth there is much intelligence for us, because we ourselves came forth thence. Each like thing understands its like. But the understanding of the four elements flows out from philosophy. This, again, is like, flowing out as it does from the same source as that from which philosophy afterwards rises. Nevertheless, as the element of earth has procreated its sig nature, so have also the other elements procreated theirs. Just as we have stones, so are they not lacking in the other elements, though those stones may not be made according to the same form as ours, but in their own peculiar form and with their own special properties. The other elements have also their minerals no less than we have. The celestial firmament puts forth its floral growths as well as its mineral products, and these we reckon among miracles. Here, however, we make the greatest possible mistake, because we reckon natural processes among prodigies, and give it out like prophets that this or that appearance of the firmament portends something special, when we ought to understand that such things happen only in the ordinary course of events. But if such thing does occasionally happen we ought to believe that such was our course or condition. In the meantime, if anything suffers from the error of the elements, the other things grow uncertain too. All ought to proceed with a perfect and unimpeded motion, and though the other three elements serve us for nutriment, just in the same way do their other three serve the firmament, the air, the water, and those who dwell in them. One derives its nutriment from the other, just like many trees in a garden. And the defects and errors of the firmament can be observed by us no less than the firmament observes our defects. And the same judgment may be passed as to the rest.

TEXT XI

That philosophy, then, is foolish and vain which leads us to assign all happiness and eternity to our element alone, that is, the earth. And that is a fool's maxim which boasts that we are the noblest of creatures. There are many worlds : and we are not the only beings in our own world. And the ignorance becomes more marked when we fail to recognise those human beings who spring from our own element, as the nocturnals, the gnomes, etc. Though they do not live in the open light of day, nor use the brightness of the firmament, but hate what we enjoy and enjoy what we hate, and, moreover, are like us neither in shape, nor in essence, nor in their sustenance ; still, that is not a subject for our wonder. They were made up as they are in the Great Mystery. We are not the only beings made ; there were many more whom we do not know. We ought to conclude, then, that not one simple single body but many bodies were included in the Great Mystery, though there existed only in general the eternal and the mortal. But in how many forms and species the elements produced all things cannot be fully told. But let all doubt be removed that eternity belongs to all these. In this way, certainly, many things which are now unknown will be investigated and in different ways become cognisable by us, not concerning those existences only which have the eternal element in them, but of those which sustain and nourish that eternity. For eternity may be understood in two ways. One is that of kingdom and domination ; the other that of ornamentation and decoration. It is opposed to all true philosophy to say that flowers lack their own eternity. They may perish and die here ; but they will re-appear in the restitution of all things. Nothing has been created out of the Great Mystery which will not inhabit a form beyond the æther.

TEXT XII

The universal procreations indicate that all things must have four mothers, neither more nor fewer. Not, however, that this can be understood from the basis of the general demonstration concerning the present Great Mystery as it appears in the beginning according to its properties; but rather the Great Mystery is known and understood by the ultimate mysteries, and by the procreations which have gone forth having their origin in the first. It is not the beginning but the end which makes either a philosopher or a magistrate. The knowledge of a thing according to its perfect nature is only found at the end of its essence. Perhaps there might have been more elements than have been given to us. In the latest knowledge of all things, however, only four are found. And although we can, indeed, suppose that it would have been easy for God, who only created four, to have created more by the same operation ; still, when we see all mortal things made up of these four only, we shall not do ill to believe that more than these would not have been well. It is reasonable to think, perhaps, that after the destruction of the four elements spoken of, certain others will come into existence essentially unlike those before mentioned ; or, that after the passing away of the present creation a new Mysterium Magnum may supervene, and we have a better and fuller cognition of it than we had of that which shall then have passed away. We base nothing on this idea, but to every one who would know the origin of the universe we say, that he must understand the world to have sprung from the elements ; and as there are four elements, so there are four worlds, and in each exists a peculiar race, each with its own necessities.

TEXT XIII

Though all things exist in the four elements whereof we have spoken, we could not, however, insist that there are four elements in all things, or that four elements abide in all things. The reason is this. The world which is separated and procreated from the element of fire needs neither air, nor water, nor earth. By the same argument, the world of air needs neither of the other three. The same is equally true of the earth and the air (*sic ?* water). For the doctrine of the elements does not lay it down that the world must be sustained by the four elements, but rather that everything is conserved by one element, namely, by that from which it is sprung. And although you may not deny that the firmament nourishes the world by means of its elementary forces, which are all igneous, as they descend to the earth, still that nutrition is not a matter of necessity. Neither is the world going to perish by itself, but suffices for its own sustentation, just as also any other world nourishes itself without the help of the earth. For example, the earth confers nothing on the water as respects its own proper essence, nor, on the other hand, the water on the earth. The same is the case with the air. Yet still we do not take the course hereof to be that each world exists solely, or *per se*, in its own element, but rather that the light from heaven is, as it were, something drawn from the four elements, and most noble in its full and perfect properties. But let no one on this account imagine that the sun received his brightness and his motion from the element of fire, since neither the planets did this, but rather from an arcanum. The splendour of the firmament irradiating the world emanated, not from the element of fire, but from the arcanum. The earth of itself gives Tronum, the water Turas, and the air Samies. These proceed, not from the element, but from the arcanum, and are present in the element. In the arcanum the four worlds agree in this way, that they are useful to each other in turn, afford mutual nutriment, and sustain one another; not, however, from the nature of the elements, for they are themselves elements.

TEXT XIV

That man lives, sees, hears, and the rest, is due not to the elements, but rather to the arcana, and especially to the monarchy. Such is also the case with other creatures. The sole entertainment and nutrition is elementary. Understand that whatever is eternal proceeds from the arcanum, and is the arcanum. Dogs die, and their arcanum remains. A man dies. His arcanum remains, and still more, his soul, which causes him to be so many degrees more worthy than the dog. The same is, in like manner, the case with all growing things. Hence has sprung up an error that in the new *Mysterium Magnum* all created things which have ever existed, will ultimately appear, not essentially, now, but in an arcane manner. We do not say that the arcanum is an essence like the immortal, but that it *is* this in its perfection. The element of fire has an arcanum in itself. From this there are added, or there flow out to the other three, light, brightness, influence, increase, but not from the element. These arcana could exist without the element itself, just as the element without the arcanum. But mark, moreover, that the element of air has in itself an arcanum from which nutriment is supplied to all the other three and to itself, not as an element from itself, but as an arcanum by means of the element. The element of earth has in itself the arcanum of permanence and fixation, which imparts to the others the virtues of duration and generation, so that nothing may perish. The element of water has the arcanum of sustentation for all the elements, and conserves whatever is in them, so that it escapes destruction. In this respect, too, there is a difference between an element and an arcanum. The one is mortal or perishable from the elements, the other permanent in the ultimate *Mysterium Magnum*, wherein all things will be renewed, yet no other things will be produced save what have been.

TEXT XV

We come to the conclusion, then, that all the elements are not joined together, but that they are altogether aërial, or igneous, or terrestrial, or aqueous, solely and without admixture. This, also, is settled, that every element nourishes itself, or that does which is in it or its world. For this reason a medicine drawn from the element of water is of no avail for those who are from the element of earth, or any other, but only nymphs, sirens, and the like. So, also, a terrene medicine does not benefit the three other worlds, but only the animals of its own world. The same is to be said of the air; for in the air there are diseases and physicians—learned and unlearned—just as in our own world, each having its own peculiar mode. So with the fire. If it sometimes happens that nymphs copulate with terrestrials, and children are born, it is open to question whether this can happen from the possibility of ravishment. By this power other aërial beings, such as melosinæ, have intercourse with terrestrials. So, too, the trifertes are borne from the fire to the earth. Now, if three of these unfamiliar beings should come from their world to ours, as we have said, they would be looked upon as gods compared with us, on account of the vast distance between us, and the entirely foreign essence which they possess. But if a man can be carried to them, there is a corresponding power of being borne to them (*sic*). So, then, the elements have no need one of the other; only one is the ark or receptacle of another. As water and earth are mutually separated, so, in like manner, the air and the fire each occupies a special position without any contact of the four elements, save that which is like the party-wall of a house.

TEXT XVI

If there is at length to follow a conjunction and a gathering together in which all things shall return to their primal essence, then that will be an arcanum, and, indeed, according to the aspect or appearance of the elements. For there nothing corporeal can appear from generation, but the appearance and present exhibition of all the generations contained will supply that place, and so all those things will be known to everyone which have before been made, or shall afterwards be made, which also will be as well known as if they were seen before the very eyes ; yet, still herein is a hidden perception of the ultimate Great Mystery. And that will be known or made certain not by Nature, but from a knowledge as to the causes of the final separation of the elements and of all created things : where each one will give an account of his own death. That is the cause of the perishable, of the living, and of the permanent. For there will be one Judge who has power from the eternal, and from age to age there has been this one Judge. This, also, is the cause or origin of religions, and of the different mystæ who serve the gods, all of which customs are false and erroneous. For no other judge has ever existed except the one God, who has been the Judge from eternity. It is too impious a piece of folly to wish to worship one who is mortal, perishable, and liable to decay, in place of the Author of creation and the Ruler of the eternal. Whatever dies has no power of governing and of ruling. There is only one way and one religion, nor should others be rashly adopted.

TEXT XVII

If, therefore, whatsoever things are created return to that unto which they were predestinated from the beginning, in that place an arcanum will be produced. For predestination is, as it were, the ultimate matter, which will be without element, and without present essence ; but that which is temporated and that which is incorrupt will the more certainly follow. For these things are not understood by the spirit, but from Nature, with this evidence that the eternal follows the mortal. For, if an insensible plant perishes, then something eternal succeeds to its place. There is no frail or fading thing in the whole world which does not substitute in its place something which is eternal. Nothing has been created void, nothing that is mortal without a succession of what is eternal. After the end of all created objects these eternal things will meet together, and be collected not only as nutriments, but as a magistery of Nature, both in the perishable and in the eternal. And thus the eternal is a sign of the dissolution of Nature, and not the beginning of created things, and the end in all things which no nature is without. And although, indeed, the fatal beings, such as melosinæ and nymphs, are to have something eternal behind them, yet we will not here treat of their corruption. For a manifold decay is here involved, inasmuch as there are four worlds. For decay is terrestrial, aërial, igneous, and aqueous. Each of these, with those created with it, is turned and led downwards to decay with the eternal which is left. Nevertheless, these four decays will reduce their eternal portion to one similitude, notably and visibly, not with their works, but with their essence. For each eternal thing is a single habitation, prevailing, however, in many differences.

TEXT XVIII

We must not pass by in silence the evestrum, according to its essence ; it is either mortal or immortal. The evestrum is like a shadow on a wall. The shadow grows and originates with the body, and remains with it up to its ultimate matter. The evestrum has its origin with the first generation of everything. Everything animate and inanimate, sensible and insensible, has conjoined with itself an evestrum, just as everything casts a shadow. The trarames is understood to be, as it were, a shadow of the invisible essence. It is born with the reason and the imagination of both intelligent animals and of brute beasts. To philosophise about the evestrum and the trarames belongs to the very highest philosophy. The evestrum gives prophecy, and the trarames gives acumen. To prophecy about what shall happen to a man or animal, or even to a piece of wood, belongs to the shadowed evestrum. What will be the method comes from the trarames. So the evestra either have a beginning or have it not. Those which have a beginning carry dissolution with them, together with something eternal surviving. That which has no beginning possesses the power of sharpening the traramium in the intellect. The mortal evestrum knows the eternal. This knowledge is the mother of prophecy. For the foundation of all intellect is from the evestrum as it is extracted or elicited by the light of Nature. Thus the prophet "evestrates," that is, he vaticinates by means of the evestrum. If the spirit prophecies, that is without the light of Nature. So it is for us fallacious, liable to imposture and uncertain, as also certain and true. And in the same way the trarames also, as the shadow of reason, may be divided.

TEXT XIX

Moreover, in the dissolution of all things the evestrum also and the trarames will be dissolved, but not without some relics of the eternal. So, then, the evestrum is exactly as it were the firmament in the four worlds. The firmament is fourfold, according to the four worlds, divided into four perfect essences, each world perfectly regarding its own creation, because it is itself of that nature : one in the earth from the firmament, one in the water, air, and fire respectively. But the firmament which is in the evestrum is dispersed, not stars which are visible, but these are firmaments to the nymphs which are not stars. Neither do they use stars, but they have a peculiar and proper firmament as fates; the igneous have each their private heaven, earth, abode, dwelling, firmament, stars, planets, and other like things, which are not in the least the one like the other. As water and fire, the substantial and the impalpable, the visible and the invisible, stand mutually related to one another, so is it with matters of this kind. In these the evestrum is divided with respect to fatal things, and remains a shadow of itself after its essence from dissolution ; and the evestrum after fire adheres to the igneous man, as another to the aqueous, and another to the terrene. This evestrum falsifies and deludes the world, shadowing itself by fraud from one world to another, shewing visions, splendours, signs, forms, and appearances. Hence arise the evestrum of comets, the evestrum of impressions, the evestrum of miracles. These three are the prophetic evestra and the shadowed evestra.

TEXT XX

It is, however, firstly and chiefly, necessary that we should know this prophet evestrum. For turban is the great essence of this kind, presaging all things which are in the four worlds. For whatever is about to happen by way of prodigy, and against Nature, or against common opinion and life, is known by the prophet evestrum, which is taken, and shadows itself forth from the great turban. It is most necessary that the prophet should know the great turban. It is most difficult to be understood, and is united with reason. It is, therefore, possible for a mortal man to know the great turban even to its extreme resolution. By this all the prophets spoke ; for in it are all the signs of the world. From it all evestra are born, from it the comets are shadowed, prodigious stars which arise contrary to the usual course of the heavens. From turban all impressions get their beginning, not from the firmament or from the stars. As often as something unheard of and rare is about to happen, outsiders sally forth and messengers in advance, by whom the coming misfortune is announced to the people. These presages are not natural, but they come from the prophetic evestrum. All epidemics, wars, seditions, have their presages, which arise from turban. Whoever has knowledge of the properties of evestra, he is a prophet and a seer of futurity. For the Most High does not parley with men, nor does he send down angels from His throne to announce these things in advance. They are fore-known and understood from the great turban which many pagans and Jews have worshipped as God, being blinded in their senses and intellect as to who the true God was.

TEXT XXI

Since, then, a shadowed evester is born with every created thing, and arises therefrom, it is well to know that by means thereof may be prognosticated the fortune and life of that thing whose evestrum it is. For example, when an infant is born there is born at the same time with it an evestrum constantly expressed in that same child, so that from the cradle to the hour of death it prophecies concerning the child, and points out what is going to happen. If it is to die, death does not arrive until the evestrum has announced it beforehand, either by knocking, shaking, or falling, or by some other warning of this kind. Whence, if the evestrum be understood, it can be presaged that this is a sign of death. Moreover, the evestrum is united with the eternal. After the death of a man his evestrum remains upon earth and gives signs whether the man is in happiness or misery.

Nor should one say, as simple folks do, that this is the spirit or soul of the man, or that the dead himself is walking about. This is the evestrum of the dead man which does not yield its place until the final end, in which all things will meet. This evestrum works signs. The gods, by their evestrum alone, have wrought miracles; and as the sun in its splendour puts forth his heat, and his nature, and his essence, so it is in our case with the presaging and prophetic evestra, in which our trust should be placed. These evestra regulate sleep and sleeplessness, the prefigurations of future events, the nature of things, reason, concupiscence, and thought.

TEXT XXII

In this way when future events are predicted in the elements by that in which evestra dwell, there will be evestra in water, some in mirrors, some in crystals, and some in polished surfaces; some will be understood by the motion of waters, some by songs and animum; for all these things can—so to say—"evestrate." The great and good God has a mysterious evestrum, in which are seen His essence and attributes. By the mysterial evestrum are known all that is good and all that is illuminated. On the other hand, the Damned Spirit himself has an evestrum in the world by which evil is known, together with all which violates and corrupts the law of Nature. But though these two evestrate, still they do not not in any way affect our life. Only by our evestrum shall we know ourselves. In all creatures are evestra, which are in all respects prophets, whether rational or irrational, whether gifted with sense or destitute of it. The evestrum is the spirit which teaches astronomy (astrology). Not that it is known by prognostications and nativities drawn from the stars; but its essence, if I may so say, is in the evestra, and its being is in these, as an image in a mirror, or a shadow in water or on the ground. And, just as growing things increase or diminish, so is it with the stars; not because from their own nature they have such a course, and damp and cold arise from the earth, but because the essence of the earth is such. So it is shadowed forth in heaven and over different parts according to the evestrum, and not as a power.

TEXT XXIII

Evestra of this kind will be corrupted, yet, nevertheless, they will not perish without the eternal. Nor will the evestra themselves be regarded, but they will dwell near those to whom they belong. Whence each will be its own adviser, so that before all else it may exhort and know itself. The nature and number of evestra is infinite. These guide the sleepers, foreshadow good and evil, search into the thoughts, and accomplish the labours or the works without any bodily motion. So, then, the evestrum is a marvellous matter, and the mother of all things in prophets, in astronomers (astrologers), and in physicians. Unless the understanding thereof be sought through the evestrum there is no knowledge of Nature. As theft betokens punishment, as a cloud foretells rain, and lotium bespeaks disease, so the evestrum shews everything without any exception. By it the sibyls and prophets spoke, but, as it were, in a kind of sleep-wakefulness. In this way the evestra are in the four worlds, the one ever communicating to the other presage, image, and miracle ; and by their regeneration they will be found much more wonderful. Nor let us here omit to say that the evestrum is a remnant of the eternal, the sustainer of religions, and the mode of operation for celestial things. For happiness alone and blessedness, and the chief good, and the last judgment, move us, and instigate us the more keenly and more profoundly to search and to investigate what difference there is between these two, the true and the false ; and this must be weighed and learnt not spiritually, but naturally.

THE PHILOSOPHY ADDRESSED TO THE ATHENIANS

BOOK THE THIRD

TEXT I

EVERYTHING in existence necessarily has a body. The mode and manner may be understood as being like a smoky spirit, which indeed has substance, but is not a body, nor is it tangible. But, though this be the case, still both bodies and substances can be produced from it. This may be understood from fuming arsenic; since after the generation of a body, nothing more is seen of the fume of the spirit, just as if it had been all reduced to a body. But this is not so. Something of a most subtle nature still remains in that place of generation. Thus by the process of separation there are produced something visible and something invisible. By this method and in this mode all things are propagated. Wood has still surviving the spirit from which it has been separated. So have stones, and so all things, without any exception. Their essence still survives just as it was separated from it. Man, in like manner, is nothing else than a relic and a survival from the separated fume. But mark this, that a certain spirit existed, and from that man is made up, and is most subtle in spirit. This spirit is the index of a twofold eternity, one being the caleruthum, and the other the meritorium. The caleruthum is the indication in the first eternity. It seeks or makes for the other, that is, God. That is a natural cause because all things affect or tend towards that from which they proceeded, or those natures which have been in contact with it; for whatever anything when building up used in the process, that the thing when it is built up desires after and pursues. And this should be understood, that a thing which has been built up does, by Nature, or by natural instinct, tend not towards its builder, but towards that from which it has proceeded. So the human body longs for the matter from which it has been separated, and not for God, since it was not taken out of Him. And that matter is the life and the habitation in which the eternal meritorium abides. So everything returns to its essence.

TEXT II

Since, then, everything has an appetency for its source, that is, for the mystery whence it sprang, it must now be further understood that this is eternal life, and that what comes forth from thence is mortal. None the less, however, there remains in the mortal something eternal, that is, the soul, as may be learnt elsewhere. And if a perishable thing is to return to its pristine condition, that can only be done by a conjunction of what is permanent ; then at length there is a collocation and a union of things. The form and substance, however, both of transitory and of non-transitory things, proceeds from that spirit of fume, just as hail or lightning emerges from a cloud. These are corporeal, and that matter from which they have proceeded remains invisible. So, then, it may be laid down that all things spring from the invisible, yet, without its suffering loss, for the matter has always the power of regenerating and recuperating that loss. Hence, also, it happens that the whole world will pass away like snow and return to the same essence of the spirit of smoke, and then will come together or coalesce apart from all tangible essence. In this way, it can be again re-born as at first. Hence, also, it is known that no created thing exists which has been born, but only as it has been built up or created. So, the chief good is constituted in the beginning of all things that anything shall thus proceed from the invisible and become corporeal, and then shall afterwards be separated again from its body, and once more become invisible. Then all things are again joined together and united and reduced to their primal matter. And although, indeed, they may be united, yet still they involve some distinction and difference one from the other. One is the abode of the other ; that other is the inmate of the abode. For that is the habitation of all things ; sensible and insensible alike must all return to that condition and to that place. For, whether rational or irrational, nothing is free from this change, but will return to its habitation, from which it has been separated, and there appear.

TEXT III

So, too, every body, or every tangible substance, is nothing else but coagulated smoke. Hence it may be assumed that such coagulation is manifold. One kind refers to wood, another to stone, a third to metal. But the body itself is none other than smoke, breathing forth from the matter or the matrix in which it is present. What grows from the ground is a smoke brought forth from the liquid of Mercury, which is various, and emits a manifold smoke for herbs, trees, and the like. But that smoke, if it issues forth from its primal, or as soon as it expires from the matrix touches foreign air, is thereupon coagulated. So this smoke constantly and persistently evaporates. As long, therefore, as it is driven and disturbed, so long the thing grows, but when the ebullition ceases the smoking also ceases. This terminates, too, both the coagulation and the growth. Wood is the smoke from Derses. Therein is latent a specific from which wood is produced. Nor is it only produced from this smoke ; it may be produced also from other dersic matter. In like manner, leffas is boiling matter, from the smoke of which all herbs are gendered. For the only predestination of herbs is leffa ;

there is no other. God is more wonderful in specifics than in all other natures. Stannar is the mother of metals, furnishing the first matter for metals by its fume. Metals, in fact, are nothing but the coagulated fume from stannar. Enur is the fume of stones. In fine, whatever is corporeal is nothing but coagulated smoke, in which there is latent a specific predestination. All things, too, will ultimately be resolved like smoke ; for the specific which coagulates has no power save for a definite time. The same may also be said of coagulation. All bodies will at last pass away and vanish in smoke, and will be terminated only in smoke. This is the consumption of everything corporeal, both living and dead.

TEXT IV

Man is coagulated smoke. Only from the boiling vapours and spermatic members of the body is the coagulation of spermatic matter produced. Man, too, will be resolved into a vapour of this kind ; so that death may be like birth. Moreover, we see in ourselves nothing else than that man is coagulated smoke formed by human predestination. Whatsoever, too, is taken or given forth is merely the coagulated fume from liquids. And so whatever is injected is consumed by the life on the same principle, so that the coagulation may be again dissolved and liquefied, as ice is liquefied by the sun, that it may afterwards vanish into the air like smoke. Life consumes everything ; for it is the spirit of consumption in all corporeities and substances. Here, too, attention must be given to the preparation of the digested mystery ; for if everything is due to return to that state from which it originated, and so anything is given forth, then it is consumed together with the life. This, however, happens only in those things which are not transmuted. Transmutation is not driven back or repressed ; and some transmutation is produced by means of life. Thus, then, is transmutation altered into the frailty of the body ; but, nevertheless, it is again separated from the body. For in its putrefaction, transmutation has no further power, and in putrefaction the digested mystery ensues as a consequence. In the meantime, there are mutually separated all the properties which man had in himself from herbs and other things, each returning to its own essence. Separation is, in fact, like that process by which, if ten or twelve things are mixed they are again disseevered, so that each regains its own special essence. Thus, eating is nothing else than a dissolution of bodies. Hence the materials of bodies are separated in vomitings and dejections from the bowels, which are simply fœtid smoke mixed with good. Nature, indeed, seeks only the subtle, avoiding what is dense. Stones, metals, and earths—in a word, all things—are dissolved by life ; nor is there any other dissolution of them by the body than that which is brought about by its life.

TEXT V

Moreover, it is equally necessary to understand the process by which each separate thing regains its own essence. This cannot be more fitly compared to anything than to fire, which is elicited from a hard flint, flaming and burning beyond all natural knowledge. For, as that hidden fire takes its origin and proceeds to work its effects, in the same form and appearance also is the essence led to its nature. And here reflect that in the beginning there existed only one thing, without any inclination or speciality, and from this afterwards all things issued forth. This origin exactly resembles some well-tempered colour, purple for instance, having in itself no inclination to any other colour, but conspicuous in its just temperature. Yet, notwithstanding, in that colour all colours are existent. For the other colours cannot be separated from it— red, green, blue, clay colour, white, black. Each of these colours, again, brings forth other blind colours, while yet each one is by itself entirely and properly tinted. And although many and various colours are latent in them, still, nevertheless, they are all hidden under one colour. In the same manner, too, everything had its essence in the Great Mystery, which the Supreme Architect afterwards separated. The crystal emits fire, not from a fiery nature, but on account of its hardness and solidity. It also hides in itself other elements, not essentially, but materially, ardent fire, blowing air, moistening water, and earth which is black and dry. Besides all these things it possesses in the composition of its qualities all colours, but hidden within itself, as the fire lies hid in the steel, betraying its presence neither by burning, nor by shining, nor by casting a colour. In this respect, all colours and all elements are present in everything. But how all things arrive at and penetrate to all other things, if anyone cares to know, let him believe that all these matters are brought about and cared for by Him alone, who is the Maker and Architect of all things.

TEXT VI

Although, as has been said, Nature lies invisibly in bodies and in substances; nevertheless, that invisibility is led to visibility by means of those bodies themselves. According as the essence of each is situated, so is it seen visibly in its virtues and in its colours. Invisible bodies, however, have no other method than this corporeal one. So mark, then, that the invisibles contain within themselves all the elements, and operate in every element. They can send forth from themselves fire and the virtue of its element; and so, too, do they send forth air, as a man sends forth his breath, or water, as a man sends forth urine. They are also of the nature of earth, and sprung from the earth. Know this, too, that the liquid of the earth always boils, and sends forth on high, beyond itself, the subtle spirit which it contains in itself. From this are nourished the invisibles and the firmament itself, and this could not be done without vapour. Incorporeal as well as corporeal things need food and drink. For this reason stones come forth from the earth from a like spirit of their nature. Each one attracts its own to itself. From the same source come spectres, fiery dragons, and the like. If, therefore, invisible as well as visible are each in its own essence, this is due to the nature of the

Great Mystery, as wood acquires ignition from a light or a taper, though this suffers no loss. And though, indeed, it be not corporeal, still it needs something corporeal in order to escape death, which is produced by the wood. In the same way, all invisibles need to be sustained, nourished, and increased by some visible thing. With these, indeed, they will at length perish and come to an end, still, however, having their activity in them without any waste or loss to other things, that is to say, to the corporeal and the visible, although this is brought about by the invisible, and apprehended by the visible.

The rest (for doubtless the author advanced further) has not come down to us.

HERMETIC ASTRONOMY

PREFACE TO THE INTERPRETATION OF THE STARS

WRITTEN BY THE DOCTOR THEOPHRASTUS OF HOHENHEIM

I have accounted it a thing at once convenient and reasonable that, seeing there is an order of the whole of astronomy, myself to explain the method by which it may be taught and known. It is the more needful that I should come forward because all science is completely corrupted and polluted by shameful notions, and imperatively calls for illumination by its true and genuine sense. Yet, having regard to the present times, I do not doubt that my labour will appear absurd and useless to many. I write these things by reason of those very persons—for the detection of their ignorant and random judgment. Such is the blindness of the world that it invariably prefers the name rather than that which the name signifies.

Astronomy contains in itself seven faculties or religions; he who does not know them all is unworthy to be called an astronomer. Let each man remain in that religion or faculty wherein he occupies himself; let the astologer deal with astrology; he who treats of the religion of magic, let him remain exclusively a magus; he who is concerned with divination, let him remain a diviner; he who regards nigromancy, a nigromantic; he who studies signatures, a signator; he who is devoted to the uncertain arts, an incertus; he who investigates matter, a physicist. Let not the astronomer deny that magic is astronomy, nor yet refuse the name to divination, nigromancy, and the rest. All these things are comprehended under astronomy as much as is astrology itself. They are natural and essential sciences of the stars, and he who is acquainted with them all, he is worthy to be called an astronomer. But albeit these sciences are sisters, they have heretofore been ignorant of their relationship, which it is important to recognise, so that one may not be despised by the other.

As, however, astronomy does not lead to the life which is eternal, though it may be called the highest wisdom of mortals in the light of Nature, it is not the highest wisdom of men. Beyond this wisdom there is another given from on high, which transcends the created and surpasses by far all mortal sapience. But, you will reply, the Father made the Light of Nature and also man himself. Blessed is the man who walks in that light for which he was formed, seeing that it derives from the Father! But what follows? The Son gave the Light of Eternal Wisdom to man that he might also walk therein. Can one contaminate the other? Happy is he that walks in the Father, happy is he that walks in the Son! It is right to live in both—in one to that which is mortal, in the other to that which is eternal. For whereas the Father is not angry with the Son, neither the Son angry with the Father, how then can the Light of Nature be separated from that which is eternal? One remains in the other. Nevertheless, these twain are separated by him who hands over to another what does not belong to himself, and speaks from his mouth that which he knows not in his heart. Each one on earth has his special predestinated gift, and it is lawful for him to work therewith. This gift in the light of Nature has regard to the neighbour. But besides this there are the gifts of the Holy Spirit—that is to say, prophetic and apostolic. Those who hunger after the goods of their neighbour possess not the divine gift, and those who speak hypocrisy have not the gifts of the Holy Spirit.

I freely confess that I have seen no prophets or apostles. I have, however, seen their writings, which dictate an eternal wisdom, and for this reason I by no means prefer the light of Nature thereto, but tread it under my feet, for the prophets have prophesied such things as no astronomer could have done. The apostles have healed sufferers whom medicine could never have restored. Therefore the relation of the physician to the apostle is the same as that which obtains between the astronomer and the prophet. What physician can restore the dead to life? Can any astronomer prophesy as David? Medicine is fallible, not so the apostles. But I only teach concerning ourselves as mortals in the light of Nature, with this limitation, that the wisdom of God is before all. The astronomer is acquainted with the figure, form, appearance, and essence of the heaven. The magus operates on the old and new heaven. The diviner speaks from the stars. The nigromancer controls sidereal bodies. The signator is versed in the micro-cosmic constellation. The adept in uncertain arts rules the imagination. The physicist composes. Now, those who give light on earth as torches in the Light of Nature shall shine, through Christ, as stars for ever. Wherefore let every one so consider these my sayings that he may gain the more from them than is written. The seed is cast into the earth, but it is another who giveth the increase. I, indeed, offer you the seed; He who brings forth seed-time and harvest, may He so conduct you to the end that ye may rejoice in an abundant vine!

THE INTERPRETATION OF THE STARS

IN order that, before advancing to definitions and proofs, I may communicate to you my full scheme, I would have you know somewhat concerning the stars. The whole machinery of the universe is divided into two parts, a visible body and an invisible body. The visible and tangible is the body of the universe, consisting of three primals, Sulphur, Mercury, and Salt. This is the elemental body of the universe, and the elements themselves are that body. The body which is not tangible, but impalpable and invisible, is the sidereal heaven or firmament. The firmament which we see is corporeal, visible, and material. This, however, is not the firmament itself, but its body. The firmament no one has ever seen, but only its body, just as the soul of man is not visible. The whole universe is thus divided into two parts, into body and firmament. Moreover, the firmament consists of two parts. One is in heaven among the stars; the other in the globe of the earth. Hence two essences of the firmament are built up. One is peculiar to the firmament of heaven, and the other peculiar to the element of this globe and sphere. The firmament of the globe or sphere is of such a nature that out of it grows whatever the body of the earth or of the elements gives or appoints. Thus from the ground the firmament of the globe brings forth the fruits, which could not be accomplished without the firmament. And the same is the case with all things that are produced from the ground. The other firmament has its special operation in heaven, that is, it relates solely to man. Now, although both star-systems, the upper and the lower, are linked together, conjoined, united, and run one with the other, still there is this difference, that the upper stars govern the higher senses, and the lower govern growing things; that is, the upper system arranges the animal intellect, and the lower those things which grow, springing forth from the sphere itself.

Beyond what has been already said, I shall enter upon no discussion as to the firmament of the globe, save so far as concerns its fruits and its growths. These are its philosophy. I shall put forward so much, however, with regard to the sense-producing star as will enable us to know that man is divided in himself; namely, into the body of the globe and the body of the senses, that is into a visible, palpable body and a body that is invisible and impalpable; or, in other words, into an elementary body of the three primals, Salt, Sulphur, Mercury, and into a sidereal body. So far as relates to the body of man, he is merely flesh and blood. That which is impalpable in him is called spirit. Thus man is made up of flesh, blood, and spirit. Moreover, the flesh and blood are not the man, but the spirit existing in himself. The spirit of man is wisdom, sense, intellect; and these are the man. The body is mere brute matter. The spirit is subjected to the stars, and the body is subjected to the spirit. So the star governs the man in his spirit, and the spirit governs the body in the flesh and blood. That spirit, however, is mortal, since it is not the soul. The soul is supernatural, and I do not speak of that here, but only of that which, being created in Adam, trenches on Nature, that is to say, flesh, blood, and this spirit. Whoever, therefore, is not reborn dies, and cannot sustain that spirit, but is carried off to death. So, then, there is a certain conjunction of the star and the man, of the elements and the man. It is a single conjunction, and a single alliance, of such a nature that no partition or separation can occur. All which can happen is that the soul departs, and is separated, that it leaves what is produced by the machinery of the world, and takes to itself what is eternal. This I point out in order that the star may be rightly comprehended among the things that are above and in the globe of the world; whilst, at the same time, it may be duly understood how each constituent part has been united in man. In this way the agreement and the operation of the star, one with another, will be understood, and we shall have ascertained what effect the external stars can have on man, and also what those things which are in man do in external things. For it is true that the external stars affect the man, and the internal stars in man affect outward things, in fact and in operation, the one on the other. For what Mars is able to effect in us, that also can the man effect in himself if he restrain himself in his manly operations. Thus are the double stars related one to the other. Man can affect heaven no less than heaven affects man.

And now we have to discuss the medium between the principal stars and the body. There is one star which governs all things; in man the animal intelligence, in brutes sensation, in the elements their operation. The star is the one supreme thing created from destruction or dissolution; and it is that in Olympus which has all these things under itself. Its office is to operate in man, to operate in elements, to operate in animals, to turn and to change their senses and their mind. Now, it is impossible to do this without a medium. This same medium is and must be a star situated in those things where the supreme operates. By this medium is produced an effect on the substance and on the body. Let us illustrate the matter by an example. If Mars is to act on a man, that cannot be done without a medium, which shall serve as the material star. By means of this Mars acts. Thus, if the higher star is to act on a parrot, it is necessary that there should be in the parrot a star as a medium by which the superior star acts. Hence it is clear that there is some star in man, in birds, and in all animals; and whatever these do, they do by the impulse of the higher influence which is received from the constellation, and regulates the unequal concordance.

Moreover, there is a similar star also in the elements, as in the earth, and that an efficacious one.   That star receives an impression from the higher star, and then of itself acts on the earth, so that there is drawn forth from the earth whatever exists or lies hid in it.   The same is the case with the element of water and the rest.   So a person is first of all an astrologer from the higher star, and another from the star of men.   There is an astrologer from the star of the elements, and there is an astrologer from the star of animals.   In this way there are four astrologers of the elements, two of the stars of men and animals respectively, which make six ; and then one of the superior star, which is the seventh.   Besides these there remains yet another astrology born of the imagination in man, superior to all the rest, and standing eighth in order.

This, like the others, has been neglected and passed over by astrologers ; but whoever would be accounted an astrologer must have a perfect knowledge of all the eight.   But, although those who are skilled in particular departments ought not to be despised, yet they cannot act universally. The star is divided into eight parts ; one is effective, six are subject to it ; the eighth is in itself effective and like the first, nay, in some respects it is superior to, and more excellent than, the first, as will hereafter be more clearly shewn, when we speak of the new heaven and firmament.   But it is only right that the celestial astronomer should know also about the rest.   Now, these inter- mediate stars act, follow one another, and agree, so that nothing shall be predicted from the higher star, with the accomplishment of which the lower star interferes and produces something else, be it better or worse.

Hence, it is clear that astronomy was always highly valued by the ancients from the time of the Deluge up to the birth of Christ.

All species of Astronomy, it is well known, were highly cultivated.   In the time of Christ, however, this ceased ; and it is (? not) matter for regret that under Christ this should have occurred, because the Father had determined by His mighty love that men should confine their thoughts to Him, and that useless things should be omitted ; and yet many of these things originated and grew up under Christ and afterwards inundated the whole world.   In order to under- stand this, I would have you know that Christ taught eternal wisdom, and took care for the soul, not without a purpose, but that the image of God might be promoted to the kingdom of its Creator, and so the lower wisdom might be neglected while the higher might be more actively cultivated.   Although, therefore, in this book I write these things like a heathen man, I profess myself a Christian.   The heathen, however, can rejoice in the Father who is not opposed to the Son ; and he is not really a heathen if he walks in the light of Nature.   The wisdom of Christ is better than all the wisdom of Nature. I myself avow this, that one prophet in a single hour speaks more certainly and more truly than all the astrologers in many years, and one apostle far excels in truth all the magicians.

What could resist the school of those who spoke with tongues of fire ?   And yet, though these gifts were possessed, a certain sect rose up in the time of Christ speciously boasting an eternal wisdom which they did not possess, though it grew and spread abroad. This sect cut itself off from the noble science of astronomy, and took its place much as dung might take the place of fruit. Hence it occurred that, in course of time, not only the foundation and light of eternal wisdom, but also the true astronomy itself was obliterated, and the entire light of Nature at length corrupted and obscured. This was a lamentable evil and sin. Woe to those who sin against the Holy Spirit as do these of whom I complain! I confess that it is better to speak from God than from astronomy; it is better to heal from God than by means of herbs; better to preach from God than from false prophets, which is the sin against the Holy Ghost. What comes from God is not halt or maimed, but has, as they say, hands and feet. What comes from Nature is, for the most part, worm-eaten and decayed. All things are not from God; but some are from Nature. All are not from Nature, but some from God. If the magi, the astronomers, the signators, the necromancers, the incerti, the diviners, should give up their science and follow the prophets, the apostles, and especially Christ Himself, who could impute it to them for a fault that they aspired to the greater from the less, from Nature to Christ? Yet we cannot but lament that they did not penetrate to that school which spoke with the tongues of fire, though they had almost lost the light of Nature. Hence it happens that they thoroughly detest both kinds of wisdom.

Concerning the stars, I lay it down that they it is which confer all animal intelligence. As the body is conferred by the globe, in the same way is the intellect conferred by the star. One cannot exist without the other. I am forced to admit that it repented the Father that He had made man, whom the Son regenerates. It is therefore wiser to be in communion with the Son than with the Father, though the light of the Father must not be abandoned. For the Father is not opposed to the Son, nor the Son to the Father. Woe to him who sins againt the Holy Spirit! I acknowledge that man is dust; for he was taken from the elements. What are the elements? Nothing. What is man? Nothing. Better is it then to follow, not that which is nothing, but that which is something. But when it comes to recognising the wondrous works of God, it cannot be but that I shall feel a difficulty. For the gifts of God are given to the Prophets, are given to Apostles, and to Saints. But so also those gifts are bestowed on astronomers and on physicians. All are by God and from God. Whatever is pre-destined to Prophets and Apostles will succeed. May that too succeed which is pre-destined to astronomy and to medicine, but all by means of God and of His operations! It is not everything that regards what is eternal or that regards even Nature; everything looks to its own. What I have to say of man, of animal, of elementary body, or of wisdom from the stars, is strictly true; and since man remains as he was formed at the beginning, I describe him as such, making in this place no mention of the new birth. Still, if the old birth and the new could not coexist, I would not describe the one or the other, but would vote all things vain. In the meantime, as to the accusation that I, being a Christian, treat a heathen topic, if the Father and the Son be agreed, and the one exist in the other, I would hope that this fact need cause no strife with any person ; and unless, indeed, opposition were raised by that sect which has darkened the light of Christ as well as the light of Nature, and so brought it about that between two stools one comes to the ground, any one would readily undertake to write on these matters.   In the meantime, if the renovation of the world takes place, then will be brought to pass the saying that was uttered by the eternal Virgin, "He has filled the hungry with good things ; and the rich He has sent empty away."

THE END OF THE BIRTH, AND THE CONSIDERATION OF THE STARS

By Dr. Theophrastus Hohenheim

Concerning the Mass and the Matter out of which Man was Made

IT follows next in order to consider how it comes about that external causes are so powerful in man.

It must be realised, first of all, that God created all things in heaven and on earth—day and night, all elements, and all animals. When all these were created, God then made man. And here, on the subject of creation, two remarks have to be made. First, all things were made of nothing, by a word only, save man alone. God made man out of *something*, that is to say, from a mass, which was a body, a substance—a *something*. What it was—this mass—we will briefly enquire.

God took the body out of which He built up man from those things which He created from nothingness into something. That mass was the extract of all creatures in heaven and earth, just as if one should extract the soul or spirit, and should take that spirit or that body. For example, man consists of flesh and blood, and besides that of a soul, which is the man, much more subtle than the former. In this manner, from all creatures, all elements, all stars in heaven and earth, all properties, essences, and natures, that was extracted which was most subtle and most excellent in all, and this was united into one mass. From this mass man was afterwards made. Hence man is now a microcosm, or a little world, because he is an extract from all the stars and planets of the whole firmament, from the earth and the elements; and so he is their quintessence. The four elements are the universal world, and from these man is constituted. In number, therefore, he is fifth, that is, the fifth or quint-essence, beyond the four elements out of which he has been extracted as a nucleus. But between the macrocosm and the microcosm this difference occurs, that the form, image, species, and substance of man are diverse therefrom. In man the earth is flesh, the water is blood, fire is the heat thereof, and air is the balsam. These properties have not been changed, but only the substance of the body. So man is man, not a world, yet made from the world, made in the likeness, not of the world, but of God. Yet man comprises in himself all the qualities of the world. Whence the Scripture rightly says we are dust and ashes, and into ashes we shall return; that is, although man, indeed, is made in the image of God, and has flesh and blood, and is not like the world, but more than the world, still, nevertheless, he is earth and dust and ashes. And he should lay this well to heart lest from his figure he should suffer himself to be led astray; but he should think what he has been, what he now is, and what hereafter he shall be.

Attend, therefore, to these examples. Since man is nothing else than what he was, and out of which he was made, let him not, even in imagination, be led astray. The knowledge of the fact tends to force upon him the confession that he is nothing but a mass drawn forth from the great universe. This being the case, he must know that he cannot be sustained and nourished therefrom. His body is from the world, and therefore must be fed and nourished by that world from which he has sprung. So it is that his food and his drink and all his aliment grow from the ground. The great universe contributes less to his food and nourishment. If man were not from the great world but from heaven, then he would take celestial bread from heaven along with the angels. He has been taken from the earth and from the elements, and therefore must be nourished by these. Without the great world he could not live, but would be dead, and so he is like the dust and ashes of the great world. It is settled, then, that man is sustained from the four elements, and that he takes from the earth his food, from the water his drink, from the fire his heat, and from the air his breath. But these all make for the sustentation of the body only, of the flesh and the blood.

Now, man is not only flesh and blood, but there is within him the intellect which does not, like the complexion, come from the elements, but from the stars. And the condition of the stars is this, that all the wisdom, intelligence, industry of the animal, and all the arts peculiar to man are contained in them. From the stars man has these same things, and that is called the light of Nature; in fact, it is whatever man has found by the light of Nature. Let us illustrate our position by an example. The body of man takes its food from the earth, to which food it is destined by its conception and natural agreement. This is the reason why one person likes one kind of food, and another likes another, each deriving his pleasure from the earth. Animals do the same, hunting out the food and drink for their bodies which has been implanted in the earth. Now as there is in man a special faculty for sustaining his body, that is, his flesh and blood, so is it with his intellect. He ought equally to sustain that with its own familiar food and drink, though not from the elements, since the senses are not corporal but are of the spirit as the stars are of the spirit. He then attracts by the spirit of his star, in whom that spirit is conceived and born. For the spirit in man is nourished just as much as the body. This special feature was engrafted on man at his creation, that although he shares the divine image, still he is not nourished by divine food, but by elemental. He is divided into two parts; into an elemental body, that is, into flesh and blood, whence that body must be nourished ; and into spirit, whence he is compelled to sustain his spirit from the spirit of the star. Man himself is dust and ashes of the earth. Such, then, is the condition of man, that, out of the great universe he needs both elements and stars, seeing that he himself is constituted in that way.

And now we must speak of the conception of man, how he is begotten and made. The first man was made from the mass, extracted from the machinery of the whole universe. Then there was built up from him a woman, who corresponds to him in his likeness to the universe. For the future, there proceeds from the man and the woman the generation of all children, of all men. Moreover, the hand of God made the first man after God's own image in a wonderful manner, but still composed of flesh and blood, that he may be very man. Afterwards the first man and his wife were subjected to Nature, and so far separated from the hand of God that man was no longer built up miraculously by God's hand, but by Nature. The generation of man, therefore, has been entrusted to Nature and conferred on one mass from which he had proceeded. That mass in Nature is called semen. Most certain it is, however, that a man and woman only cannot beget a man, but along with those two, the elements also and the spirit of the stars. These four make up the man. The semen is not in the man, save in so far as it enters into him elementarily. When, in the act of conception, the elements do not operate, no body is begotten. Where the star does not operate, no spirit is produced. Whatever is produced without the elements and the spirit of the stars is a monster, a mola, an abortion contrary to Nature. As God took the mass and infused life into it, so must the composition perpetually proceed from those four and from God, in whose hand all things are placed. The body and the spirit must be there. These two constituents make up the man—the human being, that is, the man with the woman, and the semen, which comes from without, and is, as it were, an aliment, something which the man has not within himself, but attracts from without, just as though it were a potion. Such as the principle of food and drink is, such is also that of the sperm, which the elements from without contribute to the body as a mass. The star, by means of its spirit, confers the senses. The father and mother are the instruments of the externals by which these are perfected. In order to make this intelligible, I will adduce an example : In the earth nothing grows unless the higher stars contribute their powers. What are these powers? They are such that one cannot exist without the other, but of necessity one must act in conjunction with the other. As those without are, such are those within, so far as man is concerned. Hence it is inferred that the first man was miraculously made, and so existed as the work of God. After that, man was subjugated to Nature, so that he should beget children in connection with her. Now, Nature means the external world in the elements and in the stars. Now it is evident from this that those elements have their prescribed course and mode of operation, just as the stars, too, have their daily course. They proceed in their daily agreement, and at particular epochs Nature puts forth new ones. Now, if this form of operation—if the father and mother—with this concordance meet together for the work of conception, then the fœtus is allotted the Nature of those from whom it is born, namely, of the four parents—the father, the mother, the elements, the stars. From the father and mother proceed a like image and essence of flesh and blood. Besides this, from their imagination, which is the human star, there is allotted the intellect, in proportion wherein the concordance and constellation have exhibited themselves. So, too, from the elements there is allotted the complexion and the quality of the nature. So, too, from the external stars their intelligence. As these meet, the influence which is stronger than the others, preponderates in the fœtus, or else there is a mutual commingling of all. Thus man becomes a microcosm. The father and mother are made from the universe, and the universe is constantly contributing to the generation of man. In this way, there is constituted a single body, but a double nature, a single spirit, but a twofold sense. At length the body returns to its primal body, and the senses to the primal sense. They die, pass away, and depart, never to return. The ashes cannot again be made wood, neither can man from that state in which he is ashes be brought back so as to be man again.

Now we have traced the generation of man to this point as a general and universal probation of the whole of astronomy, in order that it might be understood from thence why the astronomer studies and gets to know men by the stars, namely, because man is from the stars. As every son is known by his father, so is it here; and this science is very useful if a man knows who is from heaven, from the elements, from father and mother. The knowledge of the father and mother lies at the root. The knowledge of the elements pertains to medicine. The knowledge of the stars is astrological. There are many reasons why these cognitions are useful and good. Many men are mere brutes, and yet make themselves out angels. Many speak from their mother, calling themselves Samuels or Maccabees. Many in their earthly complexion fast and pray, and call themselves divine. Many handle those things which are not really what they are said to be. Anyone who is an astrologer knows what that spirit is which speaks and is seen. It is matter for regret that many hesitate between the two lights, culling and stealing from each in order to make themselves conspicuous. The spirits are known, indeed, to each, but in a different way, and this should not be so. But though things are thus, man is the work of God, but one only is His very son, that is, Adam. Others are sons of Nature, as Luke in his genealogy recounts of Joseph, that he was the Son of Helus, which Helus was the son of Mathat, which Mathat was the son of Levi, and so on back to Adam; yet there is no mention of the son of God. Thus man is a son in Nature, and does not desert his race, but follows the nature of his parents, the stars. Now, he who knows the father and mother of the stars and of the elements, and also the father and mother of the flesh and blood, he is in a position to discuss concerning that offspring, concerning its nature, essence, properties, in a word, concerning its whole condition. And as a physician compounds all simples into one, preparing a single remedy out of all, which cannot be made up without these numerous ingredients, so God performs His much more notable miracle by concocting man into one compound of all the elements and stars, so that man becomes heaven, firmament, elements, in a word, the nature of the whole universe, shut up and concealed in a slender body. And though God could have made man out of nothing by His one word " Fiat," He was pleased rather to build man up in Nature and to subject him to Nature as its son, but still so that He also subjected Nature to man, though still Nature was man's father. Hence it results that the astronomer knows man's conception by man's parentage. This is the reason why man can be healed by Nature through the agency of a physician, just as a father helps a son who has fallen into a pit. In this way Nature is subjected to man as to its own flesh and blood, its own son, its own fruit produced from itself ; in the body of the elements wherein diseases exist ; in the body of the spirit, where flourish the intelligence and reason ; and the elements, indeed, by means of medicine, but the stars by their own knowledge and wisdom. Now, this wisdom in the sight of God is nothing ; but the Divine wisdom is preeminent above all.* So the names of wisdom differ. That wisdom which comes from Nature is called animal, because it is mortal. That which comes from God is named eternal, because it is free from mortality. These two parts, therefore, seemed to me necessary to be treated before I commented upon astronomy itself, so that from these universal proofs the whole foundation might be the more easily gathered.

The following are the numbers, religions, and faculties of the whole of astrology, which are treated naturally and artificially. Neither more nor fewer than these exist essentially and spiritually. Their names and differences are as follows :—i., Astrology ; ii., Magic ; iii., Divination ; iv., Nigromancy ; v.,

---

* In an exceedingly abstruse treatise on The Foundation of Wisdom and the Sciences, Paracelsus thus delivers himself upon the subject of sapience and knowledge : Whosoever undertakes a treatise on the foundation of wisdom, the same before all things should admonish or teach the reader concerning the origin of the sciences and of sapience. So also the physician who has decided to write concerning diseases must first explain from what foundation he writes, and the author whom he follows, as also how he teaches, in order that the same may be proved in the case of diseases. For out of these the probity and truth of his doctrine and science can be judged. Similarly, we who are about to treat of the foundation of the sciences and the arts must necessarily teach their origin, whence they have proceeded, and whence they are to be learned. Having done this, we must proceed to explain the matter itself. This book treats not of corporal matters, but of things invisible, that is, of reason itself. Hereof I have been impelled to write, seeing that many persons before me have supposed many kinds of wisdom, whereas, so far as regards man, there is only one wisdom ; for how can one carpenter differ from another carpenter when both construct the same house, both use the same axe, and both have one method of building ? There is one compass. Concerning this, it must be stated in a treatise on science that two compasses are not to be used, for the compass is one and not two. In the same fashion, the carpenter, the quarryman, and the bricklayer can use no compass different from this. Thus the house of wisdom neither can nor ought to be established except upon one foundation. And as the builder's art is defined by one circle, one number, one line, one square, so also through all methods there is one wisdom ; and as the distribution takes place from one circle into a triangle, quadrangle, etc., which, however, are all one circle, the same is to be understood concerning the distribution of sapience. So also, as heaven, earth, air, and water use one and the same line, thus according to one line all wisdom is educed and extended. And as all men and all things are numbered by one number, so there is one number of wisdom itself, nor is any other beyond it to be taken. But inasmuch as a line is drawn by a free hand, yet by no means a correct one, or a circle, also by no means accurate, is described by hand, etc., so also lines, circles, and other figures proceed from sapience, but are by no means correct. For such sapience is by no means a true circle, quadrangle, or line. Hence I have determined to treat shortly of the foundation of science and wisdom, whence they proceed, what they are, and who bestows or imparts them. Artificers are greatly in need of the knowledge

Signature; vi., Uncertain Arts; vii., Manual Arts. What will be handled in each religion, and what the religion itself is, seek in the sequel.

ASTROLOGY.—This science teaches and treats concerning the whole firmament, how it stands with the earth and with man according to the primæval order, and what is the connection between man, the earth, and the stars.

MAGIC.—This science brings down and compels heaven from above to stones, herbs, words, etc. It teaches also the change of one thing into another, as well as the knowledge of the supernatural stars, comets, etc., and what their signification is.

DIVINATION.—This science is from heaven to man without any formal institution, so that he knows how to speak of things future, present, and past, though he has never looked into those things himself, and speaks nothing save what heaven impresses upon him. This science is most of all seen among simple persons.

NIGROMANCY.—This treats of sidereal bodies, which are without actual body, flesh, and blood. This operation stands related to the necromancer as a servant to his master, the latter commanding the former.

SIGNATURE. —This science teaches one to know the stars, what the heaven of each may be, how the heaven has produced man at his conception, and in the same way constellated him.

UNCERTAIN ARTS.—These sciences are without any principles on which they rest, or from which they proceed, and are ruled by the imagination, offering a new spirit and a new firmament by which they work.

MANUAL ART.—This science teaches the preparation of instruments for all astronomy, and with slender material expresses or comprises the form of the stars, and brings heaven and earth into one figure.

For the sake of fuller understanding I will add how many species each religion has, in this way:—

of such foundation, for what else is sapience but an art or science which one man has derived from another. He who by counsel flourishes in prudence, what has such a man learned except the art of provident wisdom, which another man does not know? Thus, the artificer has a skill in regulating the fire in which the tailor is completely wanting. This, therefore, is his art. Hence in different persons there are different arts. What, therefore, is science save art? It becomes this art to proceed out of a circle, out of a line, out of a number, for this gives a mode, and thus a mode obtains in arts. From his individual art the turner derives his special mode, and this also is true—the line, the circle, etc., give a mode of wisdom, and the said mode is wisdom itself. Further, sciences are distributed after many methods, nor could they all consist in one. One man knows one thing, and one another. No person can know and accomplish everything. Who is familiar with, who performs all things? And as no one perfects two labours in one labour, but it is necessary that one thing only should be performed at one time, so the case is the same with the arts and sciences. For the sciences are so extensive and so profound that they cannot be contained and held by one brain. One part is given to one, a second to another, etc. For as in one street of a city there are many gold beaters, in another many shoemakers, and in a third many tailors, so there is an analogous distribution of the sciences. Now, all undertakings proceed from one fount; from one fount flow all works and all sciences. Their ramifications are distributed like fruit on a tree, none of which can so separate itself as to deny that it was produced with the others from the same tree. If a guide and teacher be needed in this kind of wisdom, so that the source of a writer's instruction may be known, it becomes right and just that I should divide this discourse into two parts. The sapience of man is twofold—one relating to the soul, the other to the body. Having made this distinction, it is necessary that we should understand the animal, that is, the corporeal, and also, so to speak, the mental, that is, the eternal. But as we have elsewhere described the origin of the animal man, so here we shall chiefly concern ourselves with the man which is spiritual, and here is explained the invisible source of science and wisdom.—The rest of the work, which contains several treatises, and is at the same time only of a fragmentary nature, attributes, as might be expected, all wisdom to God, for man has nothing of himself. Man, however, is the heir of heaven, and the world exists solely for his benefit. It further divides wisdom into the animal and the angelical, and affirms that men themselves are angels, and are before all heaven and all angels.

Astrology has three species, as referring to man, the inferior bodies, animals.

Magic has six species. It belongs to comets, images, gamahei, characters, spectres, incantations.

Divination has five species, dreams, brutes, the mind, speculation, phantasy.

Nigromancy, of which there are three species, material visions or spectres, astral spirits, and inanimate, phantastical bodies, that is, those assumed by the dead or by lifeless things.

Signature has three species, chiromancy, physiognomy, and proportion.

Uncertain Arts, of which there are four species, geomancy, pyromancy, hydromancy, ventinina.

Manual Art has five species, arithmetic, geometry, cosmography, instrument, sphere (a mathematical instrument).

The Interpretation of the Species according to each religion.

ASTROLOGY—THE FIRST RELIGION

This science embraces three species in which it is occupied. It operates against man, against elements, against animals. For since heaven and the lower bodies are mutually connected, the heaven teaches us to know the lower bodies by means of a figure which represents the whole heaven. From this figure is inferred the property of the inferior bodies, and what effect heaven produces in those inferior bodies.

MAGIC

OF COMETS. This species teaches us to recognise all these supernatural signs in the sky, and to understand what they signify. Of this class are comets, halos, and the other figures of the sky. This science is founded on the Apocalypse, on dreams, and on the saying of Christ, "There shall be signs in the sun, moon, and stars." Since all these signs are supernatural, they refer not to astrology, but to magic.

OF IMAGES. This science represents the properties of heaven and impresses them on images, so that an image of great efficacy is compounded, moving itself and significant. Images of this kind cure exceptional diseases, and avert many remarkable accidents, such as wounds caused by cutting or by puncturing. A like virtue is not found in any herbs.

OF GAMAHEI. These are stones graven according to the face of heaven. Thus prepared they are useful against wounds, poisons, and incantations. They render persons invisible, and display other qualities which, without this science, Nature of herself cannot exhibit.

OF CHARACTERS. These species are words which are either spoken or written. They have power against all diseases, which they also avert. They divert misfortune and all accidents, they set free prisoners so that they are loosed from their chains, and produce those effects which Nature itself is not able to bring about, but only magical science can accomplish.

OF SPECTRES. This species exhibits the likenesses of men, so that something appears which is not really present. These visions with their signs are produced by night, not by day, and lack the body, blood, flesh, soul, and spirit of man.

OF INCANTATIONS. This species teaches how to turn men into dogs, cats, etc. It teaches a man how to convert himself into all kinds of appearances and forms. It renders people invisible, changes the minds of men at the will of the artificer, impels, leads, and directs impressions and generations according to his pleasure.

DIVINATION

DREAMS. If anything is presented to a person by means of a dream, be it present, future, or past, be it knowledge, a treasure, or any other secret, it bears reference to this art. It can direct the stars to a dream, so that anything may be thereby revealed.

BRUTES. This species teaches us to distinguish the prophecies which come from animals, so that man may see and understand what the heaven does or is about to do. It operates also in fools, in animals, and in other simple beings.

THE SOUL. This species refers only to the mind of man, so that by chance and not by premeditation it is suggested to the mind of man what he ought to do. This species is of great importance, and should be studied among the very first by man, so that he may know what the mind suggests to him from its true foundation.

SPECULATION. If any one carefully weighs and speculates, and, by means of a strong imagination, finds what he seeks, it ought to be referred to this species. It arises from the stars, which are occupied about man and teach him.

PHANTASY. If any one in mere sport finds out or learns anything, this also is from the star, when it is matured. This often reveals many things such as treasures, mines, and others which are hidden, operating without any previous knowledge or investigation, and benefiting him who does not seek it.

NIGROMANCY

VISIONS. This species sees in crystals, mirrors, polished surfaces, and the like, things that are hidden, secret, present or future, which are present just as though they appeared in bodily presence.

ASTRAL SPIRITS. This species teaches how to deal with sidereal spirits separated from the body, so that they may be compelled to serve men like slaves.

Inanimates are men without a soul produced by the stars, dwelling and conversing with men and doing the same as they.

SIGNATURE

CHIROMANCY, by which the star is exhibited in man with that appearance in which the heaven was at the time of his nativity. It appears in the hands, feet, and other lines and veins of the body, shewing themselves differently in different bodies.

PHYSIOGNOMY. This species teaches how to know a man by his countenance, manners, and gestures. This also has for its cause the hour of birth, which signs a man, and by those signs forms his nature.

PROPORTION. This species judges the properties from the general habit of a man, whether he be lame, too tall, too short, etc.

UNCERTAIN ARTS

GEOMANCY. This science is practised with a free mind without foundation or certain knowledge or signs (tesseræ). It agrees with astrology.

PYROMANCY. This species is fortune telling by fire. By fire is seen what is the motion of the heaven, what its nature and condition. In this the moon is principally consulted.

HYDROMANCY. This species teaches how to see in water certain secret and hidden things, closed and sealed letters, and persons who are travelling in distant countries, whether they are living or dead. This operation proceeds from the constellation of the new firmament, by means of imagination.

VENTININA. This teaches how to determine from the wind what the future state of the heaven will be as regards man, whether good or bad, fruitful or sterile, and other similar things in the future which cannot be determined by Nature.

MANUAL ART

ARITHMETIC. This species teaches how to find the number of heaven and earth in the stars and the like.

GEOMETRY teaches how to measure the height of heaven and earth, and of the things contained in them.

COSMOGRAPHY teaches the situation and distance of all things, the manners and nature of peoples.

INSTRUMENTATION. This species teaches how to make instruments, with which is known how heaven and earth are connected.

SPHERE. This species teaches how to learn by means of an instrument what is the knowledge and correspondence of heaven and earth.

With this brief discourse I have endeavoured to describe the different species of religions that astronomy itself may thus be more rightly understood. All these make up astronomy. But how each one may be proved is afterwards described, with this view, that it may be clear that astronomy is no inconsistent or mendacious science, but that it is based on a solid foundation drawn from the light of Nature itself; which, indeed, is necessary for establishing all truth and knowledge.

PROOF IN ASTROLOGICAL SCIENCE *

Having treated of the generation of man we must now deal with his sustenance, and in this way astrology will be sufficiently proved. There is a certain congenital virtue in man which attracts into man from the external

---

* The proofs in astrological science, magic, and divination are wanting in the treatise, and the deficiency has been supplied from another work, the *Explicatio totius Astronomiæ*, which duplicates the Hermetic Astronomy (*Interpretatio alia Astronomiæ*). It has been thought advisable at this point to compress somewhat the prolixity which further obscures the original.

sphere. Now, from that which is attracted man is sustained, and he is well and ill according to that which he has attracted. The attractive virtue is twofold, one of the elementary body, the other of the sidereal body. The desire of man for sustentation is to be understood as follows : The rays of the external sphere penetrate to us ; the internal economy of man accomplishes the rest. Thus the sphere extends its fruits from the radix even to the outward locust. Hence it follows that there is a certain nature, namely, hunger and thirst, which is implanted in us and compels us to eat those fruits. So do the rays of this sphere enter us. Now, even as the food of the physical body comes to us from the elements, so is the sidereal body supplied by the constellation with all science, all arts, all prudence. Man is formed in such a manner that he should derive all his knowledge in the same way as he gathers fruit from a tree. Thus originates music, the metallic art, medicine, agriculture: whatsoever the earthly body requires, that he finds in the wisdom of the stars, and all wisdom, whether good or bad, is derived to him from the stars. Two things only, namely, justice and holy scripture, proceed immediately from the Holy Spirit. In the stars then is the whole light of Nature founded. For as man seeks food from the earth in which he was born, so also does he seek it from the stars in which he is likewise born. Thus the wisdom to which he is born is twofold—one is animal—but of the other Christ said, " For this I was born," as if He had affirmed " I was born in the Eternal Wisdom." The wisdom of earth should be employed only over carnal matters ; the other and higher wisdom should be learnt and employed according to the words of Christ.

Now, the sidereal wisdom is foolishness before God, whence comes that saying : The wise man rules the stars, in the sense that eternal wisdom governs the animal. Thus natural wisdom is given to the body and not to the soul. Those things, therefore, which concern the soul are by no means to be polluted by the light of Nature. This must only be used with Nature. By the light of Nature all arts and operations have been invented. In the mansions of the planets there are workmen who have taught all other workmen, and they, indeed, are the best of all, for they have their arts implanted from birth. These, were they men, would everywhere forge iron and handle it as if it were wax. Mortals as yet have not learnt this arcanum, but they would do so did they drink from the true fountain. So also masons dwell in the habitations of the planets, from whom all other masons learn, and if they did this fully all matter would be plastic in their hands. Thus the firmament formed by God is our perfect instructor in all the arts if we refer to their true source. Thus, too, the palmary physician is in the firmament, who is acquainted with all diseases, and even sees those things which are hidden from our eyes. God created him such that he might beget physicians on earth. Now, concerning evil sources, there are unskilled artificers in heaven even as on earth. This ignorance and clumsiness may be discerned even by the animal wisdom, which is given for this end, that the good and not the bad may be chosen. So does the natural light lead up to the higher light. Further, Lucifer in heaven made himself other than he was created, together with his companions, and the same thing can also take place in the stars. Hence contrary conceptions may arise, adverse and perverse arts. We must not, therefore, believe every spirit, since of spirits there are two kinds, even as there are two kinds of angelic intelligences—those who remained as they were and those who fell from their first estate. Astronomy is important, in that it teaches us to discern between these two kinds of spirits. This same science also contains a great arcanum, nor can anything be learnt without it. Wisdom is eternal and natural. The eternal is immutable and constant; the natural, from its mutable conception, generates a false spirit which misinterprets scripture. But if astronomy is acquainted with this, and if, indeed, nothing is so hidden as not to be revealed thereby, who shall not extol it with the highest praises?

It has already been shewn after what manner man was made, how he possesses hunger and thirst, an elementary and sidereal body, to produce an appetite for nourishment, and finally that he tends to that which was implanted in him at conception. Hence it follows that such virtue, nature, property, and condition, and finally all the concordance and constellation, can be described by the astronomer, for in this way various nativities are constituted, and hidden things are prognosticated. In all who live according to Nature nothing is hidden from the astrologer, and thus for the generation of man a figure of heaven is erected, in order to know the properties of the stars, as also the particular mode. Understand, therefore, concerning astrology that it knows the whole nature, wisdom, and science of the stars, according as they perfect their own operation in conception and constitute an animal man. The astrologer can easily describe a man or an animal by reason of such a conjunction and concordance. But if astrology be fundamentally and properly known, and the nativities of infants be erected rightly according to the mode of the influence, may evils will be avoided which would otherwise be occasioned by the unpropitious constellations.

PROOF IN THE SCIENCE OF MAGIC

In the first place let us define the nature of Magic. It is that which brings celestial virtue into the medium, and thence is able to perfect its own operation. The medium is the centre. The centre is man. By means of man, therefore, the celestial force can be transmitted into man so that in man may be found such an operation as the constellation itself can produce. Moreover, in magic there is a further operation which it performs itself while exercising its art, that is to say, while the nature itself of the constellation does that which the magus ought to do. If the magus be himself the medium and centre, and, what is more, be capable of performing the operation of the constellation for man by means of man, it is in addition given to this art to produce another medium which is to be understood as a subject, by which subject that operation is just as well performed as by man, who is the true medium. Thus in magical science there exist two operations, one which Nature herself produces, selecting man as the instrument, and as the recipient of her influence, whether bad or good, the other operates by means of arbitrary instruments, such as statues, stones, herbs, words, also comets, similitudes, halos, and any other supernatural generation of the constellation. Thus Nature herself is able to prepare her magical powers and perform her own operations by their means, as, for example, when something extraordinary takes place amidst a rude populace and is referred to miraculous agency, whereas it is only Nature who has worked magically.

Whatsoever Nature is able to accomplish in a foreign body, the same also can man accomplish, if he direct his operation so that conception can be attained, namely, the image, having neither flesh nor blood, and being like to the comet, so that the words and characters possess their own virtues equally with medicaments. It is, in like manner, possible to bring about such a condition in herbs and gamahei, that they become like to the planets and the dwellers therein. Now, it is no matter for astonishment that man accomplishes such things, for if it be true, as the scripture says, that ye are gods, we shall certainly be superior to the stars. If the stars as a fact are found to govern the majority of men, that is because men have abdicated their power as gods ; few, indeed, are those who have exercised gifts such as those of the apostles and saints. The difference between the saint and the magus is this, that one operates by means of God and the other by means of Nature. Magic is a sublime science, and by reason of its operations is very hard of attainment. We must have regard to the word of Christ, which passes not away, when He said ; " If ye believe, ye shall accomplish more things than these." Now, if we can exceed that which is accomplished by Christ, we can also exceed that which Nature accomplishes, seeing that she was created on our account and is therefore in our power. The wise man rules Nature, not Nature the wise man. For the same reason we can accomplish more than the stars. In us, then, should abound so great a wisdom that we shall thereby control all things, not only firmamental virtues, but also living animals which yet are much stronger than man. The will of man extends over the depth of the sea and the height of the firmament.

Nature herself is a magus. If about to announce anything, she creates for herself messengers, such as comets and other celestial signs. The magus man is comparable to the physician. The physician knows the hidden virtues of herbs, but the magus the hidden potencies of the stars. The physician extracts the virtues of herbs, and produces a remedy which is small in weight but represents the powers contained in a whole field of vegetation. The magus can transfer the powers of a whole celestial field into a small stone, which is called the gamaheus. As the physician infuses herbal virtues into the sick man, and so heals his disease, so the magus infuses into man the heavenly virtues just as he has extracted them. Medicines are renewed yearly, but the stars have their exaltations in place of a summer. The sun is the highest grade of diurnal light, plus the congenital heat which belongs to it. How shall this light and heat be brought downward by means of man into a subject, so that its light will be intolerable to the eyes and sense shall scarcely be able to endure its heat? This takes place in the sphere of the crystal, which then is termed Beryl.

If the Magus can draw down virtues from heaven and infuse them into a subject, why should we be unable to make images conducive to health or disease? If poison, and the rest, can arise from earth, it can issue also from heaven. But why should not similar things take place in the case above, whether the subject be images, herbs, stones, or woods? The birth corresponds to what is sown in the constellation, and it is not man alone who operates such things; Nature also variously exercises herself. But if it be possible to Nature, why not also to man? Let Nature be an example to us. As she works we must follow in imitation. Herein lie hidden medical science, all artifices, all arts, all animal industries. It frequently happens that Nature advances some person beyond the knowledge he can derive from man, who also by skill and industry surpasses all the rest. Such a man is born like the comet, which differs from other stars. Thus it becomes possible that the Magus also, by means of magic science, may produce such an industrious man like a comet. These are the mysteries and the great things of God. The firmament, by means of the magi, exhibits the glory of God. By means of the magi out of Satra and Tharsis, by the ascendant of Christ in Bethlehem, is made manifest whatsoever the firmament and heaven do reveal in the Arcana of God.

Proof in the Science of Divination

Astronomy creates herself, and from herself performs astronomical operations which do not require art and industry. This mostly takes place among those who are of a good and honest disposition, as also temperate. The ancients preserved both their bodies and souls from pollution, so that they might more successfully perform operations of this kind in themselves. This is divination. When men, having no knowledge of astronomy, perform such operations, they are considered miraculous, and the operators are regarded as gods. The operation is revealed by dreams, by the soul, by speculation, and by animals. Divination was of much importance among the ancients. It is a part of astronomy, but it is not a science, for the operation occurs spontaneously. It is often said in common parlance: " My angel told me this." Here the operation is called an angel, as if it took place by God; it is ascribed to the angel, as if to a medium between God and the man. At the same time, the whole operation is merely celestial. Now this is the origin of divination. Man possesses a sidereal body united with an external constellation. These two communicate when the sidereal body is not affected by the elementary. In sleep, when the elementary body is quiescent, the sidereal body performs its functions. Hence arise *insomnia*, according as the constellation operates them, and as the constellations are badly or well disposed, so also are the *insomnia*. When the constellation and the sidereal body are favourably co-ordinated, future things are truly predicted. In this manner, also, many remedies have been discovered which prevail over different diseases, also hidden treasures and other concealed things, so that scarcely anything can be compared to this very great science. The firmament foreknows all future things, nor does anything escape its knowledge, whether of things past or things present. If a sidereal body of this kind be found suitable by the constellations, and if the constellations be prepared, many marvels are manifested, both present and past. In this manner old men and women, unendowed by any knowledge, as it were by their simplicity and fatuity, have often made prophecies which the event marvellously verified.

In the same way, also, many have become learned men, who, having attained a suitable sidereal body, have sedulously exercised themselves in their native influence. Hence it happens that they at last draw down upon themselves the influence of their native constellation, just as rays from the sun. So an admirable science, doctrine, and wisdom are discovered, yet is the whole animal alone, not from on high, but taken from the stars alone. Heaven being thus constituted, and producing for itself a sidereal body, there arise many great minds, many writers, doctors, interpreters of Scriptures, and philosophers, according as each is formed from its constellation. Their writings and doctrines are not to be considered sacred, although they have a certain singular authority, given by the constellation and influence, by the spirits of Nature, not of God. Operations of this kind sometimes proceed from the mind of man in a stupendous manner, when men, changing their heart and soul, would make themselves like to the saints, being made such by a drunken star; whereas wine changes man, so also these are changed. It is, therefore, worth while to understand this sort of astronomy. Intoxicated writers of this kind lead many astray; they are wanderers in the Spirit of God as well as in the Light of Nature, flitting about like dreams. Many things are done by these which yet are of no moment, nor can be understood by others.

The force and efficacy of the constellations impresses itself upon brute animals, for whatsoever lives contains in itself the sidereal spirit, and wherever the operations are, there they are manifested. So the clamour of peacocks presages the death of their owners. For no man dies without the previous indication of portents. When a man is about to die the constellation within him loses its operation, and this loss takes place by means of a sign or a great mutation. So the stars shuddered at the death of Christ. From motions taking place in Nature, the death of every man can be prognosticated. Knockings in houses will sometimes precede the death of some occupant, yet these are not the work of spectres, but are natural operations, which in this manner are accomplished in men by means of the stars. The stars singularly sympathise with man, for man has been so formed by God that the whole firmament is consensitive with man, and out of compassion gives its presages to his grief.

Proofs in Nigromancy

Regard, in due order, nigromancy, so that it may be possible to learn and judge sidereal spirits and those who have no soul. The judgment is directed to that whereof we proceed to speak. The man who buries a treasure in the earth and hides it has all his mind intent upon that treasure. If he dies, his elementated body is buried, his sidereal body withdraws from it, and walks about on the earth up to the time when its decay is complete. This body carries about with it the thoughts and the heart of the dead man. Hence, as may be inferred, it keeps itself in the neighbourhood of that place where the treasure has been buried, about which the heart of the dead man was anxious. Such sidereal spirits are constantly seen at or near the place where such treasure is. The same thing occurs in other matters about which anybody has been anxious with the whole desire of his heart, whether it has been food, or drink, or debauchery, gambling, or hunting. In all these things the spirit acts for the imagination of that heart, and it does the same thing in a shadowy way after death, until the star consumes that spirit also, as the elements have consumed the body. Hence it follows that the necromancers get to know these sidereal spirits and to ascertain for what reason they are walking about in one place or another. In the same way, they explain the nature of lemures, giants, and gnomes. Nigromancy is the philosophy of spiritual sidereal bodies, and of inanimate beings who are, nevertheless, human, as onagri, nymphs, lemures, etc. The man who busies himself about these is a necromancer. The same is the case with exorcists who adjure bodies and inanimate beings of this kind. They differ from necromancers in this respect, that the exorcists are occupied with bodies obsessed by the devil, while the necromancers find their occupation, both naturally and philosophically, with those who are not obsessed. The ignorance of men has confounded exorcism with necromancy, and taken them to be one and the same. However, their distinction has now been settled. Moreover, I have determined to say nothing about exorcists here; it will be better to relegate them to the devil, whose servants they are. But I would wish to commend necromancy to you as a remarkable natural science, which produces some marvellous effects, since by means of sidereal spirits are laid bare the very hearts of men, shewing how they are inclined, what they long for, and what their ambitions are.

It is, moreover, pleasant and delightful to rightly understand nigromancy. The knowledge of nymphs, also the discovery of lemures, gnomes, and giants, is very subtle and ingenious. Indeed, the philosophy of these four inanimate generations is a truly noble one, which many babblers oppose and prefer their own nonsense to it. Since God is wonderful in all His works, it is more than likely that, one of these days, the temerity of these people will be brought out into the light of day, and, in God's own good time, branded openly. Moreover, necromancers use beryls because, in respect of astral spirits, they have some familiarity with magicians in the way of visions, but the magi do not admit these. The causes of this fact will be noticed in these treatises. In the present discourse it has been made sufficiently clear what nigromancy is, and what is its subject-matter.

PROOF IN THE SCIENCE OF SIGNATURE

God has enriched the light of Nature with such ample gifts that even one who is not addicted to the light can know all things that are therein. Is not this a great thing which external signs offer to man's knowledge? And God has arranged it so. Possibly you wonder how this can be done. Let the following example put an end to your wonder. The carpenter is the seed of his house. Whatever he is, such will be his house. It is his imagination which makes the house, and his hand which perfects it. The house is like the imagination. Now, if such be the property of imagination that it makes a house, Nature also will be an imagination making a son, and making him according to its imagination. So the form and the essence are one thing.

Whatever anything is useful for, to that it is assumed and adapted. So if Nature makes a man, it adapts him to its design. And here our foundation is laid. For everything that is duly signed its own place should properly be left; for Nature adapts everything to its duty.

If any lord or prince builds a city he so builds and arranges all the walls, towers, citadels, and the rest, that they shall as closely as possible suit his design. If man does this, how much more shall Nature, which is higher than man? It makes one man lame, because it is going to use such an one for lame purposes. It makes another blind, he being destined for blind purposes. In one word, whatever it requires any one to be, such an one it produces.

This, then, being the custom of Nature, that it produces such a man as it wishes, those vestiges will be clear and plain in the man. By these vestiges is meant whatever Nature is going to use such a man for.

Since Nature, therefore, works thus openly and puts forth its work in public, it is right and convenient that some one should be met who sees what sort of a person Nature has in each case prepared and produced, that is, how it opposes a rascal to an honest man, and sets a man-wolf over against a shepherd.

A signature, then, is that which has to do with the signs to be taken into consideration, whereby one may know another—what there is in him. There is nothing hidden which Nature has not revealed and put plainly forward.

Rightly, therefore, should its proper place be given to signature, because it is a part of astronomy, for this reason, that the star builds the man up at its own pleasure, with the marks belonging to him. What is going to be tinged with black Nature makes black, what blue, it makes blue; that which is going to sting is made a nettle, and what is to purge is made an equisetum, what is to be used for smoothing and polishing is made a smiris. In fine, to everything is assigned its own form, by which it may be known for what purpose that thing is made by Nature.

Whatever is in anything according to its properties, quality, form, appearance, etc., is revealed in herbs, seeds, stones, roots, and the rest. All things are known by their signature. By the signature those who are instructed trace what lies hid in herbs, seeds, stones. But when the signature is obliterated and trifles are substituted for it, then it is all over with everything, even philosophy and medicine being at fault.

The cry goes everywhere that I burn with hatred of learned men, doctors, magistrates, bachelors, senators, consuls, and the like. What is the reason? Nature has signed them too clearly. I can see what they are made of; and I hate every house that lets in the rain.

In like manner, I am accused of disliking physicians and surgeons. Why? Just because they are not signed for their profession, but as rogues and impostors. The same is true of others also. I know plenty of them, if it were only safe to speak out.

How can I favour a man who is branded with so many stigmata and disgraceful marks as the Consul of Astorza, Niger, and of Nuremberg, Muffel? And how many others are there like them? Of course they detest this art, because it too clearly betrays bad men.

PROOF IN UNCERTAIN ARTS

Nature puts forward a way and clear order in which man should consider what belongs to Nature and its properties. Thus astrology teaches us to know the nature of the sun by the accustomed order of the stars. So what the moon is and what her nature, the astrologer learns from her course, which he sees to be regular. The same judgment is to be passed on the other stars. In like manner, philosophy is learnt from that which appears, how Nature stands related to the earth; hence it is ascertained that the method of philosophy ought to be the same. Thus all things have their own proof and comprehension. And so all arts, such as medicine and the rest, are conceived in a natural order. Without this order nothing can be done or brought to a perfect end.

Moreover, the uncertain arts, of which four are shewn in the table, have not this order and process, which can be materially proved and demonstrated; but this differs from the order spoken of. With regard to this it ought to be understood that there are many things which do not indeed square with the same order, but still are not opposed to Nature. They only differ from the order of material nature, as God has settled it. But what there is besides in this order ought to be understood from the Uncertain Arts in the following manner. The firmament and new heaven are constituted by the imagination; and it should be known that this imagination is effective, and produces many things, being marvellous in its operations. It often happens that the imagination of the parents, father and mother, confers on the offspring born in that creation a different heaven, another figure, another ascendant besides that which astrology gives. Thus it often happens that an offspring is begotten contrary to the star, and arranged otherwise than the figure of the heavens dictates. By the force of this imagination many learned men are often born.

Nothing, therefore, ought to be accepted beforehand in the way of proof for these uncertain arts short of the operation which takes place through the imagination by means of a new heaven, new ascendant, and firmament. In proportion as this is good, strong and just in operation, so the judgments fall. Let us take an example. Speculation is the wishing to know this or that thing. This speculation produces imagination; imagination begets operation; and operation leads to judgment and opinion. Now imagination is concerned, not with the flesh and blood, but with the spirit of the star which exists in every man. This spirit knows many things: future, present, and past, all arts and sciences. But flesh and blood are crude and imperfect, so that they cannot of themselves effect what the spirit wishes. But if flesh and blood are subject to the senses, and are purged by them, then the spirit acts thereby, if only the body be consentient. These senses are supreme in the uncertain arts. It is for this reason they are called uncertain arts; for who can know what imagination is in them? What does the spirit which is given to them imagine and effect? Yet, nevertheless, the art itself is certain. But the artist who uses it may be unfit for the creation of new heavens and the generation of a firmament. Because, therefore, there is the element of doubt on these points, credence cannot be given to opinion, but one has to wait for the issues. At last, however, the force and efficacy of these things are discovered. Moreover, it is not to fight against God if the future is explored apart from him whom God has set over the nature of the firmament. Suppose, for example, that someone is going to be stabbed with a dagger. Let this be foretold to him by some other person. Premonitions of this kind have often been found true, though there might have been strong opposition. Now, if this happens by uncertain arts God himself suggests the prophecy, the prediction, and the premonition in a manifold way. So many prophets have predicted such things by dreams. It was by a dream Joseph was admonished about Mary. And since these things did not seem natural to flesh and blood, they were thought nothing of until, the event corresponding, they were believed. Now the uncertain arts, just like dreams and other revelations, are intelligible. God chooses to appear wonderful in His works. For this reason the uncertain arts are by no means to be despised, because they eventually become known by the result. God does not intend that we should always foreknow the future for certain, as can be done by the order of Nature. He wishes us to know, indeed, but sometimes to doubt; that seeing we may not see, as Christ Himself also was known, yet not known by the Jews, seen yet not seen, heard yet not heard.

It has been said above concerning the imagination that it draws the star to itself and rules it, so that from the imagination the operation itself may be found in the star. Just as a man with his imagination cultivates the earth according to his judgment, so by his imagination he builds up a heaven in his star. The imagination of the artist in uncertain arts is the chief art and head of all. But in addition to this, imagination is strengthened and perfected by faith, so that it becomes reality. All doubt destroys the work and renders it imperfect in the spirit of Nature. Faith, therefore, ought to strengthen the imagination. Faith bounds the will.

Now, faith is threefold. There is faith in God. This produces what it believes. By faith mountains are moved, the dead are restored to life, sight is given back to the blind, the lame walk. What marvels faith produces if imagination looks to God with full faith which is unbroken and unmutilated. We find an example of this in the Saints of the Old as well as the New Testament, who, according to their belief, were made to obtain their wish, so that nothing was wanting to them. There is another faith in the Devil and his powers. Whoever has this faith, to him it happens as he believes, if only it be possible for the Devil to fulfil it. Lastly, there is also a faith in Nature, that is, in the light of Nature. He who believes in this obtains from Nature as much as he believes. Now more cannot be obtained from Nature than is given to it and conferred upon it by God. It is, then, imagination by which one thinks in proportion as he fixes his mind on God, or on Nature, or on the Devil. This imagination requires faith. Thus the work is concluded and perfected. That which imagination conceives is brought into operation.

Note an example of this. Medicine uses imagination strongly fixed on the nature of herbs and on healing. Here is need of faith that such imagination may act in the physician. If this is present, imagination conceives and brings forth spirit. The physician is spirit, not body. Hence infer that the same fact holds good in all arts. Moreover, there are physicians without imagination, without faith, who are called phantastics. Phantasy is not imagination, but the frontier of folly. These work for any result, but they do not study in that school where they ought. He who is born in imagination finds out the latent forces of Nature, which the body with its mere phantasy cannot find; for imagination and phantasy differ the one from the other. Imagination exists in the perfect spirit, while phantasy exists in the body without the perfect spirit. He who imagines compels herbs to put forth their hidden nature. So also imagination in the uncertain arts compels the stars to do the pleasure of him who imagines, believes, and operates. But because man does not always imagine or believe perfectly, therefore these arts are called uncertain, though they are certain and can give true results. The other sciences of astronomy hold their own even without faith or imagination, just as a mechanic who, if he follows his own order in working, has no need of imagination or consideration, and yet finishes his work.

But, it should be remarked, that by faith water can be crossed over without drowning or wetting; and a man without faith can do the same thing if he crosses the water by a bridge or in a ship. So also healing the sick is accomplished by means of medicine without faith; but by means of faith it is found out what medicine is. Imagination takes precedence of all. What this discovers and gives, the other, who acts phantastically, uses.

Man is not body, but the heart is man; and the heart is an entire star out of which it is built up. If, therefore, a man is perfect in his heart, nothing in the whole light of Nature is hidden from him. Thus from one point in Geomancy his whole will is accomplished. So, too, in Austrimancy, Pyromancy, and Hydromancy. The newly-born and self-begotten spirit shadows forth its knowledge and intelligence, in a figure and by a figure, as the man imagines, and remains firm therein without any dissolution. It is in this way the spirit of those sciences is begotten which at last operates and perfects that which is sought. The first step, therefore, in these sciences is to beget the spirit from the star by means of imagination, so that it may be present in its perfection. After that perfection is present even in uncertain arts. But where that spirit is not, there neither judgment nor perfect science will be present. Hence wonderful things are now found out in future and occult things, which are laughed at and despised by the inexperienced, who never realise in themselves what is the power of Nature in their spirit, that spirit, I mean, which is born in the manner described, and given and assigned by God for this special purpose.

To believe in the Devil leads to doubtful results, and the thing is mixed up with fraud. The reason for this is to be sought from God, who has determined that all who believe in the Devil should be or become liars like himself. But this faith in God is perfect and free from all defect. It is in Nature such as its power is. So, then, the uncertain arts are sciences, but with this condition added, that a new generation of the prophetic and Sibylline spirit shall take place by which the art and hand may be ruled and guided. Who was the inventor of these uncertain arts, I have not been able to ascertain. I know this, that these arts are very old, were held in great esseem by the ancients, hidden and handed down as special secrets. They spent their time on imagination and faith, by which they tracked out and demonstrated many consummate results. At present, so much imagination and faith do not exist; but most men fix their minds on those things which minister to the pleasures of flesh and blood. These they follow; to these they give their attention. These arts, therefore, even on this account, are uncertain, because man within himself is so doubtful. He who is doubtful can accomplish nothing certain; he who hesitates can bring nothing to perfection; he who pampers the body can attain to nothing solid in the spirit. Everyone should be perfect in that which he undertakes. So the spirit will be entire, and will conquer the body, which is nothing worth. The spirit is fruitful. This a man should have perfect within him, and put aside flesh and blood.

THE END OF THE PROOF IN UNCERTAIN ARTS

Imagination has impression, and impression makes imagination. Therefore from impression descends imagination. Hence, it follows that whatever be the impression, influence, constellation, star—such is the imagination.

Hence, too, it ensues that imagination brings forth a new heaven above impression, and as the imagination, such is the figure of the heaven.

Proof in Manual Mathematical Science

Though everything in the whole of astronomy be seen and discovered, yet there must be respectively numeration, dimension, occasion, and instrument. These are the principles of all sciences, that is to say, they are those things which concur with all sciences.

It is difficult to understand how numeration can be brought to bear in the case of stars, on account of their infinite number. The greatest part of them is never seen, or seen with difficulty, yet all of these must be reckoned in their number. It is, however, impossible for a man who only uses his eyes to count these. He who uses more than his eyes can count them, but not that other one.

The same is the case with geometry, for the measuring of height, depth, breadth, etc., is much too difficult to be undertaken by all. That is not geometry which is handed down among the seven liberal arts. Our geometry is astral, not terrestrial, and is known only to him who makes his measurement magically, not elementarily, but beyond the elements.

In like manner, the work of cosmography is material. The invention of the art itself is material, not elemental, but rather connected with nigromancy and divination. They who practise it examine the state of all things in heaven and earth, in what position they are placed and constituted, and with what conjunction they are connected. These matters are found out with so much subtlety that they will be described by-and-by with reference to the globe, instrument, and sphere.

Now, if a manual mathematician be so skilled a numberer in arithmetic, a measurer in geometry, an explorer in cosmography, an experimenter with instruments, then he may with the utmost propriety give himself out as a mathematician. Of these three departments does mathematical science consist, and these four make up mathematics. In this way the invisible body of astrology, which is known to the wise men, can be deduced.

But there are other mathematics, which only concern the Magi. They are very apt at making magical instruments, such as gamahei, images, characters. For these things, too, are instruments. The art of making them has to be sought in magic. Their preparation is part of mathematics. It is necessary, therefore, that these persons should be certain and well constellated, fit for preparing these things and disposing them in their place. That is, they must be virgins.

So, also, in Nigromancy. It is mathematical so far as making its preparation goes. Divination and signature need no mathematics. In nigromancy, however, it is necessary that an instrument of certitude, as also one regalia, and other defensives be used; for spirits are very prone to obsession. It is, there-fore, necessary that all should fortify themselves well against them, since the danger is imminent. But where that kind is (if I may use the expression) obsessible, it is worth while to know.

And so with regard to the mathematics required, as has been said, for the science of astronomy, let this be settled and determined, that herein is need for the most consummate prudence and intelligence. Nothing will be done by the common method. It is requisite that a man should be one who discovers these things in a more sublime way than by the ordinary and earthly light of Nature. There is need, I say, of a higher light, that is, of one that is above the artificial.

In this way, the mathematics in astronomy are proved by means of their own instruments, which agree with the great world. These instruments are so connected and bound up with the elements and the stars that they assume the form of a microcosm, which is itself made from the greater world, but consists of a smaller body, yet one which contains the universal world in itself like a quintessence extracted from it.

Here follow certain fragments aud schedules on the same matter as the preceding.

CONCERNING THE KNOWLEDGE OF STARS

Before all else you should be taught about the stars, what they are; for the astronomer is directed to the stars only and to nothing else. It should be known, however, in this place that elementary bodies are not concerned; also, that flesh and blood effect nothing, but only stars. In order that you may thoroughly understand this, I would have you know that man's senses are apart from his body. Whatever is not corporeal is either star or ether. But of those things which have not body there are many species in man. However this may be, man's sensation is certainly not flesh and blood. The body, therefore, is one thing and the sense another. The body is flesh and blood. The sense is soul. The soul, not the body, is the subject of astronomy. But the body is ruled by the soul. So, then, the body, too, is the subject of astronomy, because the body underlies the soul, is obedient to it, and ruled by it. Moreover, the soul is not something eternal in man; it is not the *summum bonum*, but is something mortal existing in man; it is the man built up in Adam. Since, then, the soul is subject to astronomy, and astronomy acknow-ledges the star alone as its lord, know that in the star there are many essences, that is, not one star, but many. It is known, also, that one star exists higher than all the rest. This is the Apocalyptic star. The second star is that of the ascendant. The third is that of the elements, and of these there are four; so that six stars are established. Besides these there is still another star, imagination, which begets a new star and a new heaven.

But although, as is now understood, there are seven stars, still the astrologer is not so conditioned as to act the astrologer in these seven. One is an astronomer of supernatural astrology; another over the ascendant;

another over the four elements, and yet another over the imagination. Each discourses of his own astrology, each one is an astrologer, and each sufficient in himself. Now, he who is an astrologer does not rest in one thing, but is conversant with all, if he does not expound his own species with which he is conversant. But that is an intolerable error which, neglecting the different kinds of stars, deals only with the horoscope, the ascendant, and the figure of the heaven. But though the rest of the horoscope should not be understood by the astrologer, this would matter little if he only confessed that these other parts were good and belonged to astrology. For there is a star of the firmament, that is, fire, which has nothing to do with the horoscope. There is a star of the earth, because the earth, no less than the heaven, has its astrologer. So water and ether equally have their own star, and in like manner the air. Let no one think there is only one star. There are more ; but beyond all that have been mentioned there is one. Beyond the fifth, again, there is another supernatural one ; and beyond this sixth, one which is hidden in man himself, making the seventh.

I speak of the seven kinds of astronomy which make up the entire man, as has been before pointed out. Moreover, these seven kinds are not under seven stars. But I say this, that astrology alone embraces these seven in itself, and hence it is necessary to understand how these seven stars are essentially conditioned. In this way a perfect astrological judgment issues forth, which can be obtained in no other way. For there must be a medium, by which the last operates, and another after the first of the four, add also in the last. I add this with the view of making quite clear what is not sufficiently insisted upon in astrology, that Mars in the sky must be thoroughly understood, which looks there like a live coal. For besides this many another exists, and, moreover, four others in the four elements, and, lastly, one in the imagination. What sort of smith would he be who could forge a horse-shoe but not a nail ? What sort of a carpenter who could only cut his wood, and not join it ? Science ought to be perfect in all particulars, without exception. What things should be joined, let them be joined.

In this place it should be specially considered that before the Deluge our ancestors, up to the birth of Christ, devoted themselves with constant zeal and unwearied labours to the discovery of wisdom ; and now, since the advent of Christ, all this has perished and become extinct, so that it is difficult to find any of it anywhere surviving. I will tell you the cause of this. Christ offered eternal wisdom to the world. When this was offered it was only right and just that the inferior wisdom should be repudiated, and the higher acknowledged. In this respect I confess that I write like a heathen, though I am a Christian. For by right the lower wisdom gives place to the higher. The wisdom of Christ is better than the wisdom of Nature. A prophet or an apostle is better than an astronomer or a physician. A prediction from God is better than one from astronomy. A cure wrought by God is better than one by herbs. Prophets speak infallibly. The sick are healed and the dead raised by apostles ; nor is there any deceit about these things.

Although, therefore, astronomy with its light was obliterated by Christ, who will impeach that light? And thus much farther I am commanded to say. The sick have need of a physician, but not all of them need apostles. So predictions need an astronomer, but not all need a prophet. Distribution being made, one part goes to the prophets, another to the astronomers; one part to the apostles, another to the physicians. Each has his own limitations. And so, indeed, astronomy is not taken away from or interdicted to us Christians, but we are commanded to use it in a Christian way. We are created by the Father for the light of Nature, and it is only right that we should know and practise this. We are called by the Son for eternity, whence this, too, should be known. So, therefore, the light is transferred to us from the Father as if by inheritance, and the light from God the Son here in this world to eternity. Neither hinders the other—the Son the Father, or the Father the Son. By this means man is able in both ways to learn, to know, and to work.

Having made this excursus, I end my treatise on astronomy, that you may know what the stars are, and what power the astronomer or astrologer has, in what respect the one differs from the other, and how the stars are situated. I have made mention of seven, not for the moment taking thought of one other star, which is the Signed Star of the Microcosm, so that really they should be reckoned as eight. In the following explanation and proofs all these things will be found connected together so that you will understand them.

I could wish, indeed, that those who put themselves in the place of Christ would shew themselves His real disciples. Then the light of Nature would be more rightly understood, that is, the miracles of God would be more carefully looked into. As it is, mere trifles and deceits are obtruded, in which there is no juice, no marrow, no wisdom. If folly and wickedness like this are allowed to succeed, what success can there be for the noble wisdom of Nature? In this way no consideration is given either to the wisdom of Nature or to eternal wisdom, but both lose esteem together. It is the way of the world to oppose every kind of wisdom. This being so, I thought it best not to refrain from writing, but by all means to go on. For the renovation of the world will be upon us; and then at length will be found that which is now sought after; and it will be so put before us that nothing of it will perish. It is a good thing to keep for our heirs a treasure predestinated to its special purpose. This is a real treasure, which is dug up with that end in view. Let no one think that I mean here to treat of anything save of the stars, and these are sufficiently explained, that being added which so far ends our knowledge. What we have deemed necessary we have linked together. And it should be known that the medium must be rightly nnderstood; for without this nothing is done; this is so.

The higher star governs all lower things. Now, if there is no star in the earth, the higher star will affect nothing. But the star of the earth conceives the power of the higher star, and is capable of containing it, which else would not be the case. It is so also with the water and the rest, as we have said above.

ANOTHER SCHEDULE

I. This threefold operation of astrology has one mode in a figure of the heaven. By this it is understood how the heaven stands related to lower things, so that a perfect judgment is able to be made.

II. COMETS. What we understand as such are newly-begotten stars, not produced at the first creation, but freshly exhibited by God. Such were the star of Christ and others like it.

IMAGES are made from terrestrial things endued with celestial powers, by means whereof they heal diseases and turn aside wounds in the case of those whom they mark.

CHARACTERS are words which heal diseases and act like images. They are drawn from the higher stars and are artificially assumed by the lower.

GROWING THINGS OF THE EARTH are like Characters and Images. Sometimes trees are brought to such a state as to put forth flowers. Sometimes these growing things are changed into frogs, serpents, owls, scarabæi, dragons, etc.

SPECTRES are visions which sometimes appear to men. They portend wars and other future evils, like comets. They should be explained magically.

DREAMS occur if the heaven and its sidereal spirit sport and joke with men, concerning the past, present, and future.

BRUTES are used when heaven works by them and foreshadows the future, so that by them we can be informed of some impending evil and misfortune.

III. THE MIND. This is when the mind itself within man expects something, good or evil. The origin of this is from heaven, which thus sways the mind, and impresses on it its good or evil fortune.

SPECULATION is when a man speculates and imagines within himself, and thereby his imagination is united with heaven, and heaven operates so within him that more is discovered than would seem possible by merely human methods.

PHANTASY is when a fool or silly person speculates, and heaven is at the same time in connection, and so operates by him that from the phantasy of a fool heavenly influence is recognised.

IV. VISIONS are apparitions artificially produced in mirrors, crystals, nails, etc.

ASTRAL SPIRITS are those which dwell in man on the earth, separable from man, and serving him, as long as they exist.

INANIMATES are men who are produced without the seed of Adam by the operation of Nature, such as are giants, lemures, nymphs, gnomes, etc.

CHIROMANCY is a science pointing out the stars by the lines in the hands, feet and other parts of the body, as we have said above in this treatise.

SCHEDULE CONCERNING THE PROOF OF MAGIC

From what source magic proceeds, and how it interprets new signs. " There shall be signs."

Besides, how impressions from above impinge upon lower bodies. Moreover, what effect heaven has with its signs, as earth with its medicines. In order that you may understand this source from which magic draws its interpretation, attend. All sciences, all branches of human knowledge, are from God. These sciences either come from the light of Nature, or are learnt by instruction, or are secretly instilled by God.

The first mode is that in which man learns by himself without the instruction of man. For magic is not learnt by its interpretation, unless it be spoken from on high.

The magician is born, as all arts are born, as is the case with those who find out new arts, as letters, or Montanica :—

The Magi have the new spirit, not created by this or that man. That spirit is born by asking, by searching, by knocking, out of the heart and by the spirit.

NOTE.—Whatever God says, He adds an interpreter thereto. Let none, therefore, ask, whence is this? It is from God. He, for example, has said, " There shall be signs in the sun and in the moon." This needs interpretation. It cannot be explained by Nature, because it transcends the limits of Nature. The spirit must concur with what is said, and he who interprets this is a magician. The spirits are in the stars.

HERE ENDS THE TREATISE ON HERMETIC ASTRONOMY

APPENDICES

APPENDIX I

CONCERNING THE THREE PRIME ESSENCES *

CHAPTER I

EVERY thing which is generated and produced of its elements is divided into three, namely, into Salt, Sulphur, Mercury. Out of these a conjunction takes place, which constitues one body and an united essence. This does not concern the body in its outward aspect, but only the internal nature of the body.

Its operation is threefold. One of these is the operation of Salt. This works by purging, cleansing, balsaming, and by other ways, and rules over that which goes off in putrefaction. The second is the operation of Sulphur. Now, sulphur either governs the excess which arises from the two others, or it is dissolved. The third is of Mercury, and it removes that which changes into consumption. Learn the form which is peculiar to these three. One is liquor, and this is the form of mercury ; one is oiliness, which is the form of sulphur ; one is alcali, and this is from salt. Mercury is without sulphur and salt ; sulphur is devoid of salt and mercury ; salt is without mercury or sulphur. In this manner each persists in its own potency.

But concerning the operations which are observed to take place in complicated maladies, notice that the separation of things is not perfect, but two are conjoined in one, as in dropsy and other similar complaints. For those are mixed diseases which transcend their sap and tempered moisture. Thus, mercury and sulphur sometimes remove paralysis, because the bodily sulphur unites therewith, or because there is some lesion in the immediate neighbourhood. Observe, consequently, that every disease may exist in a double or triple form. This is the mixture, or complication, of disease. Hence the physician must consider, if he deals with a given simple, what is its grade in liquor, in oil, in salt, and how along with the disease it reaches the borders of the lesion. According to the grade, so must the liquor, salt, and sulphur be extracted and administered, as is required. The following short rule must be observed : Give one medicine to the lesion, another to the disease.

---

* The doctrine of the three prime principles being the foundation of the physics and philosophy of Paracelsus, it is the intention of this brief Appendix to exhibit that doctrine in connection with the origin of diseases.

CHAPTER II

Salts purify, but after various manners, some by secession, and of these there are two kinds—one the salt of the thing, which digests things till they separate—the other the salt of Nature, which expels. Thus, without salt, no excretion can take place. Hence it follows that the salt of the vulgar assists the salts of Nature. Certain salts purge by means of vomiting. Salts of this kind are exceedingly gross, and, if they do not pass off in digestion, will produce strangulation in the stomach. Some salts purge by means of perspiration. Such is that most subtle salt which unites with the blood. Now, salts which produce evacuation and vomiting do not unite with the blood, and, consequently, produce no perspiration. Then it is the salt only which separates. Other salts purge through the urine, and urine itself is nothing but a superfluous salt, even as dung is superfluous sulphur. No liquor superfluously departs from the body, for the same remains within. Such are all the evacuations of the body, moisture expelled by salt through the nostrils, the ears, the eyes, and other ways. This is understood to take place by means of the Archeus from these evacuations. Now, as out of the Archeus a laxative salt comes forth, of which one kind purges the stomach because it proceeds from the stomach of the Archeus, so another purges the spleen because it comes from the spleen of the Archeus; and it is in like manner with the brain, the liver, the lungs, and other members, every member of the Archeus acting upon the corresponding member of the Microcosmus.

The species of salt are various. One is sweet as cassia, and this is a separated salt which is called antimony among minerals. Another is like vinegar, as sal gemmæ; yet another is acid, as ginger. Another is bitter, as in rhubarb or colocynth. So, also, with alkali; there is some that is generated, as harmel; some extracted, as scammony; some coagulated, as absinth. In the same way, certain salts purge by perspiration alone, certain others by consuming alone, and so on. Wherever there is a peculiar savour, there is also a peculiar operation and expulsion. The operation is of two kinds—that which belongs to the thing and the extinct operation.

CHAPTER III

Sulphur operates by drying and consuming that which is superfluous. Whether this proceeds from itself or from others, it must be completely consumed by means of sulphur, if it be not subject to salts. Thus, a medicine for dropsy is made of the salts produced out of the liver of the Archeus to consume the putrefied and corrupt. But to remove the disease itself the strength of sulphur is necessary, to which diseases of this kind are subjected in virtue of their origin. Yet, it is not every kind of sulphur which will effect this purpose, and so it results from the nature of the element that every sickness produced by the nature of the body has its contrary from the nature of the element. This takes place both universally and particularly, and, consequently, from the genera of an element the genera of diseases may be recognised. One is always the sign and proof of the other.

The same sign occurs in the case of mercury; it assumes that which separates from salt and sulphur. Hence are produced diseases of the ligaments, arteries, joints, limbs, and the like. Hence in these diseases we must simply remove the liquor of mercury. But the ailments themselves ought to be removed by those things which are favourable and conducive to them, when proof has been obtained of the speciality of the thing in Nature.

CHAPTER IV

The physician should understand the three genera of all diseases as follows. One genus is of salt, one of sulphur, and one of mercury. Every relaxing disease is generated from salt, as dysentery, diarrhœa, lienteria, etc. Every expulsion is caused by salt, which remains in its place, whether in a healthy or suffering subject. The salt in the one case is, however, that of Nature, while in the other it is corrupted and dissolved. Cure must be accomplished by means of the same salts from which the disease had origin, even as fresh salt will rectify and purify dissolved salt. The sulphureous cure follows as a certain confirmation of the operation in salt.

All diseases of the arteries, ligaments, bones, nerves, etc., arise from mercury. In the rest of the body the substance of corporeal mercury does not dominate. It prevails only in the external members. Sulphur softens and nourishes the internal organs, as the heart, brain, and reins, and their diseases also may be termed sulphureous, for a sulphureous substance is present in them. Let us take colic as an example. Salt is the cause of this, because this predominates in the intestines. In its dissolved state it produces one kind of colic, and when it is excessively hard it produces another kind; for when it passes from its own temperature it becomes excessively humid or excessively dry. In the cure of colic by elemented salts the human salt must be rectified. But if a salt other than from sulphur be applied, you must regard it as a submersion of salt and not a cure of colic. Similarly, in the case of mercurial and sulphureous diseases, each must be administered to its counterpart, not a contrary to a contrary. The cold does not subdue the hot, nor vice versa, in congenital diseases. The cure proceeds from the same source as the disease, and has generated the place thereof.

CHAPTER V

The genera of diseases are also divided into various branches, locusts, and leaves. Yet is there one cure. The mercurial disease is an instance, for mercurial liquor separates into many branches, locusts, and leaves. So all varieties of pustules are subject to Mercury, because the disease is mercurial. But some are subject to common and others to metallic mercury, some to mercury xylohebenus, some to mercury of antimony. It is necessary, therefore, to know that liquor of mercury, which cures that which the salt of mercury dissolves, and it has also an incarnative virtue. For mercury is multiplex. In metals the liquor of mercury is like a metal ; in juniper and ebony it is like wood ; in marcasite, talc, and cachimia, it is like a mineral ; in brassatella, persicaria, and serpentina, it is like grass, and yet there is but one mercury variously manifested. What has been said of pustules must be understood also of ulcers, of which some are cured by the mercury of persicaria, some by the mercury of arsenic, and some by the mercury of xylon guaico. Consequently, the physician should know the tree of diseases and the tree of natural substances, but of these there are indeed many. There is the tree of salt, which is twofold, namely, of rebis and of the element. There is also the tree of sulphur and there is the tree of mercury. Accordingly, the physician must guard against inserting two trees into one cure ; he must remember that Mercury must be administered for mercurial, salt for saline, and sulphur for sulphurous diseases. To each malady let the corresponding remedy be applied. So are there only three medicines as there are three forms of diseases.

CHAPTER VI

In fine, the physician should classify diseases under the name of their medicine. It is opposed to the usage of art to say that a complaint is, for example, jaundice ; any rustic knows this. Let him say rather : This is the disease of Leseolus. Thus, in one word, you comprehensively express the cure, property, name, quality, disposition, art, and science thereof. For Leseolus cures jaundice and nothing else. I would persuade every one to become accurately acquainted with the trees, for he who knows not their seed is involved in fundamental errors. So also we must say that this or that is a disease of gold, and not that it is leprosy. In like manner let him speak of disease of the tincture, whence it will be evident that the complaint is one which belongs to age, for the tincture regenerates age. So also we shall have a disease of vitriol, and this is epilepsy, which is cured by the oil or spirits of vitriol. I have comprehended these matters under a theory because of the special mode from which it is first deduced and the mysteries of Nature which were hidden by alchemical authors. From these I prove my theory of the elementary in its production and the annual in its generation. Let us instance the operation and virtue of Mercury. There are many of these operations and virtues both in the elementated and the annual, which experience teaches to those who know in what things Mercury and in what things other spirits lie hidden. They also will know how to prepare that Mercury, and how to form one kind into a topaz, another into crocus Sandalius, a certain other into a spirit, and any they choose into the exaltation which best suits it. The power of flesh astringents and flesh formers for wounds proceeds from Mercury alone, in which there is no sulphur and no salt, and the same is extracted and produced into its own pure liquor. But some Mercury is quicker in operation than others, as the Mercury in resin, which is quicker than that in mumia or tartar. The same process must be followed with sulphur and salt; the physician must understand their exaltations if he would cure his patients. I know perfectly well that Porphyrius would marvel were he to hear that Mercury becomes sapphire and a noble jasper because he has not seen it or handled it.

CHAPTER VII

Ginger is diaphoretic by reason of the salt out of the body whereof it is made. But that virtue belongs to the fire, through which generations boil up, as is held in philosophy, and by reason of this boiling up it removes obstacles, and reduces or elevates the humours of sulphur, salt, and Mercury to the second, third, and fourth grade of ebullition. And as it is constituted out of the igneous nature of salt, so it also ascends a grade, by which grade the humidities distil through the pores and guttas. Thus purifiers perform their work by the sole force of salt, as, for example, honey. The balm of salt is situated in honey, which, consequently, does not putrefy. For balm is the most noble salt which Nature has produced.

Attractive force is of sulphurous nature or essence. Mastic is a sulphur thus produced, and so also opopanax, galbanum, and others. Nor must we accept the axiom of physicians, that it is the property of heat to attract. We should rather say that it is the property of sulphurs to attract. Hot things only attract in so far as they burn, but that which burns is sulphur, which is not fixed, and hence evaporates like gums. Laxatives also attract like a magnet from those places where they are not. But the reason why salts attract is that salt is impressed upon sulphur and coagulated by means of the spirit of sulphur. Therefore, it attracts from places more distant than itself. Thus there are aperients of sulphur, whether cold, or green, or purple red, of any fashion whatever. For it is the nature of aperient sulphur to operate and to drive before it every moveable thing which it reaches. Nor is it true, as physicians say, that it is the nature of cold to cause evacuation.

CHAPTER VIII

What we should know about tonics is explained by the Archeus, who is like man, and remains hidden in the four elements—being one Archeus indeed, but divided into four parts. He therefore is the great cosmos and the small man, and one is like to the other. From Archeus proceeds the force of tonics. That from the heart of Archeus acts as a tonic to the heart, as gold, emeralds, corals, and the like. That which proceeds from the liver of Archeus strengthens the liver of the lesser world. Thus, neither Mercury, sulphur, nor salt, bring out this kind of healing virtue. But the heart of the elements sends it forth ; from this does that flow. In the elements there is a force and potency which produce the tree from the seed; thence it derives the strength by which it bcomes erect and stands fast.   So, also, by an external strength which the eyes see do hay and straw grow up.   There is a like strength in animals, by virtue of which they stand and move.   Moreover, there is another strength which is not visible to the eye, but is inherent and is the principle of health in the subject wherein it abides.   This is the spirit of Nature, which if a thing have not, it perishes. This spirit remains fixed in its own body. The same strengthens man.   In this manner does the strength in all the limbs of Archeus flow down into the lower world by means of vegetables.

APPENDIX II

———

A BOOK CONCERNING LONG LIFE *

———

BOOK THE FIRST

CHAPTER I

SINCE it is becoming to Theophrastus that he should philosophize further concerning long life, it is necessary, in the first place, and worthy to be known, in my judgment, what life is, especially immortal life, which subject the ancients completely passed over, as I believe, either because it was by them unknown, or was not sufficiently understood. Hence it is that so far they have made provision only for the mortal life. Now, I will straightway define what life is. Life, by Hercules, is nothing else than a balsamite mumia, preserving the mortal body from mortal worms and from dead flesh, together with the infused addition of the liquor of salts. Moreover, our life is long, for neither spirits nor the light of Nature affirm that it is short. The life of the ignorant is short, with art it is long. What is shorter than art? What is longer than life, at least among those who are not super-stitious? Further, what is more durable, more healthful, and more vital than balsam? What is more transient, more weak, and more mortal than the physical body? Its measure varies between long and short. Why, therefore, is life long and why is it short? But that life which is of the supercelestial-physical is outside our rules. The pomp of our authority extends only over the mortal body, and is regulated, so to speak, by art unto the third terminus, unto the fourth, even unto the fifth. So much for the living body. What, then, about death, and what is death? Certainly nothing else than the domi-nion of balsam, the destruction of mumia, the ultimate matter of salts. The separation of immortal from mortal things produces a dissolution of the mortal

---

* This Appendix may be regarded as serving two purposes. The subject of long life is, of course, a highly important branch of the Hermetic Mystery, and whatever Paracelsus wrote concerning it should be included in a collected edition of his Hermetic writings. But the alternative treatise entitled *De Vita Longa* shews Paracelsus at his darkest, and, it may be added, at his worst. From beginning to end it is not only unintelligible, but almost incapable of translation. It is well that one specimen of his really arcane manner should be given to the reader, so that he may regard more hopefully the difficulties which encompass the comparatively lucid works which have preceded. The present version has been reasonably compressed, but it can only be affirmed that it interprets the original about as accurately as can be expected.

members. But this is accordingly called long life from the beginning. This also is short life, that is to say, it is said of death. Now, death is not life, but art is longer than this death. These are the dissolutions of life, also digested separations of that which is pure, of life long and healthy, both mortal and immortal, which the day of birth has united and conjoined, and that from both bodies. For every conjunction of perishable things of a diverse kind brings about dissolution, and how much more then will the conjunction of things natural with things which are beyond Nature be also followed by dissolution? For the cause of death is an empirical war, scarcely different from a duel taking place between the mortal and immortal. Disease may be compared to the javelin, and the Anthos to the breast-plate. What else is there over and above the struggle which these carry on? Herein is the fountain and origin of the generation of disease which presently death follows. Hence we understand what life is, both the mortal and immortal.

CHAPTER II

In order to the clear understanding of what has been already said, I consider that I should next speak of the physical body. The end of the physical body is the sustentation of all those things of which we have been treating. Herein we should divide our examination after the following manner. In the first place, let its parts be considered, not only the special organs but those which are distributed over the whole body, including a right understanding of the marrows, the conditions, the uses of tendons, the forms of bones and cartilages, the nerves, the properties of flesh, the virtues of the seven chief members, bearing throughout this rule in mind. In the first place, we must thoroughly know the whole *rationale* and nature both of the physical body and the physical life. Now, the body and the life of the physical body are alike mortal. But from that which is mortal nothing can be elicited in the direction of long life, and thus with regard to the arcanum and the elixir, in this our Monarchia, neither the body nor the mortal life ought to be considered. For long life is a thing outside the body, is preserved apart from the body, and the body is inferior thereto. Moreover, when the body intervenes, a dissolution of either life takes place. On this subject the Empiric Muse and the medical sophists, following the method of the Spagyrites, preserved the body as a balsam to avoid occasioning death, whereas the balsam is the mumia of life, not of the body, forgetting, meanwhile, that death was not in life, for the death of life is nothing else than a certain dissolution of the body from the immortal. When this takes place, then the body dies. This was the mistake of Hippocrates throughout all his prescriptions, namely, that he administered to the body instead of to the soul, and that he proposed to preserve the mortal by means of the mortal. The body is a creature, but not so the life, and it is indeed nothing but the daughter of death. Therefore, from Archa descended that which is immortal.

But you will say that the Hippocratic Muse is not altogether to be referred to death. Be it so, but you will find a much easier way to health, since the Magnale has descended from above. For God gave unto Hippocrates only those things which are creatures, and among these even the chief mysteries were not imparted in their fulness. To this body God has added another body which is to be regarded as celestial, that, namely, which exists in the body of life. Hereof I, Theophrastus, affirm that this is the work and this the labour, namely, lest it collapse into the dissolution which is of mortal things and belongs to this body alone. And although dissolution can take place in that perishable body, and from this dissolution may be gathered a loss of the heavenly body, yet it cannot stand in the way of long life, by reason of the restoration which must shortly take place, so that the body may be altogether without any defect, for as fire continues to live so long as wood is present, even is it the same in the case of long life, so long as the body out of Archa is present, because the body as a body is to be preserved by the intervention of a body, which extrinsically grows strong. By means of this it is preserved. For the body is nothing else than the subject wherein the long life of the eternal body flourishes. So much for the physical body.

CHAPTER III

It is needful that we should now state after what manner the matter of the same is to be preserved from all corruption. In the first place, whatsoever the body corrupts in itself the same is to be restored by a foreign body, that is to say, as the Monarchia of the Spagyrists does not admit, by the common nature of balsam which labours to preserve the body. For as it is impossible that wood should not be consumed by fire, so it is impossible for the body not to be at length corrupted by life. Therefore, those who are skilled in essential things, who think they can attain long life as a balsam, are the less to be heeded, since the nature of the balsam is rather to preserve the body from corruption lest there should be a vacuum in the body. For every vacuum is a disease of that place, or a sickness in the body, or, again, it is a certain atrophy of long life. For long life has place in the perfect body; in the imperfect it continually fails until it is dissolved in death. We know that the physical body can be sustained from death, and that by virtue of its innate mumia. These things have reference to a healthy life, not to a long one, which is a terminus for the physical body. But it will be worth while before we explain long life to first exhibit and ensure a healthy life. There are certain things which ward off diseases. It is to be observed, therefore, that corruption is to be removed from the body, and that which blazes forth in long life is to be again returned to the refrigerium. Wherefore, in this place the specifics of Nature which are prepared for this purpose must necessarily defend the body wasted by any disease, which is the duty of the physicians, but in long life nothing of the kind is required. And now concerning short life.   The specifics for given diseases have nothing to do with long life ; they are used solely to fortify the body.   It makes little difference as to long life whether provision is made against fevers, etc.   For as long as the spirit of Nature remains it preserves the celestial body, and the long life remains, together with the torture of diseases.   Death here is not ready, for as long as the body is committed to the care of the physician there is death, but not the celestial body, for out of the body flows a poison into life which so inflames it that it bursts forth altogether into bad flesh, and seeing that death takes its origin from corrosives, and a certain arsenical realgar is for all, therefore it does not cease from the nature of a poison until it has satisfied its nature and consumed the body, converting it into incinerated eschara.   Nor after the end will it cease from its malice therein.   Therefore, a double praxis is to be begun—one to preserve life, and the other to repress the body and to alleviate it from day to day by reason of the corruption which takes place daily.

CHAPTER IV

Whereas, by the nature of its creation, the body, and its physical life, passes as one part into the composition of the form, and because the physical body is the half, being the whole with the celestial, the physician ought to give more consideration to the question how the major physical man is to be preserved. For in the major life consists long life, but in the minor is the subject of mortality, and this is implanted according to predestination, both the body and the celestial life in the physical body, which, as an individual companion, accompanies this conjunction.   Now, it is to be understood concerning pre-destination that there are some things which are free therefrom, and are therefore disposed according to the Divine Will, without the violation of any law.   Further, when another conjunction of these two forms has been produced, of the natural, that is to say, and of that which is beyond Nature, into the form of Nature, and when it is completely elicited in the matrix that there are two fathers and one son, two mothers and one daughter, and that these four persons generate, this celestial seed produces, together with the mortal, that animal which appears when born, the elemental seed, and also the celestial at the same time.   For in this place the corporal seed truly works, which also ought to be preserved in the predestination of the natural channel, for that which is beyond Nature ought to be considered in the first place, in order that by these things even those which are beyond Nature may be preserved.   Thus a boy is designated as the heir of two inheritances—of the nature and of the essence which exists from Nature, according to the decree of the creation which comes forth at the same time, and of that also from him who is the parent beyond Nature, which parent rules the body and governs it.   Out of the two, that is to say, from these parents, arises a conjunction of matrimony from on high.   For Adam ob-tained nothing from his creation, nor was he made subject to ascending signs or to any other matters.

Nothing, therefore, out of the four stars can participate with man. For the stars and the homuncula are not divided, but man has received long life from that which is beyond Nature. In the first place, the physician is to be admonished in order that he may possess and use the truth, not following everywhere the figments of the unskilled, who have written most frigidly on the matter, that he should pay more attention to the things which are beyond Nature than to those which are according to Nature. Next, that he should be fairly acquainted with predestination. From this, as from a source, proceeds that monarchia which is beyond Nature rather than the specific and qualitative. Hence is spread abroad that error wherewith not a few are imbued, so that they determine to study the body, and attribute many more things thereto than of right belong to it. Wheresoever present life exists, it is not in its fulness, and therefore it exists without force. This, although it be dead, because it does not operate, is yet implanted in the body. Thus, in the case of one who holds a knife to his throat, a blow takes away long life. Whatsoever further life arises in the body is the congenital life of Nature. To this, however, attention need not be paid, but to that only which revivifies the body. Moreover, long life exists as a man with us, even as fire put among wood, whereby a man recuperates himself.

CHAPTER V

We will discourse briefly of all these things, that it may become more clear as to what has been said concerning the parent which is beyond Nature, and is engrafted definitely on the natural, also the cause of the dual life of the natural body and the parent of that which is beyond Nature. From this it is clear that man is born of a double seed. For now from the time of Adam his complexion has changed with the nature of the generations in the flesh, by means of the importunate and unseasonable operations of persons in their contrary nature. It is clear that neither the sanguine, the melancholy, the choleric, or the phlegmatic temperament is born with us. From none of these has a complete temperament ever arisen. The physicians should now, therefore, pay no heed to the four complexions, for they did not exist in Adam, and much less then in his progeny, nor can four such diverse things co-exist. For first, as by the intervention of an unseasonable birth, the respective temperaments are corrupted, and this not without loss of children (for what is the temperament? It is the nature of the parent, and that without hot or cold, black or white); thus, also, in the body which is beyond Nature there exists a certain hereditary seed, and if two human beings of the same temperament unite, yet the supernatural semen under which both wisdom and life are hidden is never truly conjoined. There is therefore a dual marriage—one which human reason counselled, the other which is the conjunction of God. The former is not properly marriage, except as the eyes in the senses of Nature permit. In this, although a man in every way considers how he may excuse himself and his children, and require the Spirit of God to unite them, and profess to be honourable, there is nothing but hypocrisy. The new change of locality is a proof. But the conjunction which is of God is properly marriage, and belongs to long life by reason of divorce which in this place cannot take place, a fact which none can understand without the intervention of children. For this reason many are sanctified in the womb of their mother; these are they whom God has joined, as she who was once wife of Uriah and afterwards of David, although this in all human judgment was diametrically contrary to a just and legitimate marriage. But God effected this union because either had attained a long life beyond that which is of Nature, as by heredity, on account of Solomon, who could not otherwise be born except from Bathsheba, by a meretricious power, with David. Therefore whatsoever is beyond Nature is as a treasure committed to God, a fact well known to those acquainted with the Spagyric art, and marvellously conducive to long life.

CHAPTER VI

The practice, therefore, being divided into two, one for physical and one for long life, the physician will diagnose from the end, so far as the use of either shall come in. But of this life which is beyond Nature, whereof we are at present speaking, it must be ascertained whether it may be possible by any means to attain it in the physical life, since its sphere is beyond the powers which are accorded to Nature, and under it lie hidden the arcana of long life. For in this place the impressions which are beyond Nature are openly produced; they flow together into the supernatural life, even as the firmament passes by influx into the body which is according to Nature, and although supernatural impressions appear, yet the knowledge of them is obscure. Hence it is that they received the name of impressions from some, from others that of incantations, from others that of superstitions, while yet further names were bestowed on them according to the rules of magical art. From these proceeds that which the Greeks term Magiria, treating exclusively of impressions, which they call incantations and superstitions, which also belong to the supernatural body. It is important to treat of the supernatural body in relation to its impressions, because the whole of Magia has been perverted to a foreign use by astronomers; it has been wrongly called superstitious, and a certain medical sorcery. After the same manner they referred necromancy and nigromancy to the same source, so that each might be regarded as an idolatry, which things, unless an influence intervenes, would come at once to silence, for although the manes may answer on every side, nevertheless this does not happen without the influence which is beyond Nature, a thing which is wrongly believed to be an imposture of Satan because it is impossible to man, whereas it can be easily produced, as you see in the case of the exorcisms of fantastic spirits. The whole of cabalistical magic is contained in the separation of the body which is according to Nature from the body which is beyond Nature, and is implanted in us as an image, to be sustained and administered, so that although absent it may establish communication between those who are widely separated, and may manifest unknown thoughts. This at the same time may be very difficult for those who are uninstructed in the cabalistic art, seeing that a great mistake has been made even by its professors, a fact which is indicated by their translations out of the Hebrew and the Canons of the Spagyrists. Hence we conclude concerning long life as follows : that out of the supernatural influence not only incantations but the arts of images and gamahei have proceeded. Philosophasters have referred this influence to the stars of the firmament, and out of the coals of heaven have feigned a Mars and Jupiter to govern that body which is beyond Nature, whereas this does not pertain to them except in regard to mortal things which have nothing to do with long life. Hence the things which are to be used for long life are to be extracted from supernatural and not from natural bodies, for the whole of that supernatural force is magic, and every magus obtains the influence which is beyond Nature, together with the body in which it inheres. The body which man bears about within him is invisible to man, as is the case with generation, etc.

CHAPTER VII

But that you may rightly understand after what manner incantations or manes came to be considered superstitions, and how they since came to be abused, so that they ought neither to be called manes nor superstitions, know that the beginning of these things was from the Protoplast, who united a supercelestial and mortal body in his own long life. Now, every phantasy and imagination is a principle and special thing in supercelestial bodies. As the mortal body preserves itself in its own special substance, so does that supernal body in the imaginative, as phantasy is of that body, and is indeed itself a body. Whosoever would, in any sense, control a supercelestial body of this kind, must be thoroughly acquainted with the method of resisting the imagination, for the more frequently that body has intercourse with mortals, with the more peril do these things accompany the body. The protoplasts clearly overcame this, but their posterity, having no solid and perfect knowledge hereof, deceived themselves like madmen and fools, nor are they unjustly considered such. Moreover, that supercelestial body is in no wise dissimilar to the stars out of a certain fire, out of which invisible things there arises a visible cloud. Such also is the property and nature of supercelestial bodies that out of nothing they clearly constitute a corporeal imagination, so as to be thought a solid body. Of this kind is Ares. Those who ignorantly perverted, and knew nothing of the foundation upon which this art is built, feigned that there were manes, which originally were called Fate, and afterwards superstitions and incantations. Out of these supercelestial bodies both nigromancy and necromancy originated, and so also geomancy, hydromancy, pyromancy, and lastly also the arts of mirrors, the divining rod, divination by key, and innumerable other things which are classed among superstitions.

CHAPTER VIII

But, that the physician may know all things fully, let us remark the examples of the elders who laboured very greatly in the said magic, that they might obtain long life, without any mixture of the Hermetic rejuvenescence, and without the art of Spagyric experience, being of the body alone. We see the age of Adam and Methuselah, with whom the art of magic began. It is vulgarly thought that the Protoplast was predestined to attain the greatest possible age, but that the smallest measure of years is allotted to ordinary men. The latter point is much insisted on in the schools, but is by no means to be approved. The source of Adam's longevity was magic, by means of which influence he always lived. The death of Adam is ever to be deplored by posterity, not so much because of the fall, but of the science which died with him, who alone retained the spirit of the highest life beyond that which was of Nature. Understand the same of Methuselah, who was next to Adam. There have been other men, indeed, not unworthy mention, who surpassed the ordinary length of human life, as Moses, who completed one hundred and twenty years, yet not according to the method of magic, but rather of physical life, to whom was joined so strong a nature that it attained a great age without difficulty. Like instances occur in our own days, and will be found occasionally to the end of the world. Some, again, by the help of magic, have lived to a century and a half, and yet some have attained to a life of several centuries, and that by the adjoined force of Nature, which exists fully in metals and in other things which they call minerals. This force lifts up and preserves the body above its complexion and inborn quality. Of this kind are the Tincture and the Stone of the Philosophers, because they are elicited from antimony, and, similarly, the quintessence. These and other numerous arcana of the Spagyric art are met with, which in all manners restore the body exhausted by age, return it to its former youth, and free it from all sickness, a fact which is well known to all acquainted with this monarchia.

CHAPTER IX

There is also another way of preserving long life, which Mahomet prescribed to his disciple according to magic, and endowed him with many years ; nor did he do this from God, but from the influence which is beyond Nature. Because Mahomet, as a magus, exercised this method for the unskilled population, not for himself, he has won an immortal name. Archeus preserved his life for several years beyond a century, a thing which was laid to his discredit, and was referred to idolatry. He was equally skilled in cabalistic art with those three Sabean magi who came, not by natural magic, but by the force of horses, to the Bethlehemites, and was acquainted not only with that which was of long life, but that which is of the intellect beyond Nature. All these things proceed from supernatural influence, which rules and governs the body. These magi were afterwards followed by those who falsely claimed for themselves this almost divine name, among whom was Hippocrates, who preferred rather than that his daughter should remain in her actual form, to transform her outward natural influence into a body alien from all Nature—an evident proof of the power of incantation. In the same way Serellus attained long life and studied the metamorphosis of Nature. In the conservation of that body which is beyond Nature the most part were equal to Methuselah, but they made great errors in the transformations; their operation passed into a fantastic body, by reason of their ignorance of physical things. There are many, indeed, whose length of life will persist up to the last day. Such metamorphoses, however, take place without long life, as we see in the case of sea-wolves, who, if restored to their pristine form, again become subject to mortality. Judge also in like manner concerning the fantastic body, on the intervention of food or the osculum of man. All these things are subject to the deltic impression, but before they pass into the deltic impression death is not present, except as far as a mixed fantastic body is admitted, which produces a narcotic form, preserving even to the last day. Moreover, many have lived upon the life of another, and that according to the rule of the Deltic Nature, among whom was Styrus, who when struggling for life is said to have attracted to himself the strength and nature of a robust young man, who chanced to stand by, so that he succeeded in transferring to himself his senses, thoughts, and even the mind itself. By this imagination Archasius is said to have attracted to himself the science and prudence of every wise and prudent man. Such is the strength of mind in which that supernatural vigour exists, that it sometimes satisfies a glowing and, what is more, a ravenous concupiscence. Hence arises that contempt of images and gamahei among those who abuse this image even to destruction. Hence are those words, characters, signs, forms, and figures of hands, imprecations and orations, which are the principal cause of incantation, and, what is more, of words which are commonly applied to wounds and other diseases. Finally, whatsoever can change into this form does so by the force of that body which, beyond Nature, is implanted in us. Further, out of those impressions which are beyond Nature arise the stars of the firmament, Venus and Saturn, and other planets, so that that influence which is beyond Nature rules and governs inferior things. Whatsoever, therefore, takes place in gamahea and imaginations, by the accession of planets and signs, all this can be transferred to the superior signs. Wherefore those bodies which are perishable can easily be set free from death by that supernatural force. Moreover, Venus and Saturn, Mars and Mercury, exercising their force in the superior firmament, have endowed the most part of mortals with immortality, and that without any human operation, by the accession of imaginations, of whom not a few exist, visible and invisible, both on earth and in the sea. Some of these have attained this point by means of Deltical impressions, not, however, the nymphs, as is the case with animal generations.

A BOOK CONCERNING LONG LIFE

BOOK THE SECOND

CHAPTER I

HAVING spoken of the several arcana which restore to its pristine health a body affected by diseases, we will begin where we last left off. To finish what we handed down in former books, and to shew how the physical body may be preserved like a balsam, the particular arcana and the matter of this second book are referred to the same body. Although, then, one and the same preparation holds good, still the practice comprised in this elixir differs from that special mode of healing. In this second book the first places are held by Flos Cheiry* and Anthos. In this is comprised the arcanum of elixirs, and that by the force and virtue of the whole quintessence. At the outset, therefore, in order that each may be the more clearly noted, I will, with this view, point out in a few words what the quintessence is. Nature procreates the four elements, from which a certain tempered essence is prepared by the spagyrist, as is expressed by the Flos Cheyri.*

Now, here I think it matters little what the art of Lully teaches on this matter, since he wanders more than sixteen feet from that universal Monarchia which the Archidoxies prescribe. One thing is Extraction, another Confortation, another Melioration, to adopt the terminology of these men, of which Raymond makes mention in that treatise which is entitled "The Art of Lully," and from these he has made a false estimate of the quintessence. Since these are mere trifles rather than truths, we will pass them by in silence. But the Flower of Gold, the Flower of Amethyst, and lastly, whatever is of a transparent nature, pearls, sulphurous bodies, cachymiæ, and whatever belongs to the aluminous zerebothini, including all the genus of other things which the water produces, such as carabæ and corals,—these, I say, are all capable of forming quintessences according to the rate of temperation which is wont to be produced by the spagyrist through the intervention of a corruption of the elements.

CHAPTER II

Moreover, the sum total of the whole matter lies in this (since what is said in the book on the Elixir must each and all be referred to the subject of long

---

* Thus differently spelt in the original.

life), that universal Nature is reduced to the spagyric mixture, or temperation, which is nothing else than the goodness of Nature, in which is nothing that is corruptible, nothing of an adverse character. And yet, by another method and a different one, the same goodness of Nature is found in the tincture, according to the prescription of Nature, which exists in the Philosophers' Stone, in antimony according to the Nature of the crow, in sulphur according to the effect of the Lunary, and in the same way in other cases. Nevertheless, in all these there is one and the same temperation which among metals lurks under Mercury (I mention Mercury, which is in all metals), among gems under the crystal, among stones under the zelotus, among liquids under carabe, among herbs under valerian, among roots under sulphur-wort, among bitters under vitriol, among flints (say rather among marcasites) under antimony. Moreover, as Mercury is in all metals, so is antimony in all flints (or rather marcasites), vitriol in salt, and melissa in herbs. These are names of the tempered elixir. It should be remarked, too, that in elixirs following upon the sulphur of those substances which certain people call minerals, there is a quintessence, the Mercury of the Metals, from which is extracted the nature of the body. Cheyri prevails in Venus, Anthos in Mars; and the force and nature of these are not only that they drive away diseases, but that they preserve that body for a long life which is dependent upon the lower influence. With this view we will further say that the Elixirs of Long Life shall be embraced under many and various names, since the force of them all is one and the same. To us (if, perchance, you wonder at this mode of treatment) it has seemed good in the meantime to play with words.

CHAPTER III

Of all elixirs, the highest and most potent is gold. We will, therefore, treat of this first. If you understand the principle of this, you will understand that of other substances which are separated from their bodies. The rest, which are not separated from the body, will be indicated below when we come to mention wine. Concerning the Elixir of Gold, then, so far as relates to practice, act thus: Resolve gold, together with all the substance of gold, as a corrosive, and continue this until it becomes identical with the corrosive. Nor let the mind revolt from this method of treatment; for the corrosive excels gold, so far as it is gold, and without the corrosive it is dead. The quintessence of gold, therefore, without the corrosive, we assert to be useless. It follows, then, that the resolution must be renewed anew by means of putrefaction, although the corrosive adheres somewhat closely. For if the force of gold is so great that it preserves the body and renders it free from all sickness, nor allows it to be corrupted, how much more itself, and that without any infection? It corrects and purifies everything that is not pure. The corrosive, therefore, in the case of gold, ought not really to be called a corrosive at all. For the force of the arcanum overcomes all poison. All realgar dies in the elixir of gold, and goes off to the tincture which excels in medicine. And thus it is in this way that Potable Gold is produced after putrefaction. The common practice of the Spagyrists prescribes this dose, or rather a certain harmony. Lastly, you will notice about the elixir that wherever an elixir is brought to bear on anything, it so transmutes it that it remains fixed in a form similar to itself.

CHAPTER IV

Concerning Pearls

Now, in order to give greater clearness to what we have said about the quintessence, it should be remarked that nothing is nearer to gold than pearls. You must, therefore, reduce to temperation the four elements which are in pearls, whereby exists a quintessence without any loss of substances. Moreover, if you wish to transmute pearls into a quintessence, according to prescribed rule, act after the method of a quintessence. Do not change anything except the principle, in which it is necessary there should be joined the ultimate matter which exists as first matter in finishing the quintessence of Sol. This is extracted by a prescription of the following kind: First of all reduce to liquid a lemon newly re-elevated, in which pearls have been calcined, dried, and resolved; this serves for a resolution into the element, in which resolution is no complexion whatever. There is herein an universal force like a quintessence. I cannot in this place advise you to admit that method of extraction which Archelaus prescribes, nor any other spagyric separations of that kind. The mode of transmutation given above not only restores to their former power those members which are weak, but also keeps in the same vigour those which are strong and robust. So there is much more in pearls than in other sperms; and among these I consider the most excellent are those which come from the oysters.

In this place, too, the homunculus treated of in the Archidoxies bears no small part. The necromancers call it the Abreo; the philosophers name such creatures naturals, and they are commonly called Mandragoræ. Still, error prevails on this subject through the chaos in which certain persons have involved the true use of the homunculus. Its origin is in the sperm. By means of complete digestion, which takes place in a venter equinus, a homunculus is generated like in all respects, in body, blood, principal and inferior members, to him from whom it issued. We will, however, in this place pass by its virtues, because the subject has not been dealt with, as that of pearls has, by those who are acquainted with this matter.

CHAPTER V

Concerning the Extraction of the Quintessence from Herbs

We have made mention above of that Quintessence which should be produced without any extraction, and it is necessary to regard this subject in connection with our present opinion. The quintessence cannot be got from herbs without extraction, on account of the diversity of those essences which are included under one substance. These must be separated, so that the herb shall remain a herb and the quintessence a quintessence. Although in every herb there are four duplex elements, still the quintessence is not duplicated, but one part only. The other part, which belongs to the substance, we relegate to those arts that are special, and will treat of what belongs to the elixir. This is made quite clear by the example of melissa. Digest melissa for a philosophic month in an athanor; then separate it so that the duplicated elements appear separately, and immediately there will shine forth the quintessence, which is the Elixir of Life. Such is the case, too, with generous wine, and differently in other instances. In nepita it is bitter; in the tare, like clay; in tincium, blackish; in the hop-plant, slender and white; in the Cuscuta, harsh. In other cases it must be judged in like manner according to the prescriptions of experience.

Moreover, when this spirit has been extracted and separated from the other, behold the wine of Health! The philosophers have strenuously tried for ages to attain this; but they have never succeeded. A good part of them, followers of Raymund, have emptied several casks, in order to extract the quintessence of wine, but they arrived at nothing save burnt wine, which they erroneously used for spirit of wine. All that is necessary on this subject will be found elsewhere, in the "Philosophy of Generations." Enough to have warned the Spagyrist under what form the quintessence exists in herbs, and what it is worth while to investigate in them.

CHAPTER VI

Concerning Antimony

As antimony refines gold, so, in the same way, and under the same form, it refines the body. There is in it an essence which allows no impurity to be mixed up with that which is pure. No one, even though he be skilled in the Spagyric Art, can apprehend to the full extent the power and virtue of antimony. In the beginning of things antimony was developed, and was so related to the metals, which were produced by the water, that, when the Deluge was over, its genuine force and virtue remained after such a manner that it directs itself under the form of influence, and has never lost anything of power or virtue. With due cause, therefore, we assign to this alone everything which is attributable to minerals, whereof antimony includes within itself the chief and most potent arcanum. It purifies itself, as well as other things which are impure. Nay, more, if there be nothing wholesome present, it still transforms an impure into a pure body. This has been dealt with in the exposition of leprosy; and spagyric practice makes everything clear and comprehensible. But, not to digress at undue length, let us come at once to the mode of preparing the virtue of antimony (one jot or tittle of which is better than all the texts in your possession). First of all, take care that the antimony be not corrupted, but that the total, whatever it be, remains entire, without any loss of form, for, under this form lurks the arcanum of antimony, which should be impelled through the retort without any *caput mortuum*, and be reduced anew in a third cohobation to the third nature. Then the dose will be four grains of it given in the quintessence of melissa. To this the Archeus of the earth assigns nothing further.

CHAPTER VII

Concerning Sulphur

It is specially difficult, yet worthy of all celebrity, to realise the power and nature of the earth which procreates balsam, the characteristic whereof is that it suffers nothing to putrefy. But think of the resins, whereof the principal ingredient is sulphur, and there is nothing which deserves greater praise. In sulphur there is a balsam which none who study the different arts should fail to remember. In it are the balsamic liquids which do not allow wine or any-thing dead to putrefy, but do so conserve the body that there can attach to it no evil influence, natural corruption or any impressed on it from without. None need be surprised that so great a power is in resins, or that we speak its praises beyond the balsam which grows on the earth, and but, as it were, illustrates the force and virtue of this balsam. In those which are occult much more is found than in those which are manifest. And so, too, much more is found in sulphur than in the other departments of resins. In the case of sulphur, in order that we may arrive at the method of treating it, proceed thus: Elevate sulphur by colcothar in the spagyric manner. Do this so long as the fire does not get the mastery, as colcothar is wont to do in the case of sulphur. This same fixed spirit is the balsam of the earth, concerning which we write very little in this treatise. Its virtue is made clear by experience; and, though certain gums and resins, and other substances of this class, have the same nature as balsam, still, I think that among these sulphur is the first and the best.

CHAPTER VIII

Concerning Mercury

The Elixir of Mercury, prepared in the same way as that in which it is used for transmuting metals, avails in the very highest degree for driving away disease. Its rust, which the followers of Lully falsely call its flower, is nothing but death. As death consumes and wears away the body, so does rust affect the metal. In whatever way, then, this tincture affects it, the result will be that it ministers to long life, and the more efficaciously and powerfully in proportion as (let the expression be allowed) it reaches the grade of a poison, and the more actively and subtly its preparation has been repeated. Let no one be alarmed by those fables of Rupescissa, who, as his custom is, has written at once rashly and frigidly on this subject, namely, that, in the tincture of the body, you should altogether avoid gold and substances of that kind, which belong to Mercury, and, lastly, whatever is prepared from the spirit of salt or of arsenic. Albertus and Thomas have approached more nearly to the tincture of mercury (the virtue whereof is of subtle sharpness, though it derives its nature from the Archeus), but in their excessive coagulation, and also in the degree of repetition wherein they have overwhelmed the whole affair, they are entirely wrong. In preparing the tincture they verge on the true tincture, as in the following opinion: As metals are transmuted and fully fixed, so also is the body in the following manner: Reduce mercury in elevation until it assumes the form of a fixed crystal; then digest it to the point of resolution and coagulation; join it with gold so that this shall produce its ferment. Then proceed according to the prescript of Hermes, and continue to the completion of the stone. The dose thereof is one grain. Its power and virtue preserve the whole body in its entirety.

CHAPTER IX

Concerning the Spirit of Wine

When I mentioned the essence of herbs above, I pointed out that it is nothing but wine, which I would have you thus understand. The spirit of wine proceeds from its substance. Wine is a subjection of this just as marrubium is of proper and native wine. In order, therefore, to get the spirit of wine as an essence, which is truly an elixir, understand thus: As a pound of persicaria sends forth Ɔij. of wine, so a pound of wine takes not more than one scruple. The rest is the phlegma of wine which has no bearing on the present elixir. Let the preparation of this essence proceed in the following manner: Digest in horse dung wine which has been poured into a pelican. Continue this for a period of two months, and you will see a thin, pure substance, like a sort of fat, which is the spirit of wine, spontaneously evolved on the surface. Whatever is below this is a phlegma possessing none of the nature of wine. The fat, put by itself in a phial, and separately digested, is of the utmost power for long life. And not only does it avail for long life, but this preparation can also be adapted to other purposes by the intervention of cinnamon, xylobalsamum, myrobolani, and other things of this kind, in the following manner: Mix, and by the use of digestion so join these ingredients that with the addition of the above-mentioned elixir and of gold, a medicine shall be prepared which removes all contractions and gives free play to the limbs.

CHAPTER X

The Extraction of Mumia

The extraction of the virtues out of mumia is made magisterially (if I may use that expression) by its mixture with the essence of wine taken from chelidony. Digest it for ten days, and distil for five. Moreover, let it be once more digested afresh until the mumia turns into a liquid. When this takes place above as well as below, these portions being separated from the middle, add the sixteenth part of balsam from woods, and a twelfth in weight of the sealed earth of Pauludadum with the same quantity of liquor Horizontis. Digest this for its month, then shut it up and reverberate it. In this way it ascends to its highest degree. Of all those preparations which are dominated by poisons, this is the most powerful and efficacious.

CHAPTER XI

The Extraction of Satyrion

Whatever has to be extracted from satyrion must be procured by means of separation. In satyrion lurks a Saturnian power which, as it were, secretly steals away and weakens the virtue which satyrion possesses, and so its exaltation reduces it by thirty grains. Hence it not unfrequently happens that when satyrion is used it fails in its effects. It is worth while, therefore, to consider how not its form but only its virtue shall be separated. This must be done in the following way : Let satyrion be digested with panis siliginis in a venter equinus for a month. When this is over, take it away from the bread, and throw away the dregs. Then let the blood of the satyrion be digested thoroughly and allowed to effervesce. When this effervescence has subsided you have obtained a medicine which leaves far behind all others for every purpose which relates to conception.

CHAPTER XII

The Extraction of the First Metal

The most complete and perfect conservation of the body is attained by the First Metal; and this is so efficacious, not by the nature of its own strengthening power, but rather by virtue of the minerals which it contains. For, in order to conserve long life, it is necessary to use the prince of minerals, since minerals make up the physical body. This is the temperament which singly and alone resists corrosives, and Ares operates as much chemically as by means of the Archeus. Moreover, it blends the strongest and the weakest body in one degree. Strength, indeed, is that which exceeds the strength of Ares, and weakness is that which falls below it. That which is taken away from the stronger is conferred upon the weaker, and so each is reduced to a mean. It is done in the following way : Take the liquor of coral, in its most purely transparent form, to which add a fifth part of vitriol, which is from Venus. Let these be digested in a bath of Mars for a month. In this way the wine of the First Metal separates itself to the surface, and the vitriol of Venus lays hold of whatever dregs there may be. Thus the First Metal becomes a clear, transparent, and ruby-red wine, whereof the special virtue and power is that of all the minerals over the whole physical body.

A BOOK CONCERNING LONG LIFE

BOOK THE THIRD

CHAPTER I

LEST anything should be omitted which concerns Long Life, it is proper to observe that within the testa and over and above that quintessence, there is enclosed something out of which a certain conjunction, both of the corporal and of that which is beyond the body, outside of that quintum, produces the body into long life. Concerning this understand that it is absolutely nothing and invisible. But in the body there is something exquisite which not only confers long life upon the microcosmic body, but even preserves Dardo itself whole even to the thirtieth year, and guards the anthos and the great cheyri up to the third age. This microsmic thing sustains both the anthers and the leaves which ought to remain in their own conservation throughout the whole anatomy of the four elements. Wherefore at this point the physician must note that the whole anatomy of the four elements can be contracted into a single anatomy of the microcosm, yet not out of the corporal, but from that rather which preserves the corporal. Indeed, the superquintessence sustains the quintessence itself as well as the other four. If it be proper to give it a just and true nomenclature, I may rightly call it the balsam itself out of which life is preserved, which rightly separates itself from the balsam of the body, and is such a balsam as to surpass Nature herself. This surpassing of Nature is by a corporal operation.

CHAPTER II

But of that balsam whereof we have now spoken, which ought to produce long life, a declaration takes place in two ways—one which is secret and happens by accident, whence it follows that long life is dispensed to the majority, who yet have no idea what it is in itself. But the other mode takes place by arts, that is, by those who are able to obtain that conjunction, nor can it take place without a medium. For herein is situated the point of the matter, because in the Iliaster both long and short life are found. For that which is adjoined to herbs has its terminus; similarly, also, there is a terminus to that which is of the water of minerals; in the same way tereniabin, and so also nostoch.   Besides all these things such and so great is the strength and power of the conjunction itself, that everything which is produced out of the four elements is conserved above its first terminus, and that is the terminus of Iliaster, by which, indeed, we wish overcome that subtle man who says that a terminus cannot be crossed over, which, if it does not take place, and passes over, is for this and the other reason.   For there are two of them, one of which cannot pass over, because the terminus is placed in the nature of the microcosm ; one is in the nature of the elements, the other in that of the quintessence, moreover, also, the other is out of the last Iliaster.   For these termini consist in the power of the physician, who in these can change what he wishes according to his will, except only the fixed, where he ought to expect the end together with the mutation of himself.

CHAPTER III

Understand this Iliaster as follows, since here three virtues are found besides the quintessence.   For there is the Iliaster of sanctity, the Iliaster of the *Paratetus*, and finally that great Iliaster.   Of the first understand that such sanctity imparts long life, according to the industry of him who uses it ; the second dispenses it by favour ; the third, being bruised, consists without harm in long life.   Hence consider the Iliaster comprehended in long life. All three are together subject to the microcosm, so that it may reduce them into one gamonynum ; but the other is in no wise controlled, for it is acquired according to favour.   With the third the case is exactly as with the Enochdiani and the Heliezati, just as it is clearly the case with Aquaster. In the first place, therefore, it has its origin from the elements, as the testa shuts up, and the superquintessence is attributed to the arcana themselves. The second is ascribed to the Magnalia, the third is out of its own specifics. Hence it follows that the dwellers in the earth, the nymphs, the undines, and the salamanders receive their long life in an alien essence.   For there is a death, a time, and a will of that third Iliaster, and he it is who grants to the ear of corn that it should bear more than it would by Nature, as also the fragrance of myrrh, and the strength in Leris.   This being so, the physician must consider that a conjunction of this kind takes place in a similar manner as the tree of the sea when once fixed and reduced to bondage thereby, can also become an approved and constant cheyri ; so by a similar conjunction in the microcosm the same thing comes to pass.

CHAPTER IV

But concerning that first Iliaster, understand that it exceeds a thousand species, not that one excels another, but rather for this reason, that every microcosmus has its peculiar and, what is more, perfect conjunction and virtue : so great is the virtue and potency of Iliaster that by it a dead body is preserved alive, for this reason, because that first terminus is transmuted.

This conservation of long life transcends our powers, but not those of the higher powers. Its sustentation takes place as follows—that it confers long life, yet without the expulsion of the disease. Life, indeed, it affords, but not good health, yet sometimes it affords both, being long life of the kind which proceeds from that Iliaster. It lasts for years and is extended, as, for example, to the tenth year in the case of one who ought to have died in the fifth—a thing which takes place both by reason of the superior and inferior conjunctions.

CHAPTER V

With regard to the true Iliaster, the fact is that nothing of the kind can be reduced without signification or necessity towards the greater Iliaster, which you are to understand as follows. The greater Iliaster which is to extend long life of this kind can by no means do so where there is no place for long life. Hence observe that such a thing cannot take place without transmutation of the place as well as of the elements. That is to say, as the four mortal elements are in the testa, every moment producing a new generation, they lead forth the same to death. In another direction there is a fixed (generation) in its firmament which remains unshaken, neither causing disease nor death. Such fixed spirits suffer nothing to perish altogether, whose long life is immovable and firm even to their transmutation again into the first. The similitude is, indeed, taken out of the text because according to the fixation, as I have said, of the firmament, long life is more prolix in one confirmamentum than in another, although each arrive at their first terminus only. However, some few inhabitants accompany this kind of transmutation, so it is permissible to call it, into the tenth or even into the twelfth, whose death follows on the destruction of that great firmament, where bodies, both celestial and terrestrial, shall be shaken, yea, the supercelestial also. Notwithstanding, this takes place without any distinction of Nature, for in the first moment when such mutations happen the putrefaction of Nature commences, and that is with a still living body.

CHAPTER VI

Now, concerning Iliaster, it is necessary, in the first place, that the impure animate should be depurated without separation of the elements; this takes place without any corporal and mechanical labour, which disposition arises according as man grades himself in mind that he may be rendered like to the Enochdiani, not that he desires the Enochdianian life, for in his mind he differs diametrically from it. Wherefore it is necessary for the microcosmus in its interior anatomy to reverberate it with a supreme reverberation. Thereby the impure consumes itself, but the fixed which is separated from the impure remains without rust. Nor yet is it a fire wherein Salamandrine essence or Melosinic or Ares could be present, but rather a retorted distillation from the middle of the centre, above all coal fire. This reverberation thus being made, in its last terminus it exhibits the physical fulmen, just as the fulmen of Saturn and of the Sun separate from each other. Accordingly, whatsoever advances by this fulmen of long life pertains to that great Iliaster, and this fulmination and preceding reverberation in no wise remove the weight, but rather the turbulence of the body, and that by the method of diaphanous colours.

CHAPTER VII

Moreover, from that Iliaster of the first power long life of this kind does not result, for it affords an inferior grade. Yet he it is, however, who separates in that place, and exactly as a fixed thing can preserve a thing which is not fixed, defends the microcosmus from death, seeing that its operation is not to separate, but rather by means of those perspicuous arcana it should conquer that which is undigested, lest its perdition should follow. Just so mumia, which, together with the body, proceeds from the birth, being itself good, but the body is bad and putrid. Whatsoever life, therefore, the body lives, mumia lives also from it alone, for it is its property and nature to putrefy and revert to dung, of which it is a member, and this is its continual desire. But not so the celestial mumia, for it breaks the worthless part and guards the same by its own will, lest it should ever effect that which it attempts. Wherefore the following is the tenor of the recipe : that the super-celestial mumia sustains the microcosm more than its own mumia. For as often as there is a mumia there is also another terminus. Yet neither time nor number are found in these termini, for they continue to endure till they can no longer escape a second generation. The physician must be perfectly acquainted with the fact that every first matter expels the last. Hence the generation of worms begins where the ultimate matter of the physical body shews itself. Observe, therefore, this Iliaster, that it not only does not destroy the generation of worms, but when their matter is present it does not even impede their generation—a thing which mumia should prevent.

CHAPTER VIII

The natural mumia should be compounded out of three chief antimonies so that the foreign microcosm should govern the physical body, whether by means of the element of water or by means of its metals, salts, etc., or otherwise by means of the element of earth, as by its herbs and boleti, or in tereniabin or nostoch. For all these are mansions of the supercelestial things. Wherefore let no one be surprised that the great virtues of melissa are described everywhere. Seeing that in this a supercelestial conjunction takes place, who shall deny to it a most excellent virtue? These are the magnalia which the Bamahemi contain, and this is plainly Ilech, who, being composed out of the true Aniadus, can in no wise be removed from that elementated thing—a thing which takes place with exaltations of either world, exactly as exaltations of the nettle burn, and the colour of the flammula radiates. Yet in exaltations of this kind their virtue can be reduced into another. Therefore learn to diagnose their exaltations as follows. They are far more potent than the nettle, and also ye may collect the same in the true May when the exaltations of Aniadus commence. For exaltations of the virtues are not only situated in the matrices, but also in supercelestial things. That were a common Idæus and of no importance who knows how to fabricate a single thing subject only to the vision, nor besides tangible things can create greater things still, but he has constructed another May where supercelestial flowers attain their exaltation, in which Anachmus ought to be extracted and preserved, even as the virtue of gold lurks in laudanum. Such, indeed, are the virtues of Anachmus : then will you truly be able to enjoy long life.

A BOOK CONCERNING LONG LIFE

BOOK THE FOURTH

CHAPTER I

WE will in this place complete what has been said previously on the foundation of life, and on the life which is beyond Nature. In the first place, we exhibit to all Spagyrists the age of Adam and Methuselah, after speaking exactly of that long life which is in the hands of the highest Iliaster, according to the manner of magnalia, where more facts are to be dealt with concerning free will than we can administer out of the elements. To make these things understood more clearly we must revert to the Enochdiani. A comprehension of the nature of their influence will enable us to get at the principle of long life, even without any trouble, as was the case with S. John, whose nature comprehended not merely one age or one century. Lest I should give an opportunity to the libellous who wrest the scriptures, we will define nothing certainly in this chapter concerning the life of the highest Iliaster, whether this be present in corporal elements, or whether it lives in the quintessence where no body occurs, and where not only those live whom we have mentioned, but also those whom we thought buried in sleep. All these things I leave to be considered slowly by sublime spirits, while we have descended to these. If that highest Iliaster be impelled, or at least, if it have need of anything, he will easily attain to whatever is Enochdianic, where all our long life is collocated in its proper places in ether and in the clouds. But once for all Iliaster has satiated himself, so that henceforth he lacks nothing.

CHAPTER II

The end of long life is contained within the limit of six or nine hundred years. Concerning the source of this life which is beyond Nature, understand as follows. There are two forces in the power of man—one natural, the other of the air, wherein is nothing corporal. Having treated sufficiently of the natural, the incorporal force shall close our little book. Miserable in this respect are mortals to whom Nature has denied her first and best treasure (which the monarchy of Nature contains), to wit, the Light of Nature. But herein let us not labour vainly, but in the case of philosophy since it departs and diverges from Nature, we will remember the Aniadus, nor will we make further mention of philosophy. Having, therefore, dismissed natural things, and all which has been treated of concerning things out of the elements, as also those which are latent beneath the chaos, that is, the great Iliaster, we will have recourse to what was mentioned in the first book. In order, therefore, that we may arrive at the year of Aniadin, or even further, the following rules are to be observed. Let not what we are about to say of the nymphs offend any one. Here also shall be indicated the force and Nature of the Guarini, the Saldini, and the Salamandrini, and whatsoever can be known concerning Melosina.

CHAPTER III

But in order to make clear at the same time both the place and the body in these things, which have to be ordained and disposed according to a certain harmony, we must observe the nature of Iliaster. It preserves to a period of three hundred, or even six hundred years. Further, whatsoever out of its own nature admits also the nature of the place is brought to one conclusion, like the former century, but where they unite the nature both of the place and the body at the same time, they arrive there, and without any trouble, to the six hundredth year. Some who have reached that age might be enumerated, did not my pen hasten in another direction. There are, moreover, those who for a long time secretly and furtively are preserved to a long life, an account of whom may be omitted because they have given nothing except to Iliaster. Whatsoever does not pertain to aërial life is passed over in this place. Those, therefore, follow who have lived an aërial life, of whom some have arrived at their six hundredth, thousandth, or eleven hundredth year, a fact which can be easily understood according to the precept of the magnalia. Compare Aniadus, and that by means of the air alone, whose force is so great that the terminus of life has nothing in common with it. Further, if the said air be wanting, that which lies hidden in the capsule bursts forth. If the same shall have been filled by that which recently returns, and then is brought forward into the middle, that is to say, outside that under which it lay hidden, it still is so far hidden that as a tranquil thing it is completely unheard by anything corporal, so that there only resound Aniadus, Adech, and Edochinum. These three, and that which verges into these three, are not four but one. You will attain a very long hidden life. Such is the nature of that Aquaster, which is born beyond Nature. But if it has not been able to attain that which was latent, yet here it occasions that which was extrinsically Iliaster, etc.

CHAPTER IV

The monarchia remain, and to this we are recalled by the great Zenio, for there is a life far different, whereunto we are constrained. When all things have passed away the oppressor and the oppressed remain, a fact not sufficiently understood up to this present by the Aliani. Yet a time comes when all these things which we have investigated together shall pass away, from the first even to the last. As to whether a healthy life can be conjoined with a long life, note that there is a double essence, in one of which health resides, and this essence is fixed; in the other disease is centred, and this is similarly fixed. As to the place and the mutations of these things, let us not change anything. What, however, is the use of vainly lingering among those things which the light of Nature has refused to us? Wherefore he who guides us out of the desire of the mind, does not leave us gaping at what he points out. Let us then pass over what is beyond us, namely, certain creatures of a marvellously long life, and proceed to those which have no death, among whom are the Laureus, Siconius, Hildonius, and many others, whose nativity or natural death no one hitherto has attained or heard of. Add the nature and essence of those things, and how many will you find who have written anything at all about them?

CHAPTER V

I make no account of him who by the arts of Lully vaunts himself as a Necrolicus, and inveighs against what is contained in the four Scaiolæ, announcing me as the highest Scaiolus, in order that I may commence Necroleous arts according to the manner of the cedurini. But I envy even the hydra together with the envious Scaiolæ. What shall I say in this place of those things which the sagacious muse embraces in her canons together with the matrix of the four Scaiolæ, which sleep in you, and render your temples anodynic? I occasion so great an astonishment in you that you shall come even to take heed of a poppy. But I confine myself to the cosmographic life, where both the place and the body of Jesihach appear. Further, the things I prescribe I do prescribe beyond the forces of the body and the place. Whosoever understands these things the same has a lawful claim upon the title of a spagyrist. There is no mortality in the Scaiolæ. He who lives according to their manner, he is immortal; this I prove by means of the Enochdiani and their followers. Aquaster will not invade this place. But if I be inserted among the Scaiolæ according to the manner of the Necrolii, there will be something that I might take out and lead, a thing which the Great Adech antiverts, and leads out our proposition but not the mode, a thing I leave to theoretical discussions. And in this manner Melusine departs from the nymphadidic nature, by the intervention of the Scaiolæ, to remain in another transmutation, if that reluctant Adech permit, who is both the death and the life of the Scaiolæ. Moreover, he permits the first times, but at the end changes himself, from which I gather that supermonic figments in Cyphanta open the window. But the doings of Melusine prevent these being fixed, which, being of this kind, we dismiss. But as for the nymphadidic nature, in order that it may be conceived in ourselves, and that we may thus arrive immortal at the year of Aniadus, we take the characters of Venus. If ye recognise these things, nevertheless ye have put them to little use. But we have completed it, so that we may securely attain this life in which Aniadus dominates and reigns, and remains with that at which we ever do assist. These and other arcana are absolutely in need of nothing. In this fashion we leave and conclude long life.

HERE ENDS THE TREATISE ON LONG LIFE

APPENDIX III

A SHORT LEXICON OF ALCHEMY

EXPLAINING THE CHIEF TERMS USED BY PARACELSUS AND OTHER HERMETIC PHILOSOPHERS

THE materials for the following vocabulary are derived partly from Paracelsus himself, that is, from other writings not included in the present translation; in part from such alchemical authors as Arnoldus de Villa Nova, Eugenius Philalethes, Ferrarius, Raymund Lully, and Cornelius Agrippa, who, however, was more magician than alchemist; and, finally, from the following sources :—

A. T. Pernety. Dictionnaire Mytho-Hermétique ; Paris, 1781, 8vo.

William Salmon, M.D. Dictionnaire Hermétique ; London, 1695, 12mo.

William Johnson. Lexicon Chymicum (*editio ultima*). Two parts. Frankfort, 1678, 8vo.

Dictionarium Theophrasti Paracelsi ; Frankfort, 1583, 8vo.

Rochus le Baillif. Dictionariolum, a brief supplement to the Geneva folio.

Martinus Rulandus. Lexicon Alchemiæ ; Frankfort, 1612, 4to.

Michael Toxites. Onomasticon sive Dictionarium Philosophicum, Medicum, et Synonymum ; Argentorati, 1574, 8vo.

Gaston le Doux. Dictionnaire Hermétique ; published under the pseudonym of a lover of Hermetic Truth.

The vocabulary has not only been compiled, but has, for the most part, been literally translated from these authorities, and where it speaks positively upon the mysteries of Hermetic science it must not be understood that the editor himself is speaking. While the information it contains may perhaps claim to be regarded as reasonably full, the reader must not expect to find a satisfactory explanation of all strange, *bizarre*, and unaccountable terms which are to be met with in the text. Some are peculiar to Paracelsus, and outside the sage of Hohenheim himself it would be useless to look for information. Such words as Deneas, Magdalion, and Censeturis belong to this category.

Others, less apparently recondite, are notwithstanding wholly unknown to the editor, though he may claim a wide acquaintance with Hermetic literature and its curious recipes. The Stomach of Anthion and the Aqua Caudi Magnæ Mirandæ are of this class. Some expressions and some names of substances, such as the Quintessence of Gold and of the Sun, though common to all alchemy, are not capable of more lucid interpretation by recourse to other writers, and the Index which follows this Lexicon, by collecting the references in the text, will supply all the knowledge that is likely to be gleaned concerning them. In such cases the terms have been excluded from the Lexicon, as it is obviously useless to re-dress the materials supplied by the translated treatises as lights for a vocabulary. It is useless, for example, to say that the Balm of Sulphur is the Radical Moisture of metals, unless, indeed, it were possible to explain more fully than the alchemists have elected to do what they mean by their radical moisture, especially in the metallic kingdom. It is useless to look for any real explanation of such terms ; they are part and parcel of the philosophical mystery, and unfortunately their number is somewhat formidable. The First Entity of Antimony, the First Entity of Mercury, the First Entity of Salt, the special alchemical significance attaching to *Liquor Solis*, Macerated Tincture of Silver, Mercury of Life, Oil of the Sun, Philosophic Water, Precipitated Gold, Sphere of Saturn, are all of this kind, and all these have been omitted. But wherever the peculiar use which Paracelsus makes of a term is modified or illustrated in the writings of other adepts, there the term has been included ; and wherever any commentator has volunteered any explanation of a word which is peculiar to Paracelsus, then that explanation has been given. A careful reader will, however, find that Paracelsus is in most cases his own best interpreter, and much of his coined or mysterious phraseology is explained by the passages which contain it. For example, there is only one reference to *Argentum Potabile*, but the term is accompanied by an explanation concerning it, which makes its presence in a vocabulary unnecessary. A similar remark will apply to Astral Gold, mentioned in the *Catechism of Alchemy* ; to Abrissach, which occurs in the *Philosophy Concerning the Generation of the Elements*. In a few cases, mostly regarding unknown or recondite substances, such as *Emplastra Apostolica*, *Mucilago Lumbricata*, *Compositio Caudi*, etc., the interest attaching to the information is not strong enough to warrant the research which would have been necessary to provide it. In some instances the herbs mentioned by Paracelsus have been briefly referred to, with a view to save reference unnecessarily to other sources of information. But these are generally so accessible, and the catalogue in a Herbary of Theophrastus would be so large, that those who wish to become acquainted with the virtues anciently attributed to Gladwin, Gentian, Ginger, Gallingall, Fumitory, Fennel, Dittany, Dodder, Cummin, Comfrey, and a hundred others, must have recourse to Gerard's *Herbal*, or some similar storehouse. That Clary is a plant of the sage genus requires no more erudition to announce than can be derived from a popular dictionary of the English language.   In like manner it is a matter of common knowledge that the term Crocus signifies any metal calcined to a deep yellow or red colour, and it would be childish to include information of this kind in a word-book of alchemical technology.   In like manner, it may be taken for granted that the readers of Paracelsus are acquainted with the nature of ordinary chemical vessels, such as the Alembic, the Retort, Cupel, etc.

––––

ABROTANUM, *i.e.*, Artemisia Abrotanum, the herb Southernwood, an aromatic plant.

ACETUM.   Other meanings are attached to this term in alchemy in addition to the ordinary significance of vinegar.   It is the mercurial water of the Sages, or their universal dissolvent, their virgin's milk, their pontic water; it is the vinegar of Nature, and although its elements are various they all come from one root.

ACTIS, possibly Actæa, the wall-wort, or shrubby elder of Pliny.

ADAM.   The formation of Adam by God out of the earth, as described in Genesis, is counted, by the alchemists, among the great mysteries.   The material was no common potters' clay, but another, and one of a far higher nature.   He who knows this knows also the subject of the philosophical medicine, and, by consequence, what destroys or preserves the temperament of man.   It contains principles which are homogeneous with man's life, are potent to restore his decaying virtues, and can reduce his disorders to harmony.   Arias Montanus calls this matter "the unique particle of the multiplex earth."

ADAMANT.   The gem Adamas takes origin from the element of water.   As the stone called Lasurius is a transplanted silver, or a transplanted extraction of silver, as the topaz is an extraction from the minera of Mars, and is a transplanted iron, as the sapphire is a quintransplantation of Lasurius, so the Adamas is a second transplantation of Saturn.   And they are all extractions out of the fruits of the element of water.   The term also was used by the Greeks to signify the hardest metal known to them, probably steel, and also a compound of gold and steel.

ADAMITÆ.   These are white stones of an exceedingly hard quality.   There is also Adamitus, which properly is the stone in the bladder.   It is sometimes written Adamitum.

ADECH.   Hermetic Philosophers apply this name to that portion of the human body which is commonly termed the groin; sometimes also it signifies the mind creating conceptions of things with a view to their manual imitation.   The invisible and interior man.

ÆTNEAN.   Some alchemists apply this name to their fire because it is concentrated and natural, acts perpetually, and is not always manifest.

ÆTNÆI. These igneous spirits appear to be identical with the Salamanders which inhabit the fiery region of Nature. Paracelsus considers that the names which have been given to the Elementaries, and by which they are most commonly known, are not their true designations, which, under the circumstances, is exceedingly probable.

AGRESTA. A beverage made from apples or barberries. Paracelsus forbids it to be used by persons who are afflicted with diseases of a tartareous nature, for every acid is a resolved tartar produced out of a cold coagulate, or, otherwise, from the salts of a vitriolated minera combined with alums. Agresta is verjuice.

AGARIC is described as a medicine made from flies; it destroys worms in the body, and acts as a tonic to the system in certain forms of plague. Agaricon is also a mushroom growing upon high trees; it is of a white colour, and is good for purging phlegm. It was used anciently for tinder.

AGRIMONY, called also liver-wort, once used in the preparation of a medicine which was held to be a valuable tonic.

AGRIPPINE UNGUENT, probably a prepararation of the Agrippum, a wild olive.

AIR. The philosophy of Paracelsus on this subject may be compared with that of other alchemists. For example, Eugenius Philalethes says that air is not an element, but a certain miraculous hermaphrodite, the cement of two worlds, and a medley of extremes. It is the sea of things invisible, and retains the species of all things whatsoever. It is also the envelope of the life of our sensitive spirit. The First Matter of the philosophers is compared to air because of its restlessness.

ALCHEMY. The following remarkable passage occurs in the *Anima Magica Abscondita* of Thomas Vaughan. It has often been made use of as evidence that the adepts had a higher object than the transmutation of ordinary metals :—Question not those impostors who tell you of a *sulphur tingens*, and I know not what fables; who pin also that narrow name of *Chemia* on a science both ancient and infinite. It is the Light only that can be truly multiplied, for this ascends to, and descends from, the first fountain of multiplication and generation. If to animals, it exalts animals; if to vegetables, vegetables; if to minerals, it refines minerals, and translates them from the worst to the best condition.

ALCHEMY. The monk Ferarius defines it to be the science of the four elements, which are to be found in all created substances, but are not of the vulgar kind. The whole practice of the art is simply the conversion of these elements into one another.

ALCHIMILLA, a herb, otherwise called Lion's-foot.

ALCOHOPH is possibly *Sal Alacoph*, *i.e.*, sal ammoniac.

ALCOL. Some chemists have given this name to vinegar.

ALEMBIC. The name of this alchemical vessel has been sometimes applied to Mercury, because by its means the philosophers perform their distillations, sublimations, etc.

ALKALI. In addition to its chemical meaning, this term signifies the Vessel of the Philosophers.

ALÖEPATICUM, a medicine used in complaints of the liver.

ALOPECIA is a species of scab or mange arising in those portions of the body which are referable to Jupiter, *i.e.*, the cranium. Paracelsus ascribes the disease to the presence of the spirit of Jupiter. When that spirit is separated from its natural humours and passes into its own minera, the result in the metallic kingdom is cachimia ; in trees, fungus ; and in the human body alopecia. The fish known as sea-fox was called Alopecias by Pliny.

ALUM. Common alum is distinct from the alum of the adepts ; the latter is their salt, which is a basis of alum, all salts, and all minerals and metals.

ALUMEN ENTALI is identical with Alumen de Pluma or Alumen Scariola. It is said to be Gypsum and Asbestos.

ALUMEN PLUMOSUM, *i.e.*, Alumen de Pluma.

ALUMEN SACCHARINUM is Zaccharine Alum.

AMALGAM. The Amalgam of the philosophers is properly the union of philosophic Mercury with the sulphur or gold of the Sages. This does not take place after the fashion of ordinary chemistry, by pounding in a mortar, or otherwise, a solid and a liquid substance. It is the conduct of the fire of the philosophers according to the proper regimen ; that is to say, it is the perfecting of the work by continuous coction or digestion at an equal fire— sulphureous, covered in, and non-combustive.

AMETHYST. In addition to the precious stone which bears this name, there was a grape so called by the Greeks of which the juice was said to be non-intoxicant. There is also a herb called amethyst by Pliny, having leaves of a red-wine colour.

ANACHMUS, unknown.

ANGELICA. Paracelsus considers that Spica and Angelica are not in their origin the result of a natural generation, but of that which he terms transplantation. In other words, they are hybrids, after the manner of the mule. The Angelica Ursina, Cardopatia, or Carlina, has in its roots the peculiar quality of depriving persons of their virile strength.

ANIADUS. A term of spagyric philosophy which signifies the powers and virtues of the stars, from which we receive celestial influences by the medium of fantasy and imagination.

ANTHERA. A medical extraction made from hyacinths. Paracelsus affirms that the special astrological influences which reside in the sign of Scorpio are resisted by anthera.

ANTHOS. In old botany this term signifies the flower of rosemary ; in alchemy it is the quintessence of the philosophers and their aurific elixir.

ANTIMONY. The antimony of the vulgar is to be distinguished from the antimony of the wise. They have applied this name to the sulphureous mercurial matter which forms part of the philosophical composition. By means of their antimonial vinegar an incombustible quicksilver is extracted from the body of Magnesia. This philosophical antimony is identical with the permanent water and the celestial water, in a word, with philosophical Mercury. It cleanses, purifies, and washes philosophical gold after the same manner that common antimony purifies common gold.

AQUA METALLORUM. Trevisan explains that this is Mercury, and Braccesco states that it is a single substance.

AQUA PERMANENS. According to one interpretation, this is the Catholic Magnesia, or sperm of the world. It is also the Mercury of the Philosophers, and the water of the Sun and Moon.

AQUARIUS is Salt Nitre. It is also the alchemical symbol of dissolution and disintegration.

AQUA VISCOSA. The generation of a metallic sperm is the chief object of those who wish to perform transmutation in the metallic kingdom, and this is done by first of all converting the whole body into a thick or viscous water. Indeed, Rupescissa declares that the matter of the stone itself is a viscous water which is to be found everywhere, but if the stone itself should be openly named the whole world would be revolutionised.

ARCANUM. This term is understood by Paracelsus to signify an incorporeal, immortal substance which in its nature is far above the understanding and experience of man. Its incorporeal quality is, however, only relative and by comparison with our own bodies. From the medicinal standpoint its excellence far exceeds that of any element which enters into our own constitution. The term is applied also to every species of tincture, whether metallic, vegetable, or animal. In general Hermetic science it signifies viscous mercurial matter, or Mercury animated by reunion with philosophic sulphur.

ARCANUM OF HUMAN BLOOD. Many alchemists, both before and after Paracelsus, experimented with the blood of animals, and this not so much with a view to medicine as with the hope of discovering therein that matter of which the philosophers form their magistery. This magistery, indeed, passes sometimes under the name of Human Blood. According to Philalethes the reference is to the matter at the black stage. The name is really applied to philosophical Mercury. Even as the blood of animals nourishes their whole body, and is the principle of their physical constitution, so is this Mercury the base and principle of metals. Thus the blood of the little children who were slaughtered by Herod is pictured in the hieroglyph of Abraham the Jew and is a type of the radical humidity of metals extracted from the minera of the philosophers, which is symbolised by the children, this matter being still crude and left by Nature only on the way to perfection. The Sun and Moon come to bathe in this blood, because it is the fountain of the philosophers in which their king and queen lave themselves. Flamel, foreseeing that his allegory might receive a literal interpretation from some, warns his readers not to mistake actual human blood for the material of the stone, as it would be a foolish and abominable thing.

ARCHÆUS. According to one interpretation, this name is applied by the Spagyrists to the Universal Agent specialised in each individual; it is that which sets all Nature in motion, and disposes the germs and seeds of all sublunar beings to produce and multiply their species.

ARCHIALTIS, or Archaltes, Archates, Archallem, the secret power of God by which the earth is held up in its place.

ARES. The occult dispenser of Nature in the three prime principles. It is that which gives form and difference to species. The word is derived from the Greek, 'αρης, Mars.

ARGENTUM VIVUM. The transmutation of metals, says the *Clavicula* of Raymond Lully, depends upon their previous reduction into volatile sophic argent vive. This is drier, hotter, and more digested than common Mercury.

ARLES CRUDUM. Certain little drops which fall in the month of June. Called also Hydatis.

ARISTOLOCHY. This herb, which is corruptly called Birthwort, was made into a decoction with wine and applied by Paracelsus as a healing plaster for fractures. The variety which he calls Aristolochia Acuta was used as an ingredient in the composition of an unguent for corroding ulcers. Aristolochia is supposed to promote child-birth.

ARSANECH is sublimed arsenic.

ARSENIC. The arcane sense of this term refers it to the Mercury of the philosophers, and at times to the matter of the philosophers when in the stage of putrefaction. It is stated, or supposed to be stated, in one of the Sibylline verses, that the name of the matter whence philosophical Mercury is extracted consists of nine letters. Of these four are vowels and the rest consonants. One of the syllables is composed of three letters, the rest are of two. Hence it was concluded that *Arsenicum* was the name in question, more especially as the philosophers affirm that their matter is a deadly poison. However, the matter of the stone, according to other authorities, is not arsenic, though it is the matter of which arsenic and all mixed bodies are formed. Nor can the Mercury of the Sages be extracted from arsenic, for arsenic is sold by apothecaries and the minera of Mercury is found everywhere. The name has been given by some other writers to the matter in putrefaction, because it is then a most subtle and violent poison. Sometimes it refers to the volatile principle of the sages, which performs the office of female. It is their Mercury, their Moon, their Venus, their vegetable Saturn, their green Lion, etc. The arsenic of the philosophers whitens gold, even as the common arsenic whitens copper.

ARTETIC, Arthetica, a disease which contracts the nerves, tendons, ligaments, etc., and is very enervating and prostrating.

ASCLITIS is dropsy of the stomach.

ASPHALT is, according to Paracelsus, an extraction of black succinum. It is bitumen, also a kind of petroleum, or rock oil.

ASTRUM. This term in Alchemy signifies the fixed and igneous substance, the principle of multiplication, extension, and all generation. It tends of itself to generation invariably; but it acts only in so far as it is excited by the celestial heat which is diffused everywhere. The term also represents the highest virtue, power, and property acquired by the preparation of a given substance. The Astrum of Sulphur is sulphur reduced to an oil which far exceeds the virtues of natural sulphur. The Astrum of the Sun is the salt of the Sun reduced to an oil or water. The Astrum of Mercury is sublimated Mercury. The name is also given to the alcools or quintessences of things.

ATHANOR. In exoteric chemistry this is a square or oblong furnace communicating on one side with a tower. The tower is filled with coals; when these are lighted heat is communicated to the furnace by a funnel. The same name is applied analogically to the secret furnace of the philosophers, wherein a fire is always maintained at the same grade. It is unlike the vulgar athanor; it is actually the matter itself animated by a sophic fire which is innate therein, and is developed by art.

ATRAMENT. At the period of Paracelsus this term seems to have included all varieties of vitriol, chalcanthus, flower of copper, chalcitis, misy, sory, melanteria, etc.

AURATA. The fish called gilt-head by Quintillian.

AURICHALCUM. Brass, copper, ore, etc., *æs montanum*, extracted from cuprine stone. More correctly, Orichalcum.

AURIPIGMENTUM. The body of orpiment is composed of Sulphur, its coagulation is from Salt, its brilliance from Mercury. It is of a petrine and metallic nature, yet it is neither a metal nor a stone. Paracelsus, *De Elemento Aquæ.*

AURUM POTABILE is either the Oil of Gold, or gold reduced to a liquor without a corrosive. The Golden Calf, which was ground to powder, sprinkled upon the waters, and given to the children of Israel to drink, has been regarded as an allegory of the philosophers' potable gold.

AUSTROMANCY is a method of divination by the winds. It is, apparently, a branch of the science of Aeromancy, which, says Agrippa, divines by aërial impressions, by the blowing of the winds, by rainbows, by circles about the moon and stars, by mists and clouds, and by imaginations in clouds and visions in the air.

AXUNGIA is the fat of animals which was made into an oil and used as an unguent for wounds. The axungia of human beings is said to have been most efficacious, and next thereto that of the cock and capon. The only useful axungia from fishes was that obtained from the Thymallus.

AZOC. Mercury of the Philosopers, not vulgar and crude quicksilver, simply extracted from the mine, but a Mercury extracted from bodies by means of argent vive. It is an exceedingly ripe Mercury. It is with this substance that the philosophers wash their Laton; it is this which purifies impure bodies with the help of fire. By means of this azoc there is perfected that medicine which cures all diseases in the three kingdoms of Nature. It is made of the Elixir.

AZOTH also signifies Mercury. When the philosophers say that fire and Azoth suffice for the Great Work, they mean that Mercury prepared and well purified, or philosophical Mercury, are enough for the beginning and the completion of the whole labour, but the Mercury should be extracted from its minera by an ingenious artifice. Bernard Trevisan says that everyone beholds how this minera is changed into a white and dry matter, having the appearance of a stone, from which philosophical argent vive and sulphur are extracted by a strong ignition. Azoth has many names:—Astral Quintessence, Flying Slave, Animated Spirit, Ethelia, Auraric, etc. It will be seen that Azoth and Azoch are terms used interchangeably. They also stand for the universal medicine, which is arcane to the world, the sole true remedy, the physical stone, and also, according to some writers, they signify the Mercury of any metallic body.

BALM. The quintessence of Mercury is called the external balm of the elements.

BALSAM. Paracelsus affirms that the composition of balsams was first discovered by the alchemists, and he refers the name itself to the artifice which is required in its preparation. Every balsam, but more especially for wounds, should be of a sweet nature, not corrosive, not attractive, but possessing a consolidative quality.

BALNEUM MARIÆ. The furnace of the Sages, the secret furnace, to be distinguished from that of ordinary chemistry. Sometimes the name is applied to philosophical Mercury. The term Bath is also given to a matter which is reduced into the form of a liquor. For example, when it is desired to make projection upon a metal, it is said that it must be in the bath, that is, in a state of fusion.

BALNEUM MARIS In alchemy this term seems to apply both to the vessel which holds the sea-water and to the dissolution that takes place in the vessel. The *Balneum Mariæ* is a bath of warm water, and the name is still applied to a large cooking pan. The *Balneum Roris* seems an interchangeable term for the sea-bath. The term bath is also applied to the matter itself when it is in a liquid state. Circulation in the philosophical egg is also called the Bath of the Philosophers.

BALNEUM NATURÆ. Eugenius Philalethes says that this is really the philosophical fire.

BALNEUM RORIS, see Balneum Maris.

BAMAHEMI, unknown.

BASILISK. No one, according to Paracelsus, has any conception of the form or appearance of this animal, for the simple reason that no one can look at it without dying. According to one of his explanations it is a calf generated without a cow, that is, born of the male animal.

BDELLIUM. A plant, and also the fragrant gum which exudes from it.

BEANI. This word, which is rendered *novices* on p. 154 of Vol. II., is said to have originated in the following acrostic : *Beanus est animal nesciens vitam studiosorum.*

BENEDICTA CARYOPHYLLATA, that is, Caryophyllum, ther herb Benet or Blessed Avens. Paracelsus made use of it in tatareous complaints.

BERILLISTIC ART, *i.e.*, divination in the beryl. Paracelsus regards the the stone itself as a formation out of ice, by means of the glacial stars, which have a very great power of congelation.

BITUMEN, *i.e.*, Asphalt.

BISMUTH. As zinc, according to Paracelsus, is for the most part a spurious offspring of copper, so is Bismuth of tin. It is partly fluidic and partly ductile.

BLOOD. Some alchemists have pretended, as previously stated, that human blood is the true subject and matter of the philosophers, but others say that this is not to be literally understood, and that those who experiment with this substance in athanors will have reason to deplore their error. Blood and Human Blood are, however, by no means uncommon designations for the arcane substance of the Magistery. Philalethes says that it is applied to the matter. It is an analogical term, the allusion being to the fact that the blood in animals carries the nourishment to every part of the body, and is the principle of their physical constitution. Mercury fulfils the same function in the mineral kingdom, for it is the foundation and principle of metals.

BOLETI. Boletus is a mushroom. Bolitus is the same as Bolbiton, *i.e.*, the excrement of oxen. These explanations will, perhaps, not throw much light on the use of the term by Paracelsus.

BRASSATELLA, *i.e.*, ophioglossum.

BOCIA. Bocium, according to Paracelsus, originates out of menstrua and hæmorrhoids, in the same place in which both of these fluxes join, producing a third which is peculiar to itself.

BOLUS ARMENUS on account of its great dryness was used by Paracelsus to heal wounds.

BORAX. This is said to have been melted with natural or artificial chrysocolla, and was then used as a dissolvent and purifier of metals, especially gold and silver. The best borax was supposed to come from Alexandria.

BOTIN is Turpentine.

BUFONARIA, probably Buphonon, a herb mentioned by Pliny, that is, the toadstool.

BUGLOSSUM, *i.e.*, Bugloss, or Borage.

BULLÆ. ⎫ This causes wind, says Paracelsus, and its presence often
BULLA. ⎭ indicates the beginning of colic, while in women it occasions fall of the womb. It is otherwise called Bleb, a little vesicle, or blister. Bulla is also a genus of mollusca.

CABALA. The alchemists recognise a twofold Cabala. There is that which ends always in the letter where it begins, and this is the alphabetical system, the name and not the thing; the shadow, not the substance; the mere type of the inner Cabala. Within this there is the true, ancient, physical tradition, and this also has its complement in a metaphysical part.

The greatest mystery of the Cabala, according to one account by an alchemist, is Jacob's ladder. It is affirmed further that Jacob was not asleep during his vision, except in a mystical sense, for he had passed through death. It was, however, the cabalistical *Mors osculi*, or death of the kiss, of which those who know it must not speak one syllable. The false, grammatical Cabala consists only of alphabetical rotations, and a metathesis of letters in the text, by which Scripture is wrested. The book of Abraham the Jew, which was discovered by Flamel the alchemist, is supposed to prove that the true Cabala was chemical.

CACHEXIA, a disease which, according to Paracelsus, usually supervenes upon other complaints, and in which the nutriment turns to evil humours.

CACHIMLÆ. This term is loosely applied to a variety of substances. It seems to have most generally signified the dross of metals, or an undigested metallic matter.

CAGASTRIC. There are two seeds of disease, the Iliastric and the Cagastric. The first is in the substance from the beginning, the second is generated out of putrefaction. Dropsy and gout are Iliastric; plagues, fevers, pleurisy, etc., are of cagastric origin.

CALAMUS, a sweet cane, growing in Arabia, India, and Syria. The root of the sweet flag. The resin called Dragon's Blood is obtained from a palm of the genus Calamus.

CALAMINE, a stone used in the composition of brass. Also an ore of Zinc, Cadmia.

CATAPUTIA, a plant.

CALCANTHUS and Chalcanthus, see Atrament. It is copperas, vitriol, shoemaker's black, the water of copper.

CALCATRIPPA, or Consolida Regalis, a herb used by Paracelsus for the cure of ulcers.

CALCINATION is a pulverisation and purification of bodies by means of an exterior fire, either for effecting a disunion between their component parts, or for evaporating the humidity which combines them into a solid body. Calcination, corruption, and putrefaction are sometimes used interchangeably by the Spagyric philosophers, but calcination is most commonly that process which follows the rubefaction of the Stone. *Philosophic Calcination* is performed by the moist fire, or pontic water, of the Sages, which reduces bodies to their first principles without destroying their seminal and germinative virtues. These, on the other hand, perish under the calcination which is performed by a vulgar fire, and this is philosophically termed the Tyrant of Nature. There are two kinds of *vulgar calcination*; the one is accomplished by an open fire, as over ashes, the other in a sealed vessel. In the first the sulphureous volatile parts evaporate, and the salts are deprived of a power and virtue which are conserved in the second. All salts extracted from such ashes crystallize.

CALENDULA, the herb marigold.

CALERUTHUM. The reversion of any substance towards its first matter.

CALX. In the language of the adepts this name is applied to all kinds of bodies when reduced to an impalpable powder, whether by the action of fire or of corrosive waters. Some say that it should only be applied to the ashes of metallic or mineral bodies, and that others should be called cinders.

CALX LUNÆ, *i.e.*, Calcined Silver, or Blue Flower of Silver.

CALX OF LEAD, *i.e.*, Red Lead.

CANTHARIDES, a herb praised by Paracelsus because it draws out the humours from ulcers. The term stands for Spanish fly, a beetle which infests corn, and a kind of fish.

CAPELLA, a young goat or kid.

CAPILLUS VENERIS, literally, the hair of Venus, a herb mentioned by Pliny.

CAPUT MORTUUM. The philosophers describe the so-called element of earth as an impure, sulphureous subsidence, or *caput mortuum* of the creation.

CARABA, see Carabe.

CARABE is Succinum, and is said by Paracelsus to be an extract of the resin of its element. In another place he observes that it is a resolved petroleum, for wherever the latter is found, there also is Carabe. This and other stones void excrement like animals. If it be cast into water which is isolated from the air, its evacuations will coagulate the water, and worms will be produced.

CARABIS IGNEUS, Cathabis, or Cathebis, the stag beetle, a prickly kind of crab.

CARBUNCLE, is called also jaspis by Paracelsus, and he describes it as a golden stone, that is, of an aureate nature.

CARDAMOMUM, an Indian spice, of an aromatic, pungent, and medicinal quality.

CARDUUS ANGELICUS, a thistle, which, according to the signatory art, has the magical marks indicating that it is a cure for pleurisy. Paracelsus says that it is to be reduced into ashes and then made into a lute.

CARDAMUM, garden cress, that is, nasturtium, especially its seed, which was bruised and eaten by the ancients, and above all, the Persians, after the manner of our mustard. *See* Cardamomum.

CARNIOLA. The virtues of this plant are regarded by Paracelsus as of a celestial rather than an earthly origin.

CASSATUM, an unhealthy or dead blood in the veins.

CEDUSINI. One of the arcane names of air is *Cedue*, and the cedusini are probably a class of aërial intelligences, or sylphs.

CELLA, the inner chamber of a bath.

CENIFICATED WINE, *i.e.*, Calcined Wine.

CENTAURY. The virtues of this plant prevail in complaints of the liver; it strengthens and cleanses the separative force of this organ. Its liquor destroys serpents.

CERATION. This name is given to that stage of the philosophical process during which the matter passes from the black to the grey, and, ultimately, to the white colour. This is performed solely by digestion and coction without any addition whatsoever.

CEREVISSIA, that is, ale or beer.

CERUSSA, *i.e.*, White Lead.

CHALCEDONY, a stone which is said by Paracelsus to colour the water in which it is steeped. It is classed as an extract from salt.

CHAMEPITIS, the herb ground-pine, or St. John's Wort.

CHAOMANCY is substantially identical with aëromancy. It is a revelation of the stars of the air, and prognostication by means of the air.

CHAOS. Many other definitions of the Chaos are given by Paracelsus besides those which find a place among the treatises translated in these volumes. The term is also applied to the air which contains the cause of corruption. There is also a chaos in the human body, which is the motive force of all its interior operations. The universal world is in man, and that not analogically, but actually. It is also the chaos which conserves the body. Empty, immeasurable space. The rude, unformed mass out of which the world was created. The atmosphere. Finally, the term chaos is applied analogically to the matter of the work in putrefaction, because at that period the elements or principles of the stone are in such a state of confusion that they cannot be distinguished from one another.

CHEIRI is a term of various meanings. Sometimes it stands for Mercury. Flos Cheiri is the Elixir at the white, though also the essence of gold. Flos Anthos is the Red Elixir of Gold. Generally it is the flower of any vegetable or plant. Paracelsus says of it :—There is no more powerful medicine for the liver. Its blackness is sublimed away and a white substance is left, which must be drunk mixed with wine of life. It removes all hepatic corruptions. Iliastric mysteries are contained therein. Here the reference is uncertain ; it may be to the narcissus, the violet, or the yellow gilliflower.

CHELIDONY. Paracelsus calls chelidonia a constelled remedy, and names it as a powerful specific for certain ulcers which he also calls constelled. There was, moreover, a philosophic salt of chelidonia, which was medically applied by the alchemists. The same herb was regarded as a preventive of plague, and it was used as a remedy for jaundice. The plant is better known as swallowort or celandine. Chelidonia is also said to be a secret name of gold.

CHERIO. This is described as an accident of the external elements, whether cold or hot, through which all diseases are healed. Cherionium is an unalterable nature, such as indurated crystal, which it is not possible to dissolve.

CHIMOLÆA CALCIS. This should probably read Chimoleæ Calx. Cymolea or Chymolea is a Hermetic name for sedge or reeds. Cymolia is white waste ore, white silver litharge, marl, fuller's earth.

CHIROMANCY is the beginning of Magic ; it is to that science what the alphabet is to writing. It is acquired easily, but at the same time is a most useful and illustrious art. It is the star of natural things. For example, there is a chiromancy of the lily, and in the lily a star abides which corresponds to the nature of that flower.

CHRYSOCOLLA, gold solder, borax. Sometimes the term is used in an arcane sense and interchangeably with Argot, Rebis, etc.

CICHOREA, *i.e.*, Succory. To this plant Paracelsus ascribes a certain peculiar congenital influence which it derives from the Sun, towards which its blossoms turn invariably, and when that luminary is absent it has very little virtue. At the proper moment the congenital influence can be extracted from the flowers. After seven years its root becomes transformed into a bird.

CINERITIUM. This term has two meanings. In the first place it is an ash-pan often mentioned by Paracelsus; secondly, it is an equivalent of Regale, an amalgam of gold and silver.

CINETUS, the thickness of clouds.

CINNABAR. Further information concerning this substance is scattered through the chirurgical and medical works of Paracelsus, in one of which he seems to regard it as a form of Mercury.

CIRCULATED WINE, *i.e.*, the extracted spirit of wine.

CITRINUS is a stone which occupies a middle position between the crystal and the beryl; it is of yellow colour.

CITRINÆUS and CITRONES. See Citrinus. It is a pellucid variety of quartz.

CITRINULA, see Flammula.

CLAVELLATED. The *herba clavellata* is the herb-trinity, or heart's ease.

CLISSUS, the entire essence of a substance amalgamated into one composition. Clissus is also an occult force going and returning from one place to another, as the virtue of a root which first passes to the stem and then returns to the root. (Rochus le Baillif, in his Spagyric Dictionary.) Petrus Poterius, in the Spagyric Pharmacopœia, says, " Clissus is a certain union of all virtues in any plant, which virtues consist of the three primary substances, sulphur, salt, and mercury, such substances being severally educed from the single parts of plants."

COAGULATION. The bond of union in composites, and the common attraction between their parts. It is the rudiment of fixation. There are two kinds, even as there are two solutions. One is performed by cold, the other by heat, and each of these again is duplex, the one permanent, the other transitory. The first is fixation, the other coagulation simply: metals are an example of the first, and salts of the second.

COLCOTHAR is defined by Paracelsus as a salt. He affirms that it is a signal specific for obstruction of the fluxes.

COMPLEXIONS. The ancient opinion on this subject was most explicitly rejected by Paracelsus. In man he admits only one complexion, which has a dual mode, namely, as hot or cold. He affirms that the theory which was current in his own day deserves no consideration in diseases. At the same time in his *Explanatio Totius Astronomiæ* he admits four complexions, not, however, in the body, but in the essence.

COLOCYNTH. Whosoever swallows two grains of the essence of colocynth shall be free from every impression of the moon and from all noxious properties present in the air. It was also recommended by Paracelsus as a cure for worms.

COLOQUINTH, *idem.*

COMPOSITION. It is held by some of the adepts that there is no perfect specifical nature which is simple and void of composition save God alone. Thus the soul of man itself was compounded of an excessively tenuous fire and of the most uncompounded form of light.

COPPER. In alchemy an alternative name with Laton, signifying the matter at the black.

COPPER GREEN, *i.e.*, Verdigris.

COPULATION. In alchemical terminology this is the union of the philosophical male and female, the fixed and the volatile.

CORAL. This was used by Paracelsus as a remedy for the plague, the falling sickness, and against poison.

CORPORAL MERCURY. The innumerable recipes for corporal Mercury found in the *Manual* of Paracelsus may be illustrated by a classification which is given in his chirurgical works. He observes that there are three bodies in Mercury—that out of which it is generated, before it has become perfect; that by virtue of which it is that which it is; and that which it becomes when it is prepared by art. According to Arnoldus de Villa Nova corporal Mercury is another name for the Mercury of the Philosophers, which see.

COSTUS. The plant usually known as *herba Maria*, that is, Zedoary, passes under this name.

CRISPULA, *i.e.*, cristula, crispa, the herb called cock's-spur.

CROCUS. Hermetic chemists have sometimes given the name of Crocus or Saffron to their fixed matter when it has attained the colour of red orange.

CRUCIBLE. The alchemical crucible is described as a clay melting vessel, capable of withstanding a severe degree of heat. It had a narrow base and widened out into a round and triangular body. The cupel was a species of crucible.

CUBEBÆ, a drug so-called. A small, spicy berry, something like pepper.

CUBEBS, see Cubebæ.

CYANUM and Cyanus. According to some this is a kind of blue jasper; others say that it is a turquoise or lazule. It is also the flower commonly called bluebottle. A rock bird. A blue dye or lacquer.

CUPELLA, see Crucible.

CYCLAMEN, Cyclaminus, and Cyclaminum, the herb sow-bread; a tuberous rooted plant used for garlands.

CYROGLOSSUM, hound's-tongue, a plant. The liquor of Cyroglossum is an arcanum for falling of the womb.

CYPHANTIC, unexplained.

DARDO, a term of unknown meaning which occurs only in the treatise *Concerning Long Life*.

DAURA is the same as hellebore and winter aconite. Sometimes also it is supposed to signify foliated gold.

DEALBATION, the washing of the Laton, that is, the coction of the matter until all its blackness departs and the substance remains white. It is the removal of all impurities.

DELTIC. The term Deltic impression, made use of by Paracelsus in his work *Concerning Long Life*, is another mystery in the terminology of that strange treatise. The conjecture may be hazarded that it is derived from the Greek word *Deltos*, a writing tablet, when the Deltic impression would be analogical to that made upon a wax tablet by a stylus.

DENARIUS, properly the Roman denier, seven of which were at one time equal to a Troy ounce. Afterwards there were eight to the ounce. In the Lower Empire a silver denier scarcely weighed half so much.

DENARY, see Ternarius. Agrippa calls the number ten a manifold religion and power applied to the purging of souls. It possesses a divine quality; there is no real number beyond it; just as it flows back into unity so everything returns to its proper source, even the spirit unto God who gave it.

DENTARIA, toothwort.

DERSES, a certain arcane smoke or terrene vapour, which is the principle of vegetable birth and growth.

DIAGRIDIUM, a preparation of scammony and quince-juice.

DIAPENSIA, the plant alchimilla, which see.

DIATHESIS, an innate art or nature. A physical predisposition towards a given disease.

DIEMEÆ or Dienez, said to be spiritual essences which inhabit large stones.

DISSOLUTION, according to Trevisan, is the whole mystery of the art, and it is to be accomplished not, as some have thought, by means of fire, but in a wholly abstruse manner, by the help of Mercury.

DISTILLATION. Later writers agree with Paracelsus as to the variety of complicated processes in the operation of the great work. Those who would be accounted wise must labour to find out the Mercury, so that they may reduce things to their mean spermatic chaos, and may avoid broiling destruction. Futile distillation, in particular, will be dispensed with by one who remembers that sperms are not made by separation, but by composition of elements; to bring a body into a sperm is not to distil it, but to reduce the whole into one thick water, keeping all the parts thereof in their first natural union.

DIVERTALLUM, a generation of the elements, otherwise a production from metals.

DRACHUM, *i.e.*, a drachm.

DRACANCULUS, dragon's wort, or wild adder's tongue. There is a shell-fish also so called by Pliny.

DRAGON'S BLOOD. Alchemically speaking, this is the Tincture of Antimony.

DUBELCOLEPH, or Dubelteleph, a composition of white coral and amber.

DUELECH, a dangerous and painful tartareous and porous stone which forms in the human body, especially in the bladder.

DURDALES, wood nymphs, spirits of the trees, etc.

EAGLE. This name has been applied by the philosophers to their Mercury after sublimation, firstly, on account of its volatility, and, secondly, because even as the eagle devours other birds, so does the Mercury of the Sages destroy, consume, and reduce even gold itself to its first matter. The term is also applied to sal ammoniac and to sublimated Mercury, because of their facility in subliming. Yet the reference is not to the vulgar substances, but to those of the philosophers. The *Eagle which devours the Lion* signifies the volatilization of the fixed by the volatile, or of the sulphur by the Mercury of the Sages.

EDOCHINUM, unknown.

ELECTRUM is gold, according to one interpretation, but it was not used in this sense by Paracelsus. Sometimes it is a conjunction of seven metals into one composition according to the conjunction of the planets. It is also the middle substance between ore and metal, neither wholly perfect nor altogether imperfect. It is, indeed, on the way to perfection, but Nature, having encountered hindrances, has left it. Hence the philosophers say that we must begin where she leaves off. It is called, says one account, Electrum because it is composed of two substances, and immature because it must be perfected by the operations of the artist. Properly it is the Moon of the Philosophers, sometimes called Water, sometimes Plant, Tree, Dragon, Green Lion, Shadow of the Sun, etc. Electrum is also one of the names which have been given to the Magistery at the white.

ELECTUARIES, medicinal confections.

ELEMENTS. Some of the adepts enumerate only two elements, earth, the residence of the matrix, and water, which is the mother of all things visible. Every element is, however, threefold, this triplicity being the express image of their Author, and the seal He has set upon His creatures. There is no created thing too simple, vile, or abject in the sight of man, but that it bears witness of God in respect of that abstruse mystery, His unity and trinity. Every compound whatsoever is three in one and one in three.

ELIXIR. Avicenna speaks of a duplex elixir, and the first matter from which it is produced is also duplex. The Elixir is nothing else, according to Trevisan, than the reduction of the body into mercurial water, from which water the Elixir is extracted, that is, an animated spirit. The Elixir is the second part or operation in the achievement of the sages, as Rebis is the first and the Tincture the third. There are three species of Elixirs in the

Magistery. The first is that which the ancients called the Elixir of Bodies. It is that which is performed by the first rotation. The second is performed by seven imbibitions, even to the white and the red. The third is the Elixir of Spirits and is attained by fermentation; it is also called the Elixir of Fire, and with this multiplication is accomplished.

EMBRYONATED, *i.e.*, fecundated, implanted with seed. Thus Embryonated Sulphur is a term used by the alchemists to distinguish one of their principles from the ordinary substance of sulphur.

EMUNCTORY, an opening for the escape of corrupt matter. Thus, an ulcer is an emunctorium, and so are the ordinary channels of expurgation.

ENOCHDIANI. It may be conjectured that these beings are of the race of Enoch and Elias, who were wrapt away into another world wherein they are still supposed to retain their mortal bodies. The prophets mentioned are according to mystic speculation by no means the only persons who have thus been "caught up to heaven," *i.e.*, into the unseen, without dying in the flesh.

ENS, ENTIA. The first extract of mineral natures is the *primum ens* of its kingdom. The *primum ens* of the animal world is in the blood or in the ova. It is the first matter, the seat of life and motion.

ENUR, the hidden vapour of water, out of which stones are generated.

EQUISETUM, horse tail.

ERUCÆ, a palmer, or canker-worm, also the herb rocket.

ESCHARA is dead flesh.

ESSENCE OF THE GREATER CIRCULATUM, unknown.

ESTHIOMENSIS is Lupus, St. Anthony's fire.

ESULA, the herb tithymallus, or spurge.

ETHICA, a kind of fever.

EUPHRASIA, the herb eyebright, once regarded as valuable in diseases of the eyes.

EVE. When used in a purely alchemical sense, this name signifies the mastery of the philosophers at the white stage. Adam, similarly, is the Magisterium at the red.

EYEBRIGHT, see Euphrasia.

EVESTRUM, the eternal substance of heaven. Also a prophetic spirit, which interprets the signs of coming events. Finally the double, living phantasm, or sidereal body of man.

FEL VITRI, glass gall, *i.e.*, Sandiver, a whitish salt scum cast up from glass in a state of fusion.

FERMENT. The sophic Mercury which, together with the sophic sulphur, is said by Avicenna to be the original substance from which all metals were created, is also declared to be a ferment for every body with which it is united chemically. It is the universal vivific spirit which penetrates, exalts, and develops everything. It is the grand metallic elixir, and its potencies are educed by the operation of fire. Though found in all minerals it is really a terrestrial matter, which possesses lucidity, fluidity, and a silverine colour.

Another account says that it is the fixed matter which, combined with Mercury, causes it to ferment and communicates to it the nature which it needs.

FILLA, possibly Fella, *i.e.*, a name sometimes applied to sulphur-water.

FIRE. It is said that the common chemist works with common fire, using no medium, and so he generates nothing, not working, as God does, for preservation, but for destruction. Hence he always ends in ashes. The disciple of the philosophers should use it with an intermediate phlegma, so that his materials shall rest in a third element, where the violence of fire cannot reach, but its soul only.

FIRE OF THE PHILOSOPHERS. Some adepts affirm this to be the greatest crux of the art. It is a close, aërial, circular, bright fire, which the philosophers call their sun. It causes a certain vapour to arise in the glass which contains the matter, and digests the latter by a still, piercing, vital heat. It is continuous, producing at length an alteration and corruption of the philosophical chaos. Its proportion and regimen are very scrupulous. To understand the proper degree of this fire, the generation of man or some other animal should be considered.

FIRMAMENT, a name of Lazurium. Also the upper part of the Hermetic vessel.

FIRST MATTER. The first matter, by the universal agreement of all alchemists, was existent before man and before all other creatures ; it was, indeed, the mother of them all. The philosoper, in seeking it for the special purpose of his art, must avoid all common salts, stones, minerals, vegetables, and animals. It is totally impossible to reduce any particular to the first matter, or to a sperm, without philosophical Mercury, and being so reduced, it is not universal, but the particular sperm of its own species, and does not work any effects save such as are agreeable to the nature of that species.

FIXATION is a process by which a naturally volatile substance is rendered fixed. The principle of fixation is a fixed salt, and digestion at a suitable fire. Hermetic chemists say that the perfection of fixation can only be obtained by the operation and processes of the Stone of the Philosophers, that their Matter is also susceptible to it, and that the state is attained when it is brought by coction to the ruby-red colour. It is performed by a philosophical fire of the third degree.

FLAMMULA, or Citrinula, is crow's-foot.

FLOS. This term frequently occurs in Paracelsus, and it has many meanings and many variations in alchemy. It is given to the spirits which are enclosed in the philosophical matter. These spirits are so lively that it is always recommended that the fire should be applied gently, as otherwise they may burst the vessels. The name of Flowers is also applied to the different colours which appear in the matter during the process of the work. The *Flower of the Sun* is the citrine redness which precedes the ruby redness. The *Flower of Lily* is the white colour which goes before the citrine. The *Flower of the Salt of the Philosophers* is the perfection of the stone. The *Flower of Gold* is sometimes the Mercury of the philosophers and sometime the citrine colour. The *Flower of Wisdom* is the Elixir perfect at the red. The *Flower of the Air* is dew. The *Flower of Water* is *flos salis*. The *Flower of the Earth* is the dew and the flowers. The *Flower of Heaven* is a species of manna from which an admirable liquor is extracted. Some have erroneously regarded it as the true philosophical matter. The *Flower of the Wall* is saltpetre. Philosophical *Flos æris* is the matter of the work towards the end of putrefaction, and when it begins to grow white. The *Flower of Cheiri* is essence of gold. The *Flower of Sapience* is the Elixir perfect at the red. There are also other explanations of the *Flower of the Sun and of Gold*. It is the sparkling whiteness, more brilliant than that of snow itself, which characterizes the matter at the white. It is the fixed body of the Mastery, which is not to be understood of any flowers or tinctures extracted from common gold, but of the philosophical gold only, and of the fixed portion of the composition of the Mastery by means of which the volatile part is also fixed, according to the regimen of a prudent heat and the governance of a perfect coction.

FLYING EAGLE, the Mercury of the philosophers.

FŒNO GRÆCUM, *i.e.*, Fœnum Græcum, a leguminous plant of the clover family.

FOLIATED EARTH. This is said to be the Mercurial Water in which gold is sown.

FOUR ELEMENTS. The development of things out of the four elements occurs, according to Agrippa, by the way of transmutation. Each element has two specific qualities, one of which is proper to itself, and the other is a mean by which it corresponds to that which comes after it, as, for example, fire is hot and dry, earth dry and cold. He also recognizes that each element is threefold, that so, he says, the number four may make up the number twelve ; and by passing the number seven into the number ten, there may be a progress to the supreme unity, upon which all virtue and wonderful operation depend. Of the first order are the pure elements, which are neither compounded nor changed, nor admit of mixture, but are incorruptible. Through these the virtues of all natural things are brought into act. No man can declare their virtues, because they can do all things upon all things. He who is ignorant of them will never bring to pass any wonderful matter. Of the second order are the elements that are compounded, changeable, and impure, yet such as may by art be reduced to their simplicity ; their virtue, thus reduced, perfects above all things all occult and common operations of Nature, and these are the foundation of all natural magic. Of the third order are those elements which originally and of themselves are not elements, but are twice compounded, various, and interchangeable. They are the infallible medium, the middle nature, or soul of the middle nature. Very few understand the deep mysteries thereof. By means of certain numbers, degrees, and orders, they contain the perfection of every effect in what thing soever, natural, celestial, or supercelestial ; they are full of wonders and mysteries, and are operative both in natural and divine magic. From and through these proceed the bindings, loosings, and transmutations of all things, the knowing and foretelling of things to come, the expulsion of evil and the attracting of good spirits. Let no man, therefore, without these three sorts of elements, and the knowledge thereof, be confident that he is able to work anything in the occult sciences of Magic and Nature. But whosoever shall know how to reduce those of one order into those of another, impure into pure, compounded into simple, and shall understand distinctly their nature, power, and virtue in number, degrees, and order, without dividing the substance, shall easily attain to the knowledge and perfect operation of all natural things and celestial secrets.

FULIGO MERCURII. The *fuligo Metallorum* is properly arsenic in alchemical symbolism, but it often stands for Mercury.

FULMINATION is the graduated depuration of metals. It is so called because the metals become brilliant and diffuse radiance from time to time during the process. A red pellicle forms above, and when it disappears little sparkles are manifested at intervals.

FUMUS is Fimus, *i.e.*, dung.

GALANGA, *i.e.*, Galingale or Galangal, an Asiatic plant; the roots have a hot and spice-like flavour, accompanied by an aromatic smell.

GALBANUM, a gum or liquor having a very strong smell. Derived from an umbelliferous plant.

GAMALEI, or Gemetrei, Gamathei, etc., certain natural stones which, owing to some powerful astrological influence, receive extraordinary magical impressions. There are also artificial Gamathei engraved with magical figures, and used for talismans.

GAMONYMUM, unknown.

GARYOPHYLLON, *i.e.*, Caryophyllum, the clove gilli-flower.

GEOMANCY. According to Cornelius Agrippa, who preceded Paracelsus, and was, like Paracelsus himself, a disciple of Trithemius, geomancy is an art of divination whereby a judgment may be given by lot, or destiny, to every question whatsoever. It consists in the use of certain points, arranged in figures which are in harmony with celestial figures. It can, however, shew forth no truth unless it be founded in some divine virtue. It is supposed that the hand of the operator is directed by the spirits of the earth, so that incantations and other magical rites are resorted to in connection with geomancy. The soul itself of the operator also enters actively into the process.

GLUTEN, in addition to its ordinary significance, means ox-gall. It is also the sinonium of Paracelsus which resembles the white of egg.

GLUTEN OF SULPHUR. In the three prime principles Paracelsus appears to have recognised a certain superincession, and as gluten was alchemically sometimes a name of salt, the reference may be to the innate salt which was supposed to exist in sulphur. Gluten is also Blood, Lime, etc.

GOLD, the most pure and perfect of all metals, has been called by the adepts the Sun, Apollo, Phœbus, and other names, especially when it has been considered philosophically. Gold, as applied to the ordinary purposes of society, they term Dead Gold. *Etherised Gold* is philosophic gold. *Base Gold* is sometimes a deceptive description of the living gold of the sages. *White Gold* is the Magistery of the Philosophers attained to the white grade from which subsequently develops their orange gold and perfect redness, which alone is their true gold, their ferment, and their red smoke. *Gold in Spirit* is the gold of the sages reduced into its first matter, which they also term gold reincruded and volatilised by Mercury. When the sages tell the student to take gold, the reference is not to the vulgar metal but to the fixed matter of the Great Work, wherein their living gold is concealed as in a prison. Their 24-carat gold is their gold pure and unmixed with any foreign elements. *Volatile Gold* is the fulminating gold of Crollius. *Gold of Coral* is the matter fixed at the red. *Gold of Gum* is the fixed matter of the philosophers. *Exalted, Multiplied, or Sublimated Gold* is the powder of projection. *Vivified Gold* is gold reincruded and volatilised. *Gold of Alchemy* is the sulphur of the philosophers. *Foliated Gold* is the sulphur of the philosophers in dissolution.

GRANATE, *i.e.*, Granatum, a pomegranate.

GRAND MAGISTERIUM. The operation of the *Magnum Opus*, the separation of the pure from the impure, the volatilisation of the fixed and the fixation of the volatile one by the other, because no artist will succeed by operating separately on either. The philosophers say that the principle of their magistery is one, four, three, two, and one. The first unit is the first matter whence all has been made ; the number four represents the four elements which are formed of this matter ; the number three represents Sulphur, Salt, and Mercury ; the number two is Rebis, the volatile and the fixed ; the final unit is the stone, or that which is the result of the process and the fruit of all the Hermetic labours. In exoteric chemistry there are three kinds of magisteries, of which one has reference to the quality of the composites, the second to their substance, the third to their colour, odour, etc.

GREAT ARCANUM. Roger Bacon, or, more correctly, a treatise attributed to him, affirms that the great and supreme arcanum of Hermetic philosophy is hidden in the four elements. Espagnet says that its production requires a perfect knowledge of all Nature and art concerning the realm of metals, which is obtained by analysis of metallic principles.

GREEN LION. Philosophical chemists frequently make use of this term to signify one of the matters which enter into the composition of the Magistery. Usually it is their male or their Sun, both before and after the confection of their animated Mercury. Before confection it is the fixed part, or matter capable of resisting the action of fire. After confection it is still the fixed matter, but more perfect than before. In the first instance it is the Lion simply ; it becomes the Red Lion by preparation. The Mercury is made with the first, and the stone, or elixir, with the second. The Old Lion is the fixed part of the stone, so called because it is the principle of the whole. The

Green Lion is the matter made use of for the Magistery. It is certainly mineral and derived from the mineral kingdom. It is the base of all the menstrua of which the philosophers have spoken. Of this they have composed their mineral dissolvent.

GRILLUS, or Grilla, a name of Venus, or Copper. It is also a mild species of chalcanthus used as a purgative.

GUARINI. Men who derive their life from the influence of heaven.

GUTTA ROSACEA is a red breaking out in the face.

HALCYON, the kingfisher. The fabulous bird of that name, and also the foam of the sea.

HALLEREON, the Eagle of the Philosophers.

HARMEL, the seed of the herb Rue, but it is impossible to affirm the special significance which Paracelsus may have attached to the term in his supplementary treatise *Concerning Long Life*.

HELAZATI, unknown.

HELIOTROPE, the Melissa of Theophrastus. Heliotrope, a plant which follows the sun with its leaves and flowers, *herba solstitialis*. A gem used as a lens to look at the sun.

HELLEBORE. A plant used by the ancients as a specific for many illnesses, especially for madness.

HERMES. Albertus Magnus affirms that this mysterious adept was the first who discovered the grand magisterium.

HERMODACTYLUS, the wild saffron, or, according to some, dog's-bane.

HILDONIUS, an unknown hierarchy of elementary spirits.

HIPPURIS, the herb horse-tail, or shave-grass. The hipparus is a kind of lobster.

HIPPUS, or Hippeus, is a kind of crab-fish, the sea horseman.

HIRUNDINARIA, or Hirundinina, swallow-wort.

HONEY. A name given to the philosophical dissolvent.

HOP-PLANT, that is, Lupulus.

HUMIDITY OF THE PHILOSOPHERS. Arnoldus de Villa Nova says : I tell thee further that we could not possibly find, neither could the philosophers before us, anything that would persist in the fire, save the unctuous humidity. A watery humidity easily vapours away, its earth remains behind, and its parts are separated because their composition is not natural. Unctuous and viscous humidities are hardly separable from those parts which are natural to them. A treatise attributed to Albertus Magnus assumes that all metals are composed of an unctuous and subtle humidity, intimately incorporated with a subtle and perfect matter.

HUMULUS, a plant of the hop genus. See Hop-plant.

HYDROMANCY, according to Cornelius Agrippa, performs its presages by the impressions of water, its ebbing and flowing, increase, depressions, tempests, colours, etc. To this must be added the visions that are seen in the water.

HYDROMEL. A kind of meal. Honey diluted with water.

HYPOSARCHA, *i.e.*, Hypersarcosis, fungous or proud flesh.

ILECH. This terms seems identical with Ileias and Ileadus. The *primum Ilech* is the beginning. The supernatural Ilech is the supercelestial conjunction and union of the stars of the firmament with the stars of things below. The great Ilech is the star of medicine. When we take medicine we assimilate this star. Crude Ilech is a composition of the three principles, Salt, Sulphur, and Mercury. It seems also to be elemental air.

ILIASTER, Illeias, Eliaster, Iliadum, etc., the first chaos of the universal matter, constituted of Sulphur, Salt, and Mercury. The prime principle. Called also Ilion.

ISOPIC ART, see Ysopus, etc.

JACINTH ; from this stone a medicine called Antera was extracted.

JACULATION, a darting or casting.

JESICHACH, or Jesahach, is supernatural.

KING. A name used in two different senses by the philosophers. Most commonly it is the sulphur of the sages, or philosophical gold, being an allusion to vulgar gold, which is called the king of metals. Sometimes it is a name of the matter which enters at the beginning into the composition of Mercury, and is the first fire thereof, that fixed grain which has to overcome the mercurial cold and volatility. It seems to be used in both these senses by Basil Valentine in his Twelve Keys. Subsequently he applies the name of King to perfect sulphur and even to the powder of projection. The King is also identical with the Lion. When referring to the powder of projection it is said that the King so loves his brethren that he gives them his own flesh to eat, and thus makes them all kings like himself, that is to say, gold.

KIST, an uncertain weight. It may be of fifteen grains, or four pounds. It also signifies two measures of wine, or half a gallon.

KYBRICK, or Kibric, a name given to the stone. It signifies also the father and first matter of Mercury and all fluids.

LAC VIRGINIS. A name given to the Mercury of the philosophers. It is also mercurial water, and the vinegar of the philosophers. The Mercury of the philosophers, under the form of a milky water in the humid way. It has further been applied sometimes in the dry way when the matter has been brought to the white.

LAPIS LAZULI, *i.e.*, lazulus, the azure or lazule stone.

LATERINE OIL, an oil obtained from bricks.

LATON, copper, or brass, but also a certain state of the philosophical matter when the red colour has appeared, but is not yet permanently acquired. It is, further, a name of electrum. The blood of the Laton is that dry water which is extracted from the virgin earth of the sages.

LAUDANUM. This name is said to be used by Paracelsus in another than the ordinary sense, and to signify a specific for fevers which was a compost of gold, corals, pearls, etc. Another writer says that it is a medicine beyond all praise, made out of two substances, than which nothing more excellent can be found in all the world, and whereby almost every disease is cured.

LAUREUS. This, apparently, is a race of wood spirits. Compare *Laurus*, a laurel.

LEFFAS, the occult vapour of the earth, the principle of vegetable life and growth, that is, the predestination of plants, the sap of plants. See Derses.

LENTIGO, a freckle, pimple, or spot on the face.

LENTOR, a clammy or gluish humour.

LIGUSTICUM, lovage of Lombardy.

LILI. One author says that this is in general any matter from which a good tincture can be made, as, for example, antimony. The *lilium* of Paracelsus is the extraction of a tincture from metals.

LIMBUS. The spagyric philosophers for the most part identify the Limbus with ancient chaos. It has been described as a huddle of matter wherein all things were strangely contained. Some regard it as uncreated, but if otherwise, then it was an effect of the Divine Imagination, acting beyond itself in contemplation of that which was to come, and producing its passive darkness for a subject on which to operate. It also stands for the universal world, with the four elements thereof.

LIMBUS OF MAN. At death the earthly parts of man return to the earth, but the celestial portion departs to a superior, heavenly Limbus, while the spirit goes back to God who gave it.

LIME, see Calx.

LION'S-FOOT, see Alchimilla.

LIQUOR HORIZONTIS, probably liquor of Mercury.

LITHARGE, *i.e.*, Litharge of Silver, is the matter of the work when it has arrived at the white colour under the process of the Sages. Litharge of Gold is the Stone at the red, otherwise, the Sulphur of the Philosophers.

LIXIVIUM, *i.e.*, lye.

LOCCA, or Lorcha, the essential sweetness of the locusts of trees.

LOCUSTS. This name is applied to several different plants and trees. Paracelsus seems to have used it to signify the young shoots of any vegetation.

LORINDS, water spirits.

LORIPIDES, *i.e.*, bow-legged.

LUNARIA, moon-wort, an ingredient for love-potions.

MACHAON, possibly Macha, a flying worm.

MAGISTERY. The strength of the perfect magisterium, according to Avicenna, is one part upon a thousand, that is to say, it will transmute a thousand times its own quantity of base metal into gold. This tinging power, however, varies, and even larger quantities are mentioned by other adepts. See Grand Magisterium.

MAGNALIA, great and divine works.

MAGNESIA. This term, which is occasionally used by Paracelsus in its alchemical, as distinct from its chemical sense, has received many explanations from the adepts. It is the matter of the stone, which the philosophers sometimes call their red, and sometimes their white magnesia. In the first preparation the chaos is blood-red, because the central sulphur is stirred up and discovered by the philosophical fire. In the second it is exceedingly white and transparent like the heavens. It is something like common quicksilver, but of such a celestial and transcendent brightness, that nothing on earth can be compared to it. It is the child of the elements, a pure virgin, from whom nothing has been generated as yet. When she breeds, it is by the fire of Nature, which is her husband. She is neither animal, vegetable, nor mineral, nor is she an extraction from these ; she is pre-existent to them all, and is their mother. She is a pure simple substance, yielding to nothing but love, because generation is her aim, and that is never accomplished by violence. She produces from her heart a thick, heavy, snow-white water, which is the *Lac Virginis*, and afterwards blood from her heart. Lastly she presents a secret crystal. She is one and three, but at the same time she is four and five. She is the Catholic Magnesia, the Sperm of the World, out of which all natural things are generated. Her body is in a sense incorruptible ; the common elements will not destroy it, neither does she mix with them essentially. Outwardly she resembles a stone, and yet she is no stone. The philosophers call her their white gum, water of their sea, water of life, most pure and blessed water ; she is a thick, permanent, saltish water, which does not wet the hand, a dry water, viscous, slimy, and generated from the saline fatness of the earth. Fire cannot destroy her, for she is herself fire, having within her a portion of the universal fire of Nature, and a secret, celestial spirit, animated and quickened by God. She is a middle nature between thick and thin, not altogether earthly, not wholly igneous, but a mean aërial substance, to be found everywhere and at all seasons.

MAGNET. Alchemically speaking, this is the dew of the philosophers.

MAN AND WIFE. Most philosophers have compared the confection of the Magistery to the generation of humanity. They have, therefore, personified the two parts or ingredients of the work, namely, the fixed and the volatile, as the male and female, man and wife, etc.

MANNA is, in some instances, the Mercury of the philosophers, called also divine manna, because they affirm that the secret of its extraction is a gift of God, as also the knowledge of that Mercury which is the minera whence it is derived.

MARCASITE. Many species were known to the old chemists, for all stones which contained any proportion of metal were so called, and even sulphureous stones, vitriolic stones, etc., were included under the same term.

MARRUBIUM, *i.e.*, the herb horehound.

MARTAGON is Silphium, a kind of lily.

MATTER OF THE PHILOSOPHERS. This has been compared to the sperm of an animal, for it is a most delicate substance, almost a living thing ; indeed, it possesses some portion of life, and Nature produces certain animals out of it. For this reason the least violence destroys it, preventing all generation. If it be overheated for even a few minutes, the white and red sulphurs will never essentially unite and coagulate. On the other hand, however well the work has been begun, should it grow cold for so short a space as half an hour, it will come to no good end afterwards.

MELIGIA, probably Melisea, *i.e.*, motherwort, a kind of manna or balsam, obtained by a magisterial process upon vegetables. Meliagris is frittany.

MELLILOT, the *dulcis et mellea lotus* of Ovid, called *Sertula campana* by Pliny. The herb mellilot.

MELISSA, see Meligia.

MELONA, ? Melina, a drink made of honey.

MENSTRUUM. This term is used in a very arcane manner by some alchemists, who speak of the menstruum or matrix of the world, wherein all things are framed and preserved. It is a certain oleaginous and ethereal water.

MERCURIAL LIQUOR, a balsam of things in which all health resides. It is most potent in tereniabin and nostoch.

MERCURY. According to Arnoldus de Villa Nova, the Mercury of the philosophers is an aqueous, cold, and moist element. This is their permanent water, the spirit of the body, unctuous vapour, blessed water, virtuous water, water of the wise, philosophers' vinegar, mineral water, dew of heavenly grace, virgin's milk, corporal mercury, etc. All these names signify one only thing; out of this all the virtue of the art is extracted, and, according to its nature, the tincture, both the red and the white. Mercury, by another account, is composed of a metallic and fluidic earth. There are as many Mercuries as metals, which can all combine with the fluidic earth, and this in such a manner that separation is impossible. The whole secret of Hermetic philosophy and the Great Work consists in the wonderful sympathy between those Mercuries and this earth. *Dissolving Mercury*, used by spagyric philosophers for the reduction of metals, minerals, vegetables, and all bodies to their first matter, is of three kinds : Simple dissolving Mercury ; composite dissolving Mercury, which is properly their true Mercury, and common Mercury, or that which is extracted from metals. Simple Mercury is a water extracted according to the principles of their art from a matter, the true name of which they have carefully concealed under an infinity of false descriptions. *White Mercury of the Sages* is the stone at the white. *Red Mercury* is the magistery perfect at the red. *Universal Mercury* is the animating spirit diffused throughout the universe. *Crude Mercury* is the dissolvent of the sages, not that vulgar quicksilver which is called crude mercury by chemists. *Preparing Mercury* is this same dissolvent, which prepares soluble bodies for their development into the perfection of the magistery.

MERCURY OF LUNA. Flamel directs that the Mercury of the Moon should be taken with that of the Sun, and cherished over the fire in an alembic. It must not be a fire of coals or wood, but a bright, shining fire, like the sun itself; its heat must never be excessive, but always of one and the same degree. Thus is performed the plantation of one into the other.

MERCURY OF SOL. The unctuous humidity of the philosophers is called water of silver and water of the moon, but it is really, say others, the Mercury of the Sun, and partly that of Saturn. Indeed, it is an extract of the three metals, and without these it can never be made.

MERITORIUM, a tavern.

MICROCOSM. Sir George Ripley, a celebrated English alchemist, describes the stone as a triune microcosm. Proportion must be studied in its composition. The matter itself is found everywhere. It flies with fowls in the air, swims with fishes in the sea, it is discerned by the reason of angels, and it governs man and woman.

MINERA. The alchemists mention a minera of man as well as of metallic things. It is also the First Matter of the Philosophers' Stone. It is water and not water, earth and not earth, the Damascene earth, the subject of art, and that also of which God made use in Nature.

MITHRIDATIC, pertaining to mithridate, an old antidote against poison.

MOREE, Moro, Mores, a kind of abscess.

MORPHEA, see Morphew.

MORPHEW, or Morphea, a species of leprosy.

MOTHER OF METALS, *argentum vive*, *i.e.*, live silver.

MUMIA. Whatsoever when killed has the power of healing diseases. Hence Mumia of the elements is the balsam of the external elements. *Mumia transmarina* is manna. *Mumia versa*, according to some opinions, is liquor of Mumia.

MURIA, brine or salt water. Also stinking menstruum.

MYROBOLANUM, the Egyptian bean, a fruit used by apothecaries in the manufacture of precious ointments; an oil is obtained from the kernel.

MYSTERIUM MAGNUM. This is used by the alchemical philosophers in another sense than that of Paracelsus. The operation or confection of the great work is called a mystery of the philosophers, because they discover it only to their most intimate friends. Some also refer under this name to the first matter of the work, because that is the most concealed portion of all their writings.

NECROLICUS. As Necrolia and Necrolica are medicaments which prevent death and preserve life, so it is reasonable to suppose that a Necrolicus is the administer of their medicines. But he is also a person who has written learnedly on any subject.

NENUPHAR, a generic name of the elementary spirits of the air. It also signifies a water lily.

NEPITA, the herb net, or catmint.

NEUFARINIS, a generic name of elementary spirits.

NOSTOCH, the efflux of a certain star, according to Paracelsus. It is deposited on the earth in summer, appearing like a yellow fungus, but in consistency it is like a jelly. Some say it is wax.

OGERTUM or Ogertinum, *i.e.,* Orpiment.

OIL OF THE PHILOSOPHERS. Though oil simply so called has nothing to do with the Great Work, and must not be used in confection, the name has been applied to the matter during that period when it has assumed an oleaginous colour and viscosity. This is when it is undergoing putrefaction in the philosophical egg. Oil of the philosophers also signifies the Secret Fire of the Sages. What is termed by some alchemists *Blessed Oil* is incombustible, philosophical oil, the sophic sulphur, also the stone perfect at the white, or red, because it runs and melts before the fire like butter. *Oil of Nature* is the prime salt which is the basis of all salts. It is called oil because it is unctuous, melting, and penetrating, and oil of Nature because it is the base of all the individual substances in the three kingdoms, and is their material conserver and restorer. It is the best, the most noble, the most fixed, and at the same time the most volatile before its preparation. When it is sought to be employed by art, it must be brought from the fixed to the volatile, and from the volatile to the fixed : to solve and to coagulate is the whole work. *Essential Oil* is the volatile sulphur of the philosophic metals ; it is their soul, the male, the Sun, the gold of the sages. *Oil of Saturn* is the matter of the philosophers at the black stage, because their matter in putrefaction is called lead. *Oil of Sulphur* is also sometimes the matter at the black stage. *Incombustible Oil* is the mastery at the red, so called because of its fixation. *Live Oil* is the mastery at the white. *Vegetable Oil* is oil of philosophical, not vulgar, tartar.

OIL OF IRON. The Sulphur of the Philosophers at the red.

OLIBANUM, an inspissated sap, or gum-resin.

ONAGIR, *i.e.,* Onager, the wild ass.

OPOPONAX, the juice of the panax, or herb all-heal.

PANIS SILIGINIS, or Panis Siligineus, white bread, or fine manchet.

PANNUS, a cloth or bag.

PARATELUS, of unknown signification, but the text of the treatise *Concerning Long Life* makes it barely possible that it is a corruption of Paraclete.

PART WITH PART, a composition of equal parts of gold and silver.

PEACOCK'S TAIL. The matter of the work at that moment when the colours in the tail of the peacock manifest on the surface.

PELICAN. An alchemical vessel, so called after the bird of that name. The body tapers towards the neck, which is bent round, and the tube returns into the body.

PENTAPHYLLUM, the herb cinquefoil.

PERI-PNEUMONIA, inflammation of the lungs.

PERSIAN FIRE, a scorching and spreading ulcer.

PER SCOBAM, possibly per scobem, from *scobs*, any powder that comes of sawing, filing, etc.

PERSICARIA, the herb culerage or peachwort.

PETROSELINAR, a kind of parsley growing among rocks.

PHILOSOPHERS' STONE. This is one, says Arnold de Villa Nova, and he adds that it can be extracted from all bodies, including common quicksilver. The first physical work is the dissolution of the stone in its own Mercury, so as to reduce it to its first matter. Jacob Böhme describes the philosophers' stone as dark, disesteemed, and grey in colour.

PHILOSOPHICAL EGG. While the majority of chemists have supposed that the sages applied the name of Philosophical Egg to the vessel in which they enclose their matter for coction, and have consequently shaped it like an egg, this does not represent the idea or the sense of the sages, although at the same time such a shape is the most convenient for circulation. The egg of the philosophers is not that which contains; it is that which is contained; that is, the true vessel of Nature. Therein the philosophic chicken is concealed, which the internal fire of the egg, excited by the warmth of the hen, vivifies by degrees, and gives life to that matter of which it is the root, whence there is at length born the philosophical infant which perfects and enriches his brethren. Again: the egg signifies most commonly the actual matter of the mastery, which contains the Mercury, Sulphur, and Salt, even as the ordinary egg is composed of the white, the yolk, and the pellicle.

PHILOSOPHICAL MONTH, *i.e.*, the period required for chemical digestion.

PHLEGM. When this term is made use of in connection with the operations of alchemy, it signifies a certain aquosity or vapour given off from the matter of the work, becoming white in the process of distilling. It is for this reason that the same name is applied to the stone and to philosophical Mercury in the grade of the white.

PHŒNIX, the fire in the quintessence, the physical stone.

PISA, a mortar.

POCALE, a measure of wine.

PRUNELLA, *i.e.*, Prunellus, the bullace-tree.

PUTREFACTION. The corruption of the moist substance of bodies by a defect of heat, or, otherwise by the action of a foreign fire on the matter. It is in this sense that the spagyric philosophers say that the matter of their stone is in putrefaction when the heat of the extrinsic fire setting into action the internal fire of the said matter, the two act in concert thereupon, separating the humidity which binds its parts, and after several circulations in the hermetically sealed aludel, reducing the matter into dust.

PYROMANCY is divination by the impressions of fire, by comets, by fiery colours, by visions and imaginations in the fire. Capnomancy, or divination by smoke, is an art mentioned by Agrippa as allied to Pyromancy.

QUARTA, probably a measure containing five ounces of wine and four and a half ounces of oil.

QUARTATION, an old method of testing gold. In melting, nine parts of silver were added to one of gold, and both were resolved by aquafortis, which held the silver in solution, and the gold settled at the bottom.

QUINTESSENCE. Paracelsus, though possibly the first, was by no means the last of the alchemists to describe the term quintessence as a misnomer. Eugenius Philalethes affirms that there is no quintessence, no fifth principle, except Almighty God. There is a quartessence which is a moist and silent fire, which passes through all things in the world and is Nature's chariot, the mask and screen of the Almighty. Wheresoever God is there this train of fire attends Him. It was this fire which was manifested to Moses on Mount Sinai. According to another authority, the terms quintessence, specific, magnetism, bond, seed of the pure elements, etc., are all synonyms of one substance, a subject wherein the form abides. It is a material essence, which encloses an operative and celestial spirit. Others say that it is the fifth principle of composites, comprising the finest portion of the four elements. The *Quintessence of the Elements* is the Mercury of the Philosophers.

RABELOIA, the roots of the larger ranunculus.

RAINWATER. This is the common soft water, mentioned sometimes by Paracelsus.

REALGAR is red orpiment.

REBIS. Among other meanings attached to this curious term, it is said to be the last matter of things, and this may be understood either literally, as excrement, or philosophically, as opposed to the first matter. Alchemically, it is the fixed and the volatile.

RECTA CROA, *i.e.*, flos sectæ croæ, flower of the crocus, an extract of chelidony leaves. Also flower of the nutmeg.

RED LION. The spagyric philosophers have given this name to the terrestrial and mineral matter which remains at the bottom of the vessel after sublimation of the spirits, called eagles. This Red Lion is also called Laton.

REDUCTION. This is the retrogradation of a substance which has reached a certain degree of perfection to a degree of a lower order. The reduction of metals into their first matter is their philosophic, not the vulgar, retrogradation into their proper seed, that is to say, into a Hermetic Mercury. It is also called reincrudation, and is performed by the dissolution of the fixed by its proper volatile, from which it has been made.

REGULUS, literally, the little king.

RELOLLEA, arcane virtues, called also relolleum.

RUBEDO DE NIGRO, a red substance, extracted by art from black lime.

RUBIFICATION. The process under which the philosophical matter passes from the white to the red stage.

RUBY. With the philosophers this sometimes means the magistery at the red. So also their precious ruby is the powder of projection.

SAGÆ. The saga is properly a witch, sorceress, or wise woman. The sagani are the elementary spirits.

SAGAPIN, a Persian gum-resin.

SALAMANDER. In alchemical terminology this name is sometimes applied to the philosophical matter, and then the term Blood of the Salamander signifies the red state. More usually it is the redness which appears in the recipient in the distillation of nitre and vitriol.

SALT. This substance was supposed by the alchemists to consist of a small quantity of sulphureous earth and a large proportion of mercurial water. It was the substantial matter of bodies, of which the form was Sulphur. Three chief species were distinguished, the nitrous, the marine, and the vitriolated, to which some add the tartareous. The marine was regarded as the most important. From this salt volatilised comes Nitre, from Nitre comes Tartar, and from Tartar, when cooked and digested, comes Vitriol. There was also another classification as volatile, mean, and fixed Salt. *Fusible Salt* is the Matter of the Sages cooked and perfect at the white ; it was so called because it is actually a salt which melts like wax when placed upon a plate of red hot metal. *Salt of Metals.* Some investigators, taking this expression literally, have imagined that the Matter of the Philosophers was a metallic substance reduced into a salt or a vitriol, but the reference is to the Magistery at the white, because even as salt is the principle of the vulgar metals, so that of the Sages is the root of the first matter of philosophical metals. *Red Salt* is the red Sulphur of the Sages. *Salt Alacoph* is Sal Ammoniac. *Bitter Salt* is Alkali. *Salt of Greece* is Alum. *Indian Salt* is the Mercury of the Sages. *Salt of Bread* is marine or common salt. *Salt of the Sages* is Sal Ammoniac, the natural kind, or philosophic Nitre, but the true Salt of the Sages is their Matter at the white. *Acid Salt* is philosophical Mercury. *Salt Adram* is Sal Gemmæ, also called *Salt of Cappadocia.* *Burnt Salt* is the Matter at the black stage. *Salt of the Earth* is the Mercury of the Sages. *Salt in Flower* is the Mercury of the Sages, or the Dry Water of the Sages. *Salt of Glass* is the Mercury of the Sages. *Fixed Salt* is the Sulphur of the Sages. *Honoured Salt* is the Matter of which the Hermetic Mercury is made. *Salt of Saturn* is lead reduced to a salt. *Sea Salt* is the Mercury of the Sages. *Spiritualised Salt* is the philosophers' spirit of salt. It is their Mercury prepared by Hermetic sublimation. *Universal Salt* is the Mercury of the Sages.

SAL ANATRON, salt nitre, red salt of the Indies. It is also called *Sal Andaron.*

SALDINI, also called Rolamandri, are igneous men, otherwise essences of the race of the Salamander.

SAL GEMMÆ is Hungarian salt. It is also called *Sal Nominis* and Salt of Hungary.

SALT NITRE, see Salt.

SAL PEREGRINUM is very probably another name for the alchemical Salt of Tartar.

SALT OF TARTAR, white calcined tartar.

SAMECH, tartar, or salt of tartar ; also the healing power of all wounds.

SANDARAC, a bright red colour used by painters ; it is found in mines of gold and silver ; some call it red arsenic. There is another species made of burnt ceruse ; also red lead. Another name is cerinth.

SAPPHIRE. The philosophers have given this name to their mercurial water.

SATURN, *i.e.*, Lead. This metal is said by Isaac the Hollander to be the first matter of the philosophers. It is also a name for the philosophical matter in putrefaction. It is further the Adrop of the Sages or the Azotified Vitriol of Raymond Lully. Finally, it is a name sometimes confusedly applied to ordinary copper.

SATYRION, or Orchis, the herb ragwort, or priest-pintle.

SCARIOLÆ. The fourfold spiritual powers of the mind, corresponding to the four elements and the four cabalistical wheels of fire of that chariot which took up the prophet Elias. They are soul emanations or faculties, fancy, imagination, speculation, etc. These under religious sanctification become instruments for the attainment, not merely of long, but of eternal life.

SCORIÆ. Impurities separated from minerals and metals during fusion.

SEALED EARTH, a red earth.

SEED. The seed, or first matter of the Stone, says the Persian alchemist, Rachaidibi, is outwardly cold and moist, but inwardly hot and dry. Rhodion, the instructor of Calid, says that it is white and liquid, but afterwards it becomes red. It is the flying stone, aërial and volatile. It has one virtue inwardly and another outwardly. See *Sperm*, First Matter, etc.

SEPARATION. This process, so often referred to by Paracelsus, is also mentioned by pseudo-Hermes, who has these significant words : Know, ye that are children of the wise, the separation of the ancient philosophers was performed upon water, which separation divides the water into other four substances.

SEPULCHRE, the glass vessel which contains the matter of the philosophical work; the dissolvent of the sages; the black stage of the matter.

SERPIGO, a tetter.

SERUM, whey, buttermilk.

SICONIUS, unknown, a race of elementaries.

SILIGO, a species of corn.

SMAGMA, called also both Smegma and Smigma, is a detersive substance.

SMIRIS, Smyris, a stone with which glaziers cut glass, and with which lapidaries polish gems.

SPAGYRIC SCIENCE is that which teaches the division and resolution of bodies, with the separation of their principles, either by natural or violent means. Its object is the alteration, purification, and perfection of bodies, that is to say, their generation and their medicine. It is attained by solution; success is impossible if their construction and principles are ignored, because these serve for dissolution. The heterogeneous and accidental parts are separated with a view to the intimate reunion of the homogeneous portions.

Spagyric Philosophy, properly so-called, is the same as Hermetic Philosophy.

SPATHUS, or Spatha, a pounding instrument.

SPATULA, or Spathula, a broad and flat instrument, like a slice. It is used in certain furnaces for regulating the heat. A damper.

SPELT, *i.e.*, German grain, a kind of wheat.

SPERM OF THE WORLD, that is, catholic magnesia. It is said to be very salt, extremely soft, somewhat thin and fluid, having no relation to ordinary salts, a sperm that Nature herself draws out of the elements without the help of art. Man may find it where Nature leaves it, it is not his office to make the sperm, nor to extract it; it is already made, and wants nothing but a matrix, with heat convenient for generation.

SPHERE OF THE SUN, the dual matter of the stone, the heaven, *i.e.*, the quintessence.

SPODIUM, ash of gold, but sometimes applied to pompholix and to a kind of tutty.

SPREAD EAGLE. In common chemistry this represented sublimed sal ammoniac, and, in the Hermetic sense, it was the volatilisation of the matter.

STEEL. The steel of the philosophers is a subject of frequent reference, and it has caused many persons to seek the philosophic stone in this substance, but in vain. The steel of the sages is the wine of their philosophic gold, a spirit pure above all, an infernal and secret fire, most volatile in its nature, and the receptacle of superior and inferior virtues.

STELLIO, an eft or newt.

STIBIUM. In the first place, black lead and its derivatives; secondly, a name of antimony; thirdly, a certain stone found in silver mines.

STOMACH OF THE OSTRICH. The chemical philosophers give this name to their dissolvent or philosophical Mercury.

STONE OF THE PHILOSOPHERS. The Rosicrucian philosophers say that in the impregnable fortress of truth is contained the true and undoubted Philosophers' Stone, that treasure which, uneaten by moths and unstolen by thieves, remaineth to eternity, though all things else dissolve, set up for the ruin of many and the salvation of some. To the crowd this matter is vile, exceedingly contemptible and odious, but to the philosophers it is more precious than gems or gold. It loves all, yet it is well-nigh an enemy to all; it is to be found everywhere, yet scarcely anyone has discovered it. It is the one thing proclaimed by veritable philosophers, which overcomes all, is itself overcome by nothing, searches heart and body, penetrates everything stony and solid, strengthens all things delicate, and establishes its own power on the opposition of that which is most hard. It is the way of truth, and there is no other path to life. It is the true medicine, rectifying and transmuting that which is no more into that which it was before corruption, even into something better, and that which is not into that which it ought to be. The gold of the philosophers with which the wise are enriched is not that gold which is coined.

STORAX CALAMITA.   Storax is a sweet incense or gum.

STRUTHIO, an ostrich.   Soap-wort, used for cleaning wool, a chaplet of this flower.

SUBLIMATION is the purification of the Matter by means of dissolution and reduction of the same into its constituents.   It is not the forcing of the Matter to the top of the vessel, and then maintaining it separated from its *caput mortuum*, but its subtilization and purification from all earthly and heterogeneous parts, imparting to it a degree of perfection not previously possessed, or, more correctly, its deliverance from the bonds which bind it, and hinder its operation.

SULPHUR VIVE.   This substance, which is one of the profound mysteries of alchemy, is identical with the red sulphur of the philosophers.   Some persons, led astray by Hermetic symbolism, have gone to work on native sulphur, assuming it to be the true matter of the art, but though this name is applied to it, the reference is really to that substance when it has attained the perfection of the red or white.   It is then the true philosophical sulphur, and, according to Raymond Lully, is not to be sensibly distinguished from the true Mercury, which has no connection whatsoever with common sulphur.   Mention is also made of a white sulphur, which some say is Mercury educed from potentiality into activity by the operations of the Magistery and following the principles of the Medicine of the First Order.   It may be noted in this connection that Sulphur of Vitriol is the Soul of Vitriol, and so also with Mars, Sol, etc.   Black Sulphur is Antimony ; unctuous Sulphur is another name for the Sulphur of the philosophers.   Narcotic Sulphur of Vitriol is an extract of Vitriol which is described by Beguin ; Paracelsus himself regarded Vitriol as the best of all anodynes.   Ambrosial Sulphur is a natural red Sulphur, very transparent and met with in large lumps.   Green Sulphur is the Sulphur of the Sages.   What the alchemists call True Sulphur is the fixed grain of their matter, the true internal agent which acts upon, digests, and cocts its own proper mercurial matter wherein it is enclosed.   Sulphur Zarnel is another name for philosophical Sulphur.   The Sulphur of Nature is yet another.   Some, however, apply it to the matter at the white.   One author says that it is the essential menstruum made from Mercury and spirit of wine seven times rectified, which dissolves the calx of the Sun and Moon, or at least, extracts their tincture, with which gold is nourished by a simple yet secret series of operations.   But there is some doubt on this point.   Universal Sulphur is affirmed to be the Light from which all particular sulphurs proceed.

SULPHUR-WORT, *i.e.*, Peucedanus.

SUN AND MOON.   The influence of Paracelsus on the later alchemists is shewn very clearly in their speculations concerning Luna and Sol ; as for example : The Sun and Moon are two magical principles, the one active and masculine, the other passive and feminine.   As they move so move the wheels of corruption and generation.   They naturally dissolve and compound, but properly the Moon is the instrument of the transmutation of the inferior nature. There is no compound in Nature which has not a little sun and a little moon, a Son of the heavenly Sun, and a Daughter of the celestial Moon. What is performed by the great luminaries for the conservation of the great world, is analogically accomplished for the microcosm by the lesser luminaries. The little Moon is our incombustible, eternal oil, the receptacle of the little Sun, which is a fire. These are the Sol and Luna of the philosophers, not gold and silver.

SUPERMONIC, *i.e.*, enigmatical.

SYLVESTERS, the fauns of classical mythology.

SYPHITA PRAVA, St. Vitus's Dance.

SYPHITA STRICTA, the fantastic spirit of sleep-walkers.

SYRONES, pimples or boils on the hands.

TALC. The older alchemists have often made reference to what they term an Oil of Talc, to which they have attributed so many virtues that subsequently chemists have exerted all their power to compose it. They have calcined, purified, and sublimed the matter in question, but have met with no success. The reason is that the term was used allegorically, and that the reference was to the Oil of the Philosophers, the elixir at the white.

TANACETUM, the herb tansey.

TAPSUS, *i.e.*, Tapsos, a kind of herb.

TARTAR, a name applied to the mastery at the white stage.

TAURUS ♉ . This zodiacal sign represents asphalt or bitumen in alchemical symbolism.

TENTIGO PRAVA, ? a severe stiffness, or contraction.

TERENIABIN. A variety of manna.

TERNARY. The author of *Anima Magica Abscondita* observes that some philosophers who, by the special mercy of God, attained to the Ternarius, could never, notwithstanding, obtain the perfect medicine. Elsewhere in the same treatise he says that this Ternarius, being reduced by the Quaternary, ascends to the magical decad, which is the exceeding single monad, in which state it can perform whatsoever it pleases, for it is united thus, face to face, with the first eternal, spiritual Unity.

TESTA, potsherd, tile, brick ; also a metallurgical term, signifying bloom. But it seems to have had another meaning in the treatise *Concerning Long Life*. It is said to be the skin of man's body.

TESTÆ OVORUM, egg shells.

THERIAC. Any antidote to poison.

THRONUS, *i.e.*, Thronum, a flower or herb used as drugs or charms.

THUCIA, tutty.

TIBIA, the shank or shinbone.

TIGILLUM, a funnel, a crucible.

TITHYMAL, the sea lettuce, wolf's milk, or milk thistle. Also spurge, euphorbia ; many kinds were known to the ancients. Physicians used the juice or berries as a purgative.

TRAGACANTH, probably Trajanthes, a species of Artemisia. Also gum dragant, a low shrub. The astragalus, whence the gum tragacanth.

TRANSMUTATION. The soul of man is affirmed by the alchemists to possess an absolute power in miraculous, that is, in more than natural transmutation.

TRARAMES. Spirits who are not seen but heard only, as rapping or throwing spirits.

TRIFERTES, the same as salamanders.

TRIPLICITY. Besides the triplicity, which exists in the elements and in all created subjects, there is another more obscure and mystical triplicity, which is recognised by adepts. Without this the former cannot be attained. These three principles are the key of all Nature. The first is one in one and one from one; it is a pure white virgin, and next to that which is most pure and simple. It is the first created unity, the bride of God and of the stars, through which as a medium all things were made, and are still made, both in things natural and in things of art. The second principle does not differ from the first in substance or dignity, but only in complexion and order. The third principle is not, properly speaking, a true principle, but rather a product of art; it is a various nature, consisting of superior and inferior powers. It is the magician's fire, the Mercury of the philosophers, the microcosmus and the Adam.

TRINES, possibly Triones, the constellation called Charles's Wain.

TRIPOLIS. Tripolium is the herb called turbit or blue daisy.

TRITORIUM, a vessel for grinding, mortar, or pestle.

TRONOSIA, that is, Tronossa, honey-dew.

TRONUS, same as Tronosia.

TRUPHIT, the principle of development in every metal, a secret virtue of minerals.

TURBITH, a root much used in medicine to purge off phlegm. Mineral Turbith is sweet, non-corrosive, precipitated Mercury.

TUTIA, see Thutia.

UNIONS. The term *unio* or union was applied to the pearl, because though many are found in one shell, not one is exactly like another.

UNIVERSAL MEDICINE. Most recipes of the adepts given for the preparation of the absolute elixir, are more obscure even than Paracelsus, as, for an example: Take ten parts of celestial slime; separate the male from the female, and each afterwards from its earth, but physically, mark you, and with no violence. Conjoin after separation in due, harmonic, vital proportion. Straightway the soul, descending from the pyroplastic sphere, shall restore, by a vivific embrace, its dead and deserted body. The conjoined substances shall be warmed by a natural fire in a perfect marriage of spirit and body. Proceed according to the Vulcanico-magical theory till they are exalted into the fifth metaphysical rota. This is that world-renowned medicine of which so many have scribbled, and yet so few have known.

VALERIAN, the great set wall.

VENTER EQUINUS. According to an explanation of Paracelsus the Venter Equinus is the digestive power.

VESSEL OF THE PHILOSOPHERS. Geber describes this instrument as a round glass vessel with a flat round bottom. This and all other simple explanations concerning it are supposed to be wholly deceptive, and it is regarded as one of the most profound mysteries of all alchemical art. A treatise attributed to S. Thomas Aquinas says that there is but one vase, one substance, one way, and one only operation.

VERBENA, the herb Vervain.

VERTO, a weight equivalent to the fourth part of a pound.

VESSEL OF HERMES, see Vessel of the Philosophers.

VICTORIALIS is probably Victoriola, the Alexandrian laurel, or tongue-laurel. Paracelsus refers to its magical powers, and says that the lorinds can be made to manifest by means of it.

VINUM ARDENS. According to Paracelsus the correction of Vinum Ardens is by distillation as long as it will ascend, for the removal of the aqueous part.

VINUM ESSATUM, wine impregnated with the virtues of herbs or other substances.

VIRGIN MERCURY and Virgin Sulphur. According to one explanation these are the heaven and earth of Moses.

VITISTA, *i.e.*, St. Vitus's dance. Paracelsus terms it a disease of the imagination.

VITRIOL. The philosophers apply this name to their green or crude matter. Their White Vitriol is the magistery at the white. Their Red Vitriol is their sulphur perfect at the red. Metallic Vitriol is the salts of metals.

VUA, that is, Uva. This is firstly the grape, and stands, by an extension of meaning, for the vine itself. Analogically it is any bunch or cluster. There is also a swelling of the uvula which passes under this name. Finally, there is *uva quercina*, a concretion occasionally found at the roots of oak-trees, possessing medicinal properties, especially in dysentery.

VULCAN is said by Paracelsus to be the master of the alchemists and spagyrists.

WATER. So early as the days of the Greek author Theophrastus, the origin of metals is said to have been referred to water. The same writer mentions that Callias, an Athenian, endeavouring to make gold, brought his materials into cinnabar.

XYLOBALSAMUM, the wood of the balsam-tree.

YLE, wood; timber; the matter or stuff of which a thing is made; the raw unwrought material, whether wood, stone, or metal. In chemical signification, a simple substance, a base matter; as a principle of being, first found in Aristotle, and frequently later in philosophical writers,—usually as opposed to the intelligent principle.

YSOPUS.
YSOPAIC ART. } The art of smelting and fusing.

YLIADUM, Yliadus, Yleidus, etc. The interior spirit which informs the members of every body. Outwardly it generates health, but inwardly disease in humanity. It also leads on to the crisis in diseases. Disease is the resolution of the Yliadus. The reason of this seems to be that the interior spirit contains many species of salts. The resolution of arsenic in the body causes plague ; the resolution of ogertinum, or orpiment, causes pleurisy ; the resolution of vitriolated salt causes another disorder which Paracelsus describes as a disease characterized by a paroxysm, by swellings of the throat, and ulceration of the tongue. Elsewhere Paracelsus says that Yleidus is elemental air, and that its obstruction in any part of the body occasions disease. There is an Yliadus of the elements and one also of man.

YRCUS. The male coney, or rabbit. The blood of this animal was supposed to soften glass and flints when pounded and made into a paste therewith. It had also a similar effect upon other crystalline substances. It is written Hircus by the ignorant. It should be noted in this connection that *cuniculus*, that is, the word which usually signifies coney, also stands for the long pipe attached to a still or a furnace.

ZELOTUM is Petrine Mercury.

ZINCTUM, or Zinetum. Paracelsus describes this as a little known metal, of a peculiar nature and seed. Many metals are adulterated therein. It is of itself fluxile, for it is generated from the flux of the three prime principles. It is fusible but not malleable. He states that he is unacquainted with its ultimate matter. It is akin to quicksilver, and it does not admit of combination with other metals.

ZINIAT. Ferment.

ZONNETI GNOMI. These are fantastic bodies.

ZWITTER, a species of Marcasite ; also roasted ore.

NOTE.—*Some explanations contained in the foregoing Vocabulary concerning known substances are not in correspondence with modern knowledge, and it should be understood that they represent chemical science during the period of Paracelsus and of Alchemy.*

INDEX